AFRICAN VOICES ON STRUCTURAL ADJUSTMENT

A COMPANION TO
OUR CONTINENT, OUR FUTURE

Edited by

**Thandika Mkandawire
and Charles C. Soludo**

COUNCIL FOR THE DEVELOPMENT OF
SOCIAL SCIENCE RESEARCH IN AFRICA

INTERNATIONAL DEVELOPMENT RESEARCH CENTRE
Ottawa • Cairo • Dakar • Johannesburg • Montevideo • Nairobi • New Delhi • Singapore

Africa World Press, Inc.

P.O. Box 1892
Trenton, NJ 08607

P.O. Box 48
Asmara, ERITREA

Africa World Press, Inc.

P.O. Box 1892
Trenton, NJ 08607

P.O. Box 48
Asmara, ERITREA

Copyright © 2003 Council for the Development of Social Science Research in Africa
First Printing 2003

All rights reserved. No part of this publication may be reproduced, stored in a retrieval system or transmitted in any form or by any means electronic, mechanical, photocopying, recording or otherwise without the prior written permission of the publisher.

COUNCIL FOR THE DEVELOPMENT OF SOCIAL SCIENCE RESEARCH IN AFRICA
Avenue Cheikh Anta Diop, Angle Canal IV, BP 3304 Dakar, Senegal

CODESRIA would like to express its gratitude to the Swedish Development Co-operation Agency (SIDA/SAREC), the Rockefeller Foundation, the International Development Research Centre (IDRC), the Ford Foundation, the Carnegie Corporation, the European Union, the Norwegian Ministry of Foreign Affairs, the Danish Agency for International Development (DANIDA), the Dutch Government, and the Government of Senegal for support of its research and publication activities.

INTERNATIONAL DEVELOPMENT RESEARCH CENTRE
P.O. Box 8500, Ottawa, ON, Canada K1G 3H9

Book design: Getahun Seyoum Alemayehu
Cover design: Roger Dormann

ISBN 0-86543-778-5 hard cover
ISBN 0-86543-779-3 paperback

The Cataloging-in-Publication Data is available from the Library of Congress

ACKNOWLEDGEMENT

CODESRIA would like to express its gratitude to the Swedish International Development Cooperation Agency (SIDA/SAREC), the International Development Research Centre (IDRC), the Ford Foundation, the MacArthur Foundation, the Carnegie Corporation, the Norwegian Ministry of Foreign Affairs, the Danish Agency for International Development (DANIDA) the French Ministry of Cooperation, the United Nations Development Programme (UNDP), the Netherlands Ministry of Foreign Affairs, the Rockefeller Foundation, and the Government of Senegal for their support of its research, training and publications programmes.

Contents

Preface viii

Chapter 1
Introduction: Toward the Broadening of Development
Policy Dialogue for Africa 1
T. Mkandawire and C.C. Soludo

Chapter 2
In Search of Alternative Analytical and Methodological Frameworks
for an African Economic Development Model 17
C.C. Soludo

Chapter 3
Trade Liberalization, Regional Integration, and African
Development in the Context of Structural Adjustment 73
T.A. Oyejide

Chapter 4
Industrialization and Growth in Sub-Saharan Africa:
Is the Asian Experience Useful? 103
C. C. Soludo

Chapter 5
Impact of Structural Adjustment on Industrialization
and Technology in Africa 161
S. M. Wangwe and H. H. Semboja

Chapter 6
Structural Adjustment and Poverty in Sub-Saharan Africa:
1985-1995 189
A. G. Ali

Chapter 7
The Elusive Prince of Denmark: Structural Adjustment
and the Crisis of Governance in Africa 229
A. O. Olukoshi

Chapter 8
Economic Policy Reforms, External Factors, and Domestic Agricultural Terms of Trade in Selected West African Countries 275
T. B. Tshibaka

Chapter 9
Financial Sector Restructuring Under the SAPs and Economic Development, With Special Reference to Agricultural and Rural Development: A case study of Uganda 305
G. Ssemogerere

Chapter 10
Financial Sector Reform in Eastern and Southern Africa 353
F. M. Mwega

Chapter 11
Resource Mobilization, Financial Liberalization, and Investment: The Case of Some African Countries 383
M. N. Hussain, N. Mohammed and E. M. Kameir

Chapter 12
Financial Liberalization in Africa: Legal and Institutional Framework and Lessons from Other Developing Countries 411
E.L. Inanga and D. B. Ekpenyong

Chapter 13
Financial Liberalization, Emerging Stock Markets, and Economic Development in Africa 445
T. W. Oshikoya and Osita Ogbu

Chapter 14
The Outcome of Financial Sector Reforms in West Africa 471
C. Emenuga

Appendix 500

Contributors 503

Preface

The publication of a two-volume evaluation study on "Adjustment in Africa" by the World Bank in 1994 sparked off major controversies and re-ignited the debate about the direction(s) of Africa's development. For most African scholars who live in and study these economies, the World Bank reports were yet another major disjuncture between reality and dogma. The urge to provide a critical response to the reports in order to straighten the issues was irresistible. The need for such critical appraisal of the structural adjustment program (SAP) as a development strategy provides the immediate rationale for this project.

Furthermore, the failures of SAP, the simplistic diagnosis and highly tendentious performance evaluation of the 1994 report, as well as the seemingly changed environment that is more permissive of alternative viewpoints have convinced Africans to "re-enter" the debate. There is a growing call for "local ownership" of adjustment and for Africans to assume the leading role in defining the continent's development agenda. There was thus the broader goal of Africans reclaiming the initiative and providing a framework for thinking Africa out of the current economic crisis. A careful reading of the writings of African scholars on the subject indicates a discernible trend in terms of an emerging perspective. The perspective derives from the failures of SAPs as well as lessons of experience on what kinds of policies are required, which ones can work, and what kind of issues should be addressed. It is also a perspective that is informed by the lessons derivable from the experiences of other, more successful economies, especially in Asia, and in fuller appreciation of the constraints and opportunities offered by the new and continuously evolving international environment. A second but no less important goal of the project is thus to sift such perspective in a volume as a way of initiating and advancing continued dialogue on those issues.

In pursuit of these objectives, about 30 studies (largely determined by the authors through research proposals) were commissioned. About 25 of such were by economists, making it the project with the largest participation of African economists on the issue of adjustment and the way forward. Each of the studies (covering various aspects of adjustment programs) appraises the performance of SAPs so far with respect to the particular sector/issue, and evaluates the compatibility of the policies with the requirements for long-term development. Participants in the project set out to analyze the various policies under SAP from the perspective of development broadly understood as involving economic growth, structural change, and elimination of poverty.

The results of the studies were presented at two research workshops in Abidjan in 1996. Each paper was rigorously reviewed by a discussant, and the collective peer review and extensive discussions following each presentation helped to enrich the final product. In the end, the only thing common in all the reports is the conclusion that adjustment has not worked as promised. Each paper concludes with a number of recommendations about the way forward. The Synthesis Volume (*see Our Continent, Our Future*) for this project has liberally drawn from some of the papers.

In this volume, we are constrained by space to select and publish a few (about 50%) of the papers written for this project. Coverage of the broad range of issues, rather than just judgement about the quality of the papers per se, informed the selection. Needless to say, most of the papers for the project are of very high quality. We have selected papers that touch on issues which have been accorded very little attention in the SAP framework and its evaluation studies, or ones which have generated the most controversy. The issues covered in the selected papers include: critical evaluation of the model and methodologies for performance evaluation under SAP; comparative development experiences; trade liberalization and regional integration; SAP, technology, and industrialization in Africa; poverty under adjustment; reforms, external and domestic factors, and implications for agriculture and rural development; and financial sector reforms and resource mobilization. Each of the papers should be read and appreciated in its own right, and we do not attempt to summarize their individual contributions.

Finally, we thank the International Development Research Centre, Swedish International Development Cooperation Agency (SIDA/SAREC), the Dutch Ministry of Foreign Affairs for financial support and CODESRIA for support and coordination.

—Thandika Mkandawire
Charles C. Soludo

Chapter One

Introduction:
Towards the Broadening of Development Policy Dialogue for Africa

Thandika Mkandawire
Charles C. Soludo

In this introductory chapter, we skeletally evaluate the context for the development policy dialogue for Africa, as well as summarize some of the key elements of the emerging African perspective on adjustment and development. We draw liberally from the Synthesis Volume (*see Our Continent, Our Future*) of this project.

Africa's endemic poverty and pervasive underdevelopment have defied much of the development policy experiments of the last three decades. To be sure, no amount of excuses can hide the monumental failures of public policy in the past and the complicity of Africans and the outside world in the process. In several aspects the fault, to borrow the Shakespearean expression, "is not in our stars but in ourselves." Through several acts of omission and commission, we, and more so our leaders, have short-changed the continent. In addition, our problems have been compounded by the very weak/poor initial conditions, short but peculiar history of post-independence colonial heritage, and hostile external environment. On the other hand, the international community—the multilateral development institutions (especially the Bretton Woods institutions, [BWIs]), and bilateral agencies (mostly former colonial masters)—also bear much of the blame. There has been hardly any development program in much of Africa without the tacit or explicit involvement/endorsement of the donors. In several important aspects, many of the policies/programs which have turned out to be "bad" were at their insistence. With Africans adjudged "incapable of thinking for themselves and implementing policies," a deluge of over 100,000 foreign technical experts costing over $4 billion annually to maintain have literally taken over the process of policy/project design and sometimes implementation. In what has ensued, Africa has turned into a pawn in the chessboard of experimentation for all manner of ill-digested development theories and pet hypotheses. Again, Africa is largely to blame for sheepishly following along.

Sequel to the intensification of the economic crisis since the late 1970s, the efforts to "develop" Africa in the last decade and a half have been underpinned by SAP. Throughout the adjustment years, the BWIs seized much of the initiative, and foreclosed the debate by literally insisting that it was either their way or nothing, with African scholars and policy-makers largely relegated to reactive protest. Africans recognized that their economic crisis required some fundamental adjustment, but raised serious reservations about the relevance and/or adequacy of the kind of adjustment being foisted upon them by the BWIs. After over a decade of acrimonious debates and tons of evaluation reports, there is an increasing convergence of views that SAP has not worked, and as designed, it is grossly defective as a policy package for addressing the problems of underdevelopment in the region. In its latest evaluation report on Adjustment in Africa, however, the World Bank (1994) obdurately insists, contrary to all evidence (including several of its own contradictory evidence) that "adjustment is working." Such insistence could not hide the continuing disappointing socio-economic performance of the region. More recently, such self-assurances of the past seem to be giving way to a subdued humility, expressed in such phrases as "development everywhere is a complex phenomenon...nobody has all the answers...learning from experience...rapidly changing realities." This recent "re-thinking" and admission that it does not have all the answers have provided a conducive climate for productive dialogue in search of the way forward.

The current mood and search for solutions is akin to the situation in 1979/80 when African heads of state adopted the Monrovia Declaration, and later the Lagos Plan of Action and the Final Act of Lagos.

These were Africa's own first attempts at articulating a framework for solving its problems. Subsequently, as a response to the perceived inadequacies of the orthodox SAP, and the deepening crisis, the UN-Economic Commission for Africa articulated the African Alternative Framework to Structural Adjustment Programs (AAF-SAP). Most African governments signed the documents, and to date, none has publicly dissociated itself from the ideas espoused in them. The World Bank virulently attacked those documents, and every African government that wished to have successful debt rescheduling or aid negotiations distanced itself from the principles in the documents. African governments either did not have the confidence and courage to implement their own strategies, or they were constrained by resources to implement programs they did not believe in. The years of the crisis merely saw a continuation of the surrender of national policy-making to the ever changing ideas of the "international experts." Ironically however, most of what appear today as new insights about the imperatives of poverty reduction, investment in infrastructure, education, requirements of rapid industrialization and the structural and institutional bottlenecks of Africa's underdevelopment, are nothing but a rehash of old but once disparaged ideas of African scholars and policy-makers on the subject. It has taken over a decade for the international community to agree with the UN-ECA's AAF-SAP that "adjustment alone is not adequate for long-term sustainable development" as the World Bank (1994) finally admits. The acrimonious nature of the earlier debate and dogmatic proclivities of the BWIs prevented a creative search for a "consensus model" or some convergence of views about the synthesis of the disparate proposals on African development.

Areas of Consensus

The background studies in this volume as well as current discussions on Africa's development, underscore a tendency towards some broad agreement on a set of issues. One issue about which there seems a broad agreement is that Africa's economic performance, though modest especially in 1996, has been generally disappointing. The modest performance of the regional economies in 1996 raised a short-lived sense of euphoria and a few of the countries—Uganda, Ethiopia, Côte d'Ivoire, etc.—are said to be spectacular. But, everyone agrees it is not yet "*uhuru.*" For example, a few stylized facts of these economies despite over a decade of SAP can be summarized as follows:

Macroeconomic stability has improved modestly, but many analysts doubt its sustainability given that in most economies that it has occurred, volatile external finance has been largely responsible. Moreover, stability has been achieved mostly at the great expense of domestic investment—even in basic infrastructure (physical infrastructure, and human capital formation) which have been recognized to be central for sustainable growth and development. As Nissanke (1997:6) argues,

> The present narrow base for raising fiscal and export revenues means also a continuous implementation of short-term stabilization policies on a perpetual basis, leaving the economies constantly exposed to large external shocks, as primary commodity prices in the world market exhibit not only declining long-term trends but also excessively volatile fluctuations which are detrimental to growth and development. Sub-Saharan African countries have to undertake continuous adjustments to their deteriorating, and highly unstable position in current account balance. The scale of adjustment required has often far exceeded the capacity of these economies to absorb volatilities and manage associated high uncertainty and aggregate systemic risks.

Modest recovery in output (essentially improved capacity utilization in most cases) has been uneven and fragile, with economies quickly entering and dropping out of the "good or excellent" performance list. That is why, despite the fragile but encouraging aggregate performance of African economies in 1996, even the World Bank cautions that such a "recovery" might not be sustainable in view of the structural weaknesses and heavy debt burden.

The external debt stock of SSA has more than doubled since the adjustment program and without any acceleration in economic growth to sustain its servicing in the future.

Africa's infrastructural base and human capital formation which were deemed to be fragile at the beginning of adjustment have even deteriorated further throughout the adjustment period. Africa's capacity for managing the crisis has been further eroded through massive brain drain and demoralized civil service caused by sharply declining real wages, massive retrenchment, incessant vilification for corruption and insensitive denigration of their competence by peripatetic "experts" on "capacity-building" missions. Poverty has intensified despite modest recovery in some countries, and human development indicators—life expectancy, infant mortality, and school enrolment—have worsened.

The export basket remains undiversified, and though a recovery of about 3.2% in exports has been achieved, imports have grown by an average of 6.4% (due mainly to the import liberalization). Added to the external debt burden, the external viability of these economies remains in serious doubt. The fragile industrial base has even shrunk further (de-industrialization) since the adjustment program in many countries.

The already weak African states prior to the adjustment program have become even weaker, and the requisite institutions for development are either very weak or non-existent in some countries.

A second area of broad consensus is that the goal of Africa's development should be stable but rapid economic growth with poverty eradication. Furthermore, there seems also a broad agreement on the centrality of the market in any framework for organizing economic activity. However, in the light of the stylized facts about the enduring weaknesses of these economies, there are sharp disagreements about the required strategy to attain the goals of economic policy. Specifically, there are lingering differences of opinions regarding the choice of policy instruments, timing and sequencing of implementation, the nature of institutional and governance problems and the role of the state; the role of the donor/international community, etc. These differences notwithstanding, it is evident that the initial gulf in opinions seems to have narrowed considerably. One issue of apparent consensus is that there is a need to broaden the domain of discussions about what constitute the development "fundamentals."

Broadening the fundamentals of economic development

In much of recent discussions of the prerequisites of economic development, "getting the fundamentals right" has become the new fad. What is debatable is the nature of "fundamentals" and the edifice which they are supposed to support. In the popular usage under SAP, "stabilization" issues have been elevated as constituting "the" fundamentals. In times of economic turbulence and scarce resources, concern with "stabilization" tends to attain great prominence on the policy agenda and monetary and financial fundamentals have tended to be confined to issues of stability and efficiency in the use of given resources. The crisis of the last two decades has underscored the importance of macroeconomic stability. Economic conditionality now routinely insists that countries enter into agreement with the IMF before having access to other sources of funding. There can be no doubt about the importance of macroeconomic stability for both economic and political wellbeing of a country. No sustainable growth can take place in a context of hyperinflation and unsustainable balance of payments equilibrium. And so any new policy must address issues of stability and must seek to get the financial "fundamentals" in place, not necessarily sequentially but at least simultaneously with "other" fundamentals. It is important to note how preoccupation with "development issues" of the 1960s and 1970s and consequent little regard for macroeconomic stabilization was partly responsible for the debt and balance of payment crisis of the 1980s.

The appropriate reaction to the previous neglect of stabilization issues is, however, not to elevate it to such pre-eminence as to almost completely ignore the "development issues." Stabilization has proceeded without seeking to minimize the negative consequences of these policies on investment, employment and the process of technological learning. Indeed what we note is the deterioration in the support given by the financial system to the production system and the tendency to give primacy to the exigencies of financial stabilization even when these undermine economic production and social cohesion. While perhaps addressing issues of static allocative efficiency and while perhaps appropriate for stabilization policies based on these narrow perception of "fundamentals," financial stabilization does not explicitly incorporate measures that would be addressing developmental fundamentals and do not in themselves guarantee the resumption of growth. The implication is that they have the distinct danger of producing a "poor but stable and efficient" Africa.

Between the promise of "accelerated development" of 1981 and that of "sustainable development" in 1989 and the more recent Adjustment in Africa, "development" disappears from the adjustment discourse to such an extent that the "success indicators" were confined to the movement of policy instruments rather than the real economy. In the 1989 World Bank's Long Term Perspective, there was an attempt to assert the developmental thrust of adjustment. The report recognized the importance of structural and institutional factors in explaining economic performance in Africa. It correctly pointed out that "it is not sufficient for African governments merely to consolidate the progress made in their adjustment programs. They need to go beyond the issues of public finance, monetary policy, prices and markets to address fundamental questions relating to human capacities, institutions, governance, the environment, population growth and distribution, and technology" (World Bank 1989b). It suggested that the attention to human resources, technology, regional co-operation, self-reliance and respect for African values provide the main focus of the proposed strategy.

The proposed strategy stressed the need to put in place an enabling environment for infrastructure services and incentives to foster efficient production and private initiative; enhanced capacities of people and institutions; and growth strategy that must be both sustainable and equitable. This was a significant broadening of perspective on behalf of the Bank and was in some way a movement closer to the positions that the UN-ECA had pushed over many years. Although the notion of "governance" in the 1989 report was excessively technocratic and narrow, it did permit a much less jaundiced view of the role of the state in the economy and placed issues of governance at the core of the policy dialogue. This shift was, however, short-lived and was de-emphasized in the 1994 report which used indices of policy stance that said nothing about other "fundamentals." Nothing was said about proactive policies that would be required for a dynamic industrialization program, a transformative agrarian process of the "Green Revolution" dimensions and the accumulation of both physical and human capital in a forced and speeded-up pace.

Needless to emphasize that many development-inducing policies are of a micro- or meso-level nature. The primacy given to "stabilization" fundamentals led to a confusion between microeconomic policies and macroeconomic policies, with micro-level policies guilty by association with failed macroeconomic policies (Rodrik 1995). Under the current reforms, emphasis on getting the monetary/financial accounts and "fundamentals" right has almost become an end in itself because it was assumed that most other things of a microeconomic or sectoral nature would automatically follow from any such reforms. The result was that interventionist microeconomic policies that had worked in several successful Asian economies were rejected wholesale in the African case, leaving the state with virtually nothing to do in the development process. Experiences so far in most parts of the world, including Europe, point to the need to take explicit steps to get "other fundamentals" right. For example, the quest for the European common currency had imposed emphasis on "monetary/fiscal fundamentals" with complete disregard to the "real" side of the economy. Recently, many countries are realizing that the adjustment costs are simply too enormous, and it is not surprising that many newly elected governments in Britain, France, etc. are arguing that "employment and output" should be part of the targets for monetary union.

The major point of the foregoing discussions is that the new dimensions of development dialogue for Africa should synthesize the various "fundamentals" into a coherent strategy for development. We argue that "policy fundamentals" must simultaneously address the key elements of: equity, rapid but stable economic growth, and political legitimacy. The weight attached to each of these will depend on each country's political and economic conditions. A broad spectrum of African analysts insist that development-focused adjustment should simultaneously establish "fundamentals" relating to financial, production and governance systems that constitutes a development policy-package. In more recent times, one additional set of "fundamentals" relates to environmental sustainability of economic growth.

One example of these "other fundamentals" is the "production fundamentals." These address the real side of structural variables—productivity, demand for output, firm and household behavior, institutional relations, innovation process, tangible and intangible investment, export orientation, employment creation, and economic growth. A development-oriented adjustment program will be investment driven. Consequently, a central piece of policy must include measures that increase investment and improve allocation among sectors and projects.

The presumption in all these is the view that economic policies in Africa will be judged by the extent to which they contribute to economic development broadly understood as involving economic growth, structural change and elimination of poverty. Our argument is that policy needs to simultaneously address a broad range of fundamentals-macroeconomic stabilization, proactive, supply-side (production) fundamentals as well as socio-political fundamentals. All these should be mutually reinforcing.

From the overall framework of "development fundamentals," a number of micro-macro policies might need to be re-designed in ways that are more compatible with the structures of these economies and that cohere with the objectives of long-term development. Some of these policy areas span: trade and industrial policy; human resource and technological development; resource mobilization and external financial dependence; democratic governance, and measures to address the gamut of institutional and administrative bottlenecks of the system (see Chapter 4 of the Synthesis Volume for detailed adumbration of these issues).

An aspect of the policy framework under SAP about which a serious re-thinking is imperative pertains to the requirements of rapid industrialization and the consistency of unilateral, deep trade liberalization in the process. Africa's position in the new global trade relations will largely be determined by what action is taken in two important directions: first, increasing the regional and international competitiveness of its production activities by changing the structure of exports towards more dynamic, non-traditional products (in terms of their demand prospects and their potential to effect technological change); and second by tackling the structural bottlenecks inherent in the entire system of governance and those specifically addressing export promotion and industrial development. The need to effect change in the export structure will inevitably bring discussions of policy issues relating to export diversification, the transformation of production structures, and industrialization, back on the development agenda. The policy stance under SAP of an "incentive-neutral" trade policy and the associated perfect competition and comparative advantage model that underlie the attendant industrialization strategy has come under severe attack. With respect to trade policy, an emerging consensus is that adequate preparations should be made—in terms of relevant supply-side measures and institutional arrangements—to elicit the desired export supply responses before deep liberalization is implemented. Trade reforms must serve, and be consistent with the requirements for balance of payments constraints, government fiscal viability, and industrialization objectives.

In essence, the emerging consensus is that SAP was wrong by its over-arching emphasis on "getting prices right" as well as asking governments to "get out of the way" with respect to industrial policy. Industrial policy of the "selective" kinds should be brought back on the agenda but with full lessons of the negative consequences of the kinds of import substitution strategy of the past. The new strategy presupposes an aggressive export-orientation but insists that such a strategy can best be achieved with a dose of "infant industry protection" and an active, selective industrial policy by the state.

A nascent industrializing country would need a number of policies to develop the "hard" and "soft" infrastructure required to build up a competitive industrial sector. These would involve massive investment in providing such "hard infrastructure" as roads, ports, efficient telecommunication and postal services, electricity, and water supply. Human capital development through investment in education at all levels, especially in science and technology, and research and development activities would serve to provide the requisite skills to compete in modern world. The soft infrastructure would include the institutional framework for doing business—efficient and transparent regulatory framework, enforcement of contracts and well-defined property rights, insurance and accounting services, development of the money and capital markets, forging of business-government relationship, etc.

Furthermore, a gamut of tax, credit, and labor policies would need to be designed to lower the operating costs of firms. Institutions for provision of long-term finance and procurement of information relating to technologies and markets need to be designed. Perhaps, one of the areas where closer government-business relationships would need to be strengthened in Africa relates to the processing of information relating to foreign markets and technologies.

Long-term growth prospects in Africa will depend on how well agriculture performs. In most countries, agriculture will be a source of foreign exchange and savings. It will also be an important source of inputs into industry and a major contributor to the market for some of the "infant industries." The chosen pattern of agrarian transformation will also determine the course of equity in the growth process. Because of its importance, agriculture has continued to receive policy attention, but it has proven to be the Achilles Heel of virtually every strategy of development in Africa. The conviction that getting the macroeconomics right would elicit the required response has proved naïve. It has become clear that there was a need for sectoral and micro-level policies that would directly address problems of low productivity and low technological levels in African agriculture. More specifically, there is a need for increased investment in infrastructure, in extending markets to reduce transaction costs, in increased extension services, etc. All these require a much more active state than was allowed for under SAP. There is a need to devise schemes that direct credit to rural producers in a manner that encourages technical innovation. This may involve subsidized credit or inputs. We should recall that the "Green revolution" in Asia required massive intervention in the markets for credit and agricultural inputs. It also demanded policies that protected agriculture from cheap imported grains "dumped" from countries that heavily subsidized their own agriculture.

Mobilization of domestic financial resources and their use is another area that requires more attention than it has received. African governments should find ways and means of "forcing" up the domestic saving ratio. African countries have in the past achieved higher levels of domestic savings than the current ones. There is no doubt that such levels can be attained, especially if the "debt overhang" can be relaxed, allowing the public sector to also begin saving. The deployment of efforts at increasing both private and public savings will probably have a much higher pay-off than the efforts that have hitherto been devoted to attracting foreign capital —both official and private. African governments must seek ways of mobilizing domestic resources. Forced savings schemes such as fully funded pension schemes (such as those of Singapore) and taxation on luxury consumption goods ("Kaldor Tax") should be considered. Some form of "financial repression" will also have to be tolerated to direct savings and to mobilize capital for long-term development.

Partly because of the fiscal crisis of the state and the ideological stance of public investment suggesting that they were either inherently inefficient or "crowded out" private investment, public investment has collapsed in Africa. In more recent years it has become abundant clearly that one constraint on private investment response has been the collapse of public investment. There is need to revive public investment within a framework of long-term development thinking both as a guide to the private sector on what are the national priorities but for its planning in light of well-articulated state priorities and projects.

In addition, efforts must be deployed to enhance the capacity of the state to channel public, private and external savings to finance investments by means of development banks or specialized investment funds. As argued by Fanelli and Frankel (1994), carefully administered development banks could be efficiently used to evolve screening devices for the selection of private investment projects. Such an argument is reinforced by the lack of long-term capital markets in most African countries. It is unlikely in Africa that private markets will generate a flow of financial intermediation high enough to support a substantial rate of investment in productive activities. Evidence has shown that the axiomatic scheme that associates financial liberalization to high savings and then high investment simply does not hold. Studies of stock exchange markets suggest that they are unlikely to finance long-term investments at desirable levels.

For several countries in Africa, important attention must begin to be focused on strategies to encourage flight capital to return home, and be invested. For countries such as Nigeria, Zaire, etc. with several tens of billions of dollars in foreign private bank accounts, any program that attracts back a significant proportion of such funds could unleash the required momentum for growth in some sectors. Government leadership in providing the necessary incentives, legal guarantees of property rights, and personal encouragement to these owners of funds (however acquired) would be important. For such a process of "amnesty" to be credible, the governments should enjoy popular support and empowered by the electorate to grant such "amnesty." An "amnesty" declared by those who constitute the "kleptocracy" would be morally reprehensible and ultimately incredible.

Aside from efforts directed at increasing domestic savings and campaigns for flight capital to return as investments, another potential source of resources to aid investment could be the external aid. There is an on-going but largely unsettled debate about the positive and negative consequences of aid to Africa. For many analysts, aid has simply generated a dependency syndrome, cynicism, "aid fatigue," and has been largely counter-productive. Any serious development strategy must re-examine the aid-development nexus. African governments need to define in a much more precise manner what external assistance is required on the basis of clearly defined national goals and an exhaustive mobilization of national capacities and resources. It is imperative that for most countries to move forward, both the donors and recipients of aid seriously re-think the purpose and nature of aid to Africa. No doubt, some aid plays some positive roles, but policy-makers should initiate major debate around the potentials for channeling aid money in a manner that enhances African human resource utilization and building and mobilization of domestic resources and wean away African economies from an aid dependency that simply does nobody any good.

In addition to aid, perhaps the single most important issue relating to the compact between Africa and the international community is the external debt overhang. Many analysts of the African crisis have stressed for the umpteenth time that the resumption of long-term sustainable growth in Africa would be extremely difficult, if not impossible, without addressing the debt burden. Alternative proposals for solving the problem are on the table. Increasingly, however, analysts would agree with Sachs (1996) that "the assistance should come in the form of debt cancellation. No one can doubt the dreadful policy errors of the past, nor the mutual complicity of African and donor nations. A fresh start requires a thick line drawn under the past." This solution, simple as it sounds, holds a key component to the solution to Africa's economic stagnation. All efforts to find out why stabilization or adjustment has not worked, why investment has not resumed, and why the state capacity has been further eroded, will not succeed without the inclusion of this one but dominant argument—debt overhang. The immediate solution of this crisis, or an articulation of a comprehensive reform program that also indicates how countries can sustain growth irrespective of the debt should be devised.

Another, and perhaps the hottest issue on the development policy dialogue, pertains to the questions of social-political "fundamentals." An aspect of these fundamentals is the domain of the state— its institutions, and governance. It is important to underscore the importance of such socio-political fundamentals because the failure to have them in place can completely scuttle any "sound" economic policy. Africa abounds with examples of countries which impressed donors with their sound economic policies when the political system was on the verge of collapse. Furthermore, to tie up all these fundamentals into some coherent whole one needs a theory of development which delineates the role of the state. Essentially, a government that would implement the kinds of policies suggested here would have to be "strong" both in terms of technical capacity and its political legitimacy and social anchoring. The state, to use the old cliché, must be developmental. A major challenge to the transformation of African states into developmental ones must go beyond merely enhancing its techno-bureaucratic capacity and seek to "embed" such a developmental state within democratic social institutions and governance frameworks. This major challenge requires imagination and sense of history. Indeed the constitution of "democratic developmental states" may be the single most important task on the policy agenda in Africa. Such a process is not facilitated by the current practice that removes key elements of economic policy from democratic scrutiny by placing them in the hands of an insulated "technocracy." What is required is a system of policy-making and democratic governance in which political actors have the space to freely and openly debate, negotiate and design an economic reform package that is integral to the construction of a new social contract on the basis of which Africa might be ushered into the twenty-first century.

Conclusions

The single most important development in the discourse of Africa's crisis is the healthier atmosphere of co-operation rather than the confrontational, acrimonious debates of the past. The vehement rejection of some African analysts of the orthodox SAP as a development strategy often gave the mistaken impression that Africans were against "sound economics." On the other hand, the failures of the past have foisted some long overdue humility on the advocates of SAP to acknowledge that "no one has all the answers." What emerges is the need to forge a new platform for development policy dialogue based on partnership and consensus-building. In this dialogue, Africans must have the confidence and courage to take the driver's seat. While others can help, it is ultimately the primary responsibility of Africans to think for Africa and to develop it.

The new dispensation calls for fundamental change by Africans on several fronts, in response to changing reality. While one loathes the elements of the "Afro-pessimism" and the attempts to locate Africa's poor performance on its supposedly "immutable" and "peculiar" characteristics, it is also important to underscore the need for fundamental changes in some of the current attitudes, institutional arrangements, orientation to governance and economic management. For example, the tensions and suspicion between the state and the capitalist class in many African countries that lead to massive capital outflows into "safer havens" is very unhealthy for the development of what are obviously capitalist economies. Furthermore, the subordination of national goals and development agenda to the narrow and often temporary interests of political survival and, or, ethnic loyalty, is hardly the best way to build a competitive and prosperous economy. In the end, there is no wishing away the socio-political issues which the transition to a market economy brings. Each country must, out of its own historical experiences, forge its own vision and design the requisite institutions to effectuate development. Outsiders can assist, but they can never substitute for local initiative.

We cannot overemphasize the role which the international conjuncture can play in widening or further narrowing the road ahead. Africa must learn to compete in this global arena. Such a learning process will be facilitated by regional markets but also by adopting a proactive strategy for increasing and diversifying Africa's exports. Africa's natural resources may facilitate this process but we should recognize that only a strategy that relies on our human capacities will produce a development process that can respond flexibly to a rapidly changing world. Our natural resource endowment will only contribute to development if we add intellectual value to it and if we use the revenue to transform and modernize agriculture and strengthen the development of industrial structure that is made competitive by strategically orchestrated exposure and competitiveness in international markets.

Africa should know that it cannot integrate fully into the global economy by permanently depending on aid and preferential treatment. None of these have served Africa well. Aid has produced a phenomenon of a dependency syndrome that stifles both imagination and initiative, while preferential treatment (especially under the Lome convention) has provided incentives for the perpetuation of activities that have tended to fossilize our production structures in primary commodities. What Africa needs, as it approaches the twenty-first century, is not increased aid but rather a leveling of the playing ground. An important element in this is an unconditional debt write off for all the indebted SSA countries. Tying debt reduction to perceived compliance of debtors to certain performance standard could amount to a circularity of logic. Poorly performing countries could owe their performance to the debt burden, while the high fliers could be because of the debt relief—which currently comes by way of increased official aid.

African economies are "market economies." This means that while the state may draw up the larger developmental plans the implementation of such plans will depend on the responses of private agents. Two important lessons in Africa's development experience have been that (a) failure to mobilize the resource allocative functions of the market can only contribute to the inflexibility of the economy and (b) failure to recognize the weakness of "market forces" in a number of fundamental areas can lead to failed "adjustment." Development policies will therefore have to be keenly conscious of the capacities and weaknesses of both states and markets in Africa and seek to mobilize the former while correcting the latter. Dogmatic faith in either "planning" or "markets" will simply not do.

Moving beyond "adjustment" towards growth and development is, of course, not an easy task. The issues are complex and a brief introduction such as this cannot delve into all the issues as would be most appropriate. To recap, we need to underline two of the major conclusions of this project. First, Africans must avoid a "failure complex" which leads to a tendency to adopt self-fulfilling blanket condemnations of our own reality and that makes us incapable of learning from our experiences. Second, moving African economies on to a development path will require robust state and societal institutions. This in turn will require the instituting of creative mechanism to produce a truly developmental state-society nexus able to synergetically mobilize human and physical resources and address the many contradictions that are inherent to our societies and to any processes of rapid change.

Economic development is quintessentially a political process, touching upon issues of power, intra-temporal and inter-temporal distribution of not only economic resources but also of power. It is a process that taxes the political system heavily. It involves sacrifices and commitments that can only be sustained through a sense of shared vision and common purpose. It calls for the mobilization of national capacities. We have argued that such a process must be democratic not only because of the inherent value of democracy but also because it may be in the nature of African societies (their social pluralism and artificiality of national borders) and the current political conjuncture that only a democratic "developmental state" can acquire the adhesion of a citizenry that is so diverse as one finds in African countries. For sustainable long-term development therefore, the process of policy-making and governance generally must be democratic.

Chapter 2

In Search of Alternative Analytical and Methodological Frameworks for an African Economic Development Model

Charles Chukwuma Soludo

Introduction

Right from its inception in the early 1980s, the major controversies about the Structural Adjustment Program (SAP) relate to the crisis of analytical and methodological framework and thus the discordant tunes about what SAP can or cannot do, whether it has succeeded or will ever succeed, and the efficacy of evaluation techniques. Answers to such conceptual and methodological issues continue to elude analysts even as the need for answers becomes more urgent.

The contradictions, equivocations, and largely controversial conclusions of the 1994 World Bank Study Report on adjustment in Africa have raised the urgency of a rigorous re-evaluation of Africa's development model. At the heart of the shortcomings of the 1994 Report is the persistent crisis of theoretical and methodological framework. A careful read-through of the Report as well as most other recent studies of the Bank points to a clear admission (at least in rhetorics) of the inadequacies of the first and second generations of SAPs in Africa. There have been changing emphasis on such issues that were never built into the original SAP diagnosis and policy prescriptions: admission of the need for "poverty alleviation measures," recognition of the inhibiting impacts of the external debt and external trading environment, realization that SAPs are inadequate for sustainable long-term development, recognition of some institutional and structural bottlenecks that might make implementation of SAPs difficult, etc. What is surprising however, is that such "paradigmic shifts" and "add-ons" have not been translated into any fundamental rethinking of the goals, design, and policy instruments of SAP, nor have they led to any efforts to evolve an internally consistent framework that nests the various "SAP add-ons" into a coherent and measurable development framework. On the other hand, most of the existing "alternative frameworks" to SAP have several fundamental problems which range from being excessively engulfed with sloganeering and polemics to the poverty of rigorous analytical and methodological frameworks that also speak to the changing international economic relations.

The result is that the variance between any of the existing frameworks (including SAP) and the realities of the African condition is very deep. For SAP, it is little surprise that the empirical results produced by several evaluation exercises are at every point contradicted by the evidence on the ground. In practice the choice of methodologies for evaluating performance of particular development strategies cannot be divorced from the crisis of relevant conceptual and theoretical framework. What one sees is ultimately dependent, not only on where one stands but perhaps also on what he chooses to see and how he does so—with one eye closed or with dark glasses on. Thus, any serious attempt at a rethinking of the adjustment program and general development prospects should start with a critical evaluation of the existing analytical and methodological frameworks, with a view to illuminating the pathway towards potential consensus.

Our goal in this chapter is not to resolve the crisis of a development paradigm for Africa. We intend to demonstrate that existing models are at best incomplete theories of development and the prevailing methodologies for performance evaluation as amounting to a convenient use of statistics. We sketch the issues in the search for an alternative framework as deriving from the weaknesses of the existing frameworks.

Secondly, we survey and critically review the past and current models/strategies of African economic development; thirdly, examine the crisis of methodologies and the nature of macroeconomic data for evaluating performance under SAP; fourthly, outline the issues in and the elements of the search for alternative frameworks and also conclude the chapter with some recommendations.

Past and Current Development Models/ Strategies in Africa

Diagnosis of economic problems and prescriptions of policy actions are strongly influenced, not only by events, but also by the intellectual milieu of each era. What is often missing in most discussions of SAP is the connection between changes in theoretical paradigms and the nature of policies prescribed. Analytically, two schools of thought dominate the debate about the nature of the African crisis and required remedies. On the one hand is the structuralist paradigm, while the neoclassical theory presents the alternative, albeit dominant, perspective.[1] Empirically, the polar case interpretations also boil down to a debate about the relative weights of the external shocks and structural rigidities versus domestic policies in explaining the crisis. In this section, we provide a comparative evaluation of the strategies/ models and interpretations of the African development crisis.

Structuralism and import substitution industrialization challenged

Intellectually, the last few decades have witnessed major shifts in general economic thinking and development economics in particular. Structuralism as a development paradigm had been born partly in response to the failure of stabilization packages in Chile, Argentina and Uruguay (1956-62) which ignored the effects of the "bottlenecks" and rigidities that pervaded those countries' agricultural, foreign trade and government sectors. Structuralists argue that the poor initial conditions, and a long list of structural bottlenecks make developing countries different from the industrial ones, and thus require different set of policies for development. Such bottlenecks often cited include: widespread subsistence farming and activities; shortage of trained personnel and scarcity of entrepreneurial capabilities to promote and manage development; weak institutions; small and fragmented economies; inappropriate technology and traditional production techniques marked by low levels of productivity; poor physical infrastructure and transportation networks constraining the integration of various regions of a country; weakness in political structures; excessive dependence on the foreign sector rendering the economies extremely susceptible to external shocks; rudimentary money and capital markets; colonial influence marked by "too much" dependence of African manufacturing sector on imported factor inputs such as capital, skilled manpower, technology as well as spare parts and raw materials; poor weather conditions and prolonged periods of droughts which affect the major production activity—agriculture; low price elasticities rather than instantaneous responses to price incentives, etc (see Elbadawi et al. 1992:70-73).

A bottleneck greatly emphasized by the structuralists is the asymmetrical economic power relations in the world and the secular movements in the terms of trade faced by the developing countries. The centre-periphery thesis, as well as the argument of a premature integration of the developing countries into the global capitalist system is seen as necessitating different policies by the developing countries in order to achieve "self-reliance" and "balanced development."
Much of the economic woes of the developing countries are attributed to the vagaries of the external environment which is controlled by the industrialized countries.

In the light of these structural rigidities, structuralists believe that the basic principles that should underline policies in developing countries are ones that essentially contradict the neoclassical view—class based distribution of income rather than marginal productivity based distribution, oligopolistic rather than the laissez-faire capitalist markets; increasing returns to scale or fixed proportions production functions rather than "well-behaved" production functions with decreasing returns and high rates of substitution; nonequivalent or "unequal exchange" in the world rather than competitive, comparative advantage based world system, active state intervention in economic activities through planning, controls, and direct investment, to nurture the process of development, etc (see Mkandawire 1989:3-4).

The rationale and modalities of this paradigm of development are well articulated in the literature (see Prebisch 1959, 1964; Hirschman 1958; and the Lewis (1954) Labor surplus model. Ranis (1991) provides a succinct summary of the theoretical framework of the inward-looking, import-substitution industrialization (ISI) strategy especially in relation to modernization process or path for small open dualistic and labor surplus economies such as most African economies. The process dictates that the economies pass through some evolutionary phases. For example, they begin from an early or primary import substitution (PIS) sub-phase, gradually evolve to the secondary import substitution (SIS), and at a much later stage, mature into an outward looking or externally oriented sub-phase which may be of the export promotion and export substituting variety. A key feature of the early phases is inward looking policy of protection, price controls, central planning, dependence on external financial assistance—all of which imply strong state control.

The model posits an evolutionary process of moving from the inward-looking orientation to an outward looking strategy which also entails the gradual dismantling of the instruments of controls and replaced by liberalism and domination by market forces. The attainment of a mature market economy is thus the goal, but the process involves a sequence of phases as dictated by the country's stage of development. According to Olofin (1995:12),

> A good example of the set of countries which have followed this normal evolutionary path from dualistic labor surplus economies to developed industrialized economies are the Asian tigers at one extreme of the complete success stories. Some emerging NICs such as Thailand, Philippines and Malaysia are examples of intermediate success stories. At the other extreme of the spectrum are the non-success cases of mostly Latin American and sub-Saharan African countries who instead of moving steadily from (PIS) sub-phase to (EO) sub-phase via (SIS) sub-phase, either stagnate around (PIS) or oscillate between (PIS) and (SIS) moving from control regimes, to partial and feeble liberalization and back to control and strict regulation.

The veracity of the above characterization of development in Asia is certainly debatable, and is one of the sources of raging controversy in the policy context of SAP. In Africa, the desirability of this evolutionary model, and the extent to which the successive phases have materialized or failed to do so in individual countries are germane issues for empirical examination.

While structuralism and import substitution industrialization (ISI) strategy provided the dominant intellectual impetus for the African development model in the 1960s through 1970s, the specific forms, contents, and patterns of the policies differed among the countries. Countries generally followed variants of this dominant model as dictated by the peculiarities of their respective political economies and institutional and managerial capabilities. For example, while some countries implemented the strategy under the aegis of socialism, others adopted variants of the so-called "mixed economy" models. In general however, some identifiable approaches characterized the development models in most countries: the capital formation centered approach; the economic nationalism centered approach; the socialist development approach; and the basic human needs approach (see Mongula 1994, for detailed review of these approaches). We summarize the key elements of each approach below.

Following the intellectual leadership of growth and development economists like Solow, Mirlees, Nurkse, Lewis and Rostow, growth in Africa after independence was perceived as a function of increased capital formation, and also increased investment in education and training. This "modernization" strategy relied heavily on flows of external financial assistance, private investment and aid, and also overlooked the participation of the rural/grassroot population. Related to this approach was the economic nationalism approach. By this, countries sought to take increasing control of their economies by the Africanization of the public service, legislation of partnerships between foreign investors and mostly the states, part or full nationalization of foreign owned companies, etc. The central concern was the need to overcome dependency and neocolonialism. Because of the absence of organized and significant private sector capital in most countries, the state got increasingly involved in major economic activities—increased interest towards domestic production of capital goods, development of domestic scientific and technological capacity, and domestic-resource based industrial production. State involvement was essentially seen as a strategy to serve the people rather than neocolonial commercial interests and also accelerate economic development. Aside from the deep structuralist bent of this orientation, several scholars provided strong justification for this policy in the Africa specific case. For example, Green (1972) argues that;

> Given the very late start, weak private domestic sector capacity, poor domestic savings mobilization capacity and extreme openness to trade, African economic development (whether capitalist or socialist) will require large public sectors—both government and public enterprise ... Without public-sector leadership African economic development will not take place.

An approach which some countries (for example, Mali (1960), Congo (1963), Tanzania (1967), Zambia (1967), Benin (1960s), Ethiopia (1974), etc) adopted (with varying degrees of success) to ensure state leadership in economic activities was the socialist model. This model was predicated on the perceived inherent inadequacies of "periphery capitalism," and thus regarded socialism as the best approach which could truly usher the neo-colonial countries out of the deep-rooted economic and the underlying political structures which nurture and reproduce the system of underdevelopment. The basic human needs approach, though ideologically neutral, derives from the moral justice of development and complements the tenets of the socialist model. Under it, development was seen as centered first and foremost on human beings rather than on infrastructure. This approach developed in the wake of the growing impetus of the dependency theory which criticized the world capitalist system and the periphery capitalist status of the developing countries. It led to increased government expenditures on health, education, urban food subsidies, water supply, and housing.

It could be argued that the central goal of these approaches as embodied in the structuralist/ISI paradigm was the "structuralist transformation" of these economies. With the strong support of the dominant development theory of their era, these approaches led to huge investment in socio-economic infrastructure and education, generally raised the living standards of citizens, and provided the foundation for what is today the industrial sector in most countries. The sustainability of this "transformation" strategy was soon called into question as external debt soared, income and productivity growth stagnated even below the population growth rate, the feeble external sector became overwhelmed by the declining export market shares and bouts of deteriorating terms of trade, and unsustainable twin deficits (fiscal and external) became almost permanent features of most economies.

What caused this sordid state of affairs is the source of unending debate: was the paradigm inherently wrong; was it wrongly implemented; or were external factors dominant in explaining the economic woes that soon befell these economies? There are no consensus answers to these questions. The debate is mostly couched in polar case explanations: the Bretton Woods institutions (BWI) insist that the structuralist/ISI development paradigm was wrong and hence wrong policies derived from it led to economic stagnation; while others (mostly African institutions and scholars) prefer to heap the blame on the shoulders of the "overwhelming external shocks." Befekadu (1994) provides another plausible explanation for the failure of the ISI strategy. According to him, the poor performance should be attributed to poor implementation of an otherwise potentially effective development strategy since these countries merely implemented a caricature of the true model. The result was that no African country successfully went beyond the first sub-phase, and instead of leading to vibrant self-reliant competitive economies, the ISI strategy (at least in implementation) ended up with producing highly dependent economies with manufacturing enclaves that easily rendered the economies hostage in relation to foreign exchange gap. During the period of the ISI strategy, the agricultural sector was pauperized and in the process deepened the poverty of the rural and urban informal sector (see Befekadu 1994, and Olofin 1995). For example, the overvaluation of the exchange rate and/ or the administrative lowering of prices offered for agricultural exports produced undesirable consequences for several countries. As Bates (1989:222) explains;

> Private agents then responded. They did so by smuggling, as in the case of Ugandan coffee, which was then sold as "Western Kenyan" coffee on the Nairobi auction floor, or as in the case of Ghanaian cocoa, which was marketed in the franc (CFAF) zone economies of Togo or Cote d' Ivoire. Alternatively, they responded by shifting out of the production of export crops, growing food crops instead. Or, more dramatically still, they shifted out of agriculture, quitting their farms for the cities, investing less in agriculture, or spending greater amounts of time in leisure.... And government attempts to institute price controls result in corruption and black markets, as essential commodities, such as food stuffs, are sold privately at market clearing prices rather than at officially posted prices.

The resulting socio-economic crisis raised the urgency of the search for alternative models of development.

Some attempts to rescue the ISI strategy

With the deepening of economic crisis especially in the late 1970s and early 1980s, several countries made frantic efforts to resist any change in paradigm but adopted various "rescue operation" measures whose central objective was to respond to the crisis and ensure "self-reliance." These "self-reliance formulations often seem to be smudged copies of models associated with the United Nations Economic Commission for Latin America (ECLA) of the 1950s. Such models are oddly blended with copying out the Treaty of Rome as a guide to economic cooperation based on managed markets and planned economic cooperation" (Green 1993:67). The Lagos Plan of Action (1980) and the Final Act of Lagos was an example of such models of "self-reliance" (strongly rooted in the structuralist/ISI paradigm) and was a start towards a strategy designed for a world context similar to the 1970s.

At the individual country levels, different measures were adopted under what can be referred to as National Economic Survival Programs (NESP). In Nigeria, it was specifically referred to as the Austerity Measures as embodied in the Economic (Emergency) Stabilization Act, 1982. These were essentially country programs which were prepared and sold to the Paris Club in order to solicit further aid or to avoid impending cuts, as well as "rescue" the economies from imminent collapse. A major objective was to tackle the balance of payments (especially trade) deficits, and the contents and details varied across countries depending on the extent to which the BWIs participated in their design and also on the individual country's perceptions. The instruments employed varied among countries: some countries sought to control imports more rigidly while simultaneously imposing export quotas, while others sought to restrict imports through currency devaluation and price adjustments. Other policy measures included: reduction in price controls, control over the increasing government budgetary deficits, and restriction on bank lending. In some countries, the elimination of food deficits (which was a disturbing and recurring problem) constituted another objective of policy (Mongula 1994:92).

These individual country programs did very little to stem the deepening crisis in Africa. In fact, in the late 1970s, sub-Saharan African (SSA) states requested a World Bank study on the reasons why their growth was so low. Late in 1981, the (in)famous Berg Report entitled *Accelerated Development* (AD) appeared, and had the impact of a bombshell. Appearing barely a year after the Lagos Plan of Action (LPA) and solidly rooted in hard neo-liberalism, this Report was in every respect an anti-thesis of the LPA. As summarized by Green (1993:56), the AD briefly argued that:

- SSA growth had been low throughout the 1970s (about 2%).

- the underlying cause lay in low agricultural (and especially agricultural export) growth (under 1.5% overall, somewhat below 2% for domestic food, 0% for exports).

- the basic reason for the poor performance was that SSA states had the prices wrong and restricted imports, as well as trying to do too much and interfering with markets and private enterprises.

- getting the prices right and securing a doubling of net resource inflows should yield a 3.5% GDP (and agricultural export) growth rate, especially because primary product terms of trade were set to improve (a projection the Bank reversed a few weeks before AD appeared).

For Africa, the Report represented an articulate challenge to the structuralist/

ISI orthodoxy. It was put bluntly as the authorized version of what SSA countries must do in order to get significant World Bank finance beyond a decreasing number of project loans. That most of the economic reform programs in Africa started immediately after the Berg Report indicates that the Report could be fairly regarded as providing the immediate intellectual basis for the design and implementation of the World Bank/ IMF supported structural adjustment programs (SAP).

Neoclassical ascendancy and SAP as an incomplete theory of development

The Berg Report and all the SAP policies implemented at the behest of the BWIs in developing countries have as their intellectual precursor the ascendancy of the "new classical economics" and the principle of "monoeconomics" as the dominant paradigm for diagnosing economic problems and prescribing solutions. Between 1973 and 1980, enormous intellectual transition regarding development theory and policy occurred. The neo-Keynesian consensus was breaking up quite rapidly as a consequence of the inflationary and other macroeconomic distortions in the industrial economies. Increasing interest in the classical view was re-enforced by the public's disenchantment with neo-Keynesian policies of the time and consequently the election of a more conservative group of leaders in major industrial countries: Margaret Thatcher in Britain (1979), Joe Clark in Canada (1980), Ronald Reagan in the United States (1980), and Helmut Kohl in Germany (1981). There was strong political attack on big government as being both incompetent and oppressive, and this was reinforced by the evident failure of centralized planning in the former USSR and other communist countries in the 1980s. There was increasing attention to the twin concepts of "efficiency" and "market forces" and the new classical paradigm provided the anchor.

As Mkandawire (1989:5-6) observes, "the neoclassical interpretation is based on the theoretical and empirical corpus of work that essentially derives from a set of theories on the efficacy of the market system in resource allocation." Its major principle is that of "monoeconomics" by which it insists on the universality of rational economic behavior and the existence of marginal substitution possibilities in production and consumption. The classicals vigorously criticized the range of established theoretical constructs in development economics, such as dual-gap analysis, the Lewis theory of growth and the use of input-output in planning. The critiques sought to show how such theories violated normal economic principles of response to price incentives, while offering empirical proof of rational economic behavior by the poorest peasants in developing countries (Toye 1994:22). Thus, according to the classical view, the so-called "structural bottlenecks" that could make basic economic principles inapplicable in developing countries were not empirically founded. Factor substitution was shown to be high, and commodity markets performed well wherever they were allowed to operate.

Furthermore, the classical view also challenged the argument that developing countries in Africa could not develop because of the adverse terms-of-trade shocks, limited access to foreign credit, and declining demand for African exports. Both the faster growth of low-income countries in other regions (especially South-East Asia) that have faced similar external conditions as African economies and better performance in Africa in earlier periods (with similar external shocks) point to the potential dominant impacts of domestic policies. The economic success of the four "Asian tigers" which pursued "market and outward-oriented" policies was presented as the evidence par excellence that domestic policy distortions are to blame for poor performance.

Thus, neoclassical economics sought to re-establish a presumption of a "standard economics" for policy analysis in all economies—developing as well as industrial economies. The localization of these arguments in the African case was made in the Berg Report, and more so, the World Bank and the IMF have become the major converts and protagonists of this diagnosis. The consequence is the design and implementation of SAP whose focus is the domestic policies and the central strategy is to "role back the state and get prices right." While stabilization remains the major policy thrust, some of the original concerns of the structuralists seem to be addressed in some, albeit "distorted" ways. Stabilization had failed in some Latin American countries (1956-62) because it neglected the structural issues, and structuralist policies have failed in most of Latin America and Africa, arguably because, they focused on long-run structural transformation while ignoring the imperatives of short-run macroeconomic stabilization. The "new consensus" policy package still incorporates short-run stabilization through the traditional instruments of money supply control, fiscal deficit reduction, devaluation, and removal of price controls, but it also recognizes the need to complement these with medium to long-term institutional changes. As Toye (1994:23) argues, "the resurgent monoeconomics of the 1980s is, therefore, not simply the old monetarism, but the old monetarism trying to incorporate the insights of the old structuralists into a new policy consensus." But the manner in which such insights are incorporated is such that even the structuralists have difficulty recognizing several of such policies as deriving from their development theories.

In content, the SAPs can be distinguished from the IMF programs. The Fund is primarily in the stabilization business: reducing external, fiscal, banking, exchange rate, and price imbalances, and doing so by cutting absorption of domestic production and imports into private consumption, public services, and investment. The main instruments used in practice include drops in real wages, cuts in real public spending (except on external debt service), lower real credit, sharp devaluation, and high real interest rates. By its own account, the Fund is not in the development or growth business, although it engages in limited lending. The Fund does not derive its power from its limited net lendings (which have been negative to SSA from 1987), but from the fact that a highly conditional agreement with the Fund is a precondition for a Bank SAPs, a Paris Club (official creditors) debt rescheduling and for enhanced bilateral assistance (see Green 1994:57-8).

The SAPs on the other hand seek to restructure production capacities in order to increase efficiency and help to restore growth over the medium to long-term. It is believed that SAPs should complement stabilization measures on the logic that once equilibrium is established by the Fund program, the best way to prevent imbalances from recurring is to remove structural impediments and microeconomic distortions that may impede the increase in productive capacities. The SAPs therefore take the Fund's program as a necessary pre-condition for successful adjustment. The major policy instruments can be categorized into four groups as follows (see Stewart et al. 1992:6-7):

- mobilization of domestic resources through fiscal, monetary and credit policies, and improved financial performance of public enterprises;

- improving the efficiency of resource use throughout the economy: in the public sector, measures include reform and privatization; in the private sector, price decontrol, reduced subsidies, competition from imports and credit reform, and encouragement to direct foreign investment;

- trade policies: liberalization, with import quotas removed and reduced lower tariffs; improved export incentives and some institutional reforms to support exports;

- institutional reforms: strengthening the capacity of the public sector generally and increasing the efficiency of public enterprises; improved institutions to support the productive sectors.

In program design, though the principles are the same for most countries, the specifics are shaped to the local context more, and is less monolithic—and sometimes, less internally consistent—in approach than the Fund. Over the years, the Bank has undergone some important paradigmic shifts, and in the process attempted to reform the SAPs (at least in rhetorics). For example, the first generation of SAPs were essentially the orthodox stabilization measures with scant attention to the structural issues. Over time, such issues as "sustainable growth and development," "social dimensions of adjustment," "adjustment with a human face," "poverty alleviation measures," etc have become important, albeit ad-hoc, tack-ons to the adjustment literature.

Ever since the introduction of the SAP policies in the early 1980s, controversies about their relevance to the African conditions and effectiveness in achieving stated and desired growth and development objectives have dogged the process. It is generally agreed, however, that some form of "adjustment" or "structural transformation" of the African economies was needed in response to the plethora of economic crisis that befell them (whatever the dominant causes). Thus, one major area of apparent consensus among most analysts is that what countries do or do not do in matters of policy responses to internal and external shocks will have great implications for their growth and performance. What is perhaps not too obvious, and in fact the major source of controversies, is the claim that one particular type of policies is more consistent with growth and development than others (which is what the BWIs are insisting). In broad terms, the debate relates to questions about the diagnosis of the African crisis, the policy objectives of SAPs, the choice of instruments and relevance to African conditions, and the empirical results of SAP policies.

The World Bank has published numerous performance evaluation reports and the verdict is the same: countries that have tenaciously and consistently implemented SAPs are experiencing growth and improved economic performance. On the other hand, performance assessment by individuals and other agencies have produced results that are as varied as the number of assessments. The results range from those which are cautionary with mooted success stories at best, to ones that are downrightly critical—pointing to the deteriorating economic crisis and thus to the irrelevance or inappropriateness of SAPs. Among Africans and African institutions, there is hardly any consensus about the desirability and performance of SAPs. This is reflected in the disparate views of several African economists about the adjustment process. For example, a view widely credited to the former Executive Secretary of the ECA (Prof. Adedeji) and shared by many analysts, is that the 1980s was Africa's lost decade and that SAP did more harm to the continent than all the decades of colonialism put together. On the other hand, in a recent study, the Central Bank of Nigeria (1993:121) notes with satisfaction that "On balance, the gains of the SAP have been so overwhelming that it is difficult to imagine what the Nigerian economy would have looked like without it." Most other analysts fall on either side of these extreme views.

Among the protagonists of the SAP (including the World Bank), there is a general sense of disappointment that SAP has not worked as expected in Africa. According to the World Bank (1994: 1- 2), after over a decade of SAPs in Africa, "reforms remain incomplete. No African country has achieved a sound macroeconomic policy stance, and there is considerable concern that the reforms undertaken to date are fragile and that they are merely returning Africa to the slow growth path of the 1960s and 1970s." The disappointing performance raises the debate about whether the deficiencies are caused by the design and content of adjustment policies themselves, or by non-implementation or distorted implementation of major aspects of the package. Granted that some important aspects of SAP may not have been properly implemented, several important gaps and weaknesses of the program itself are increasingly being recognized. Some of these are summarized in the Report by the Center for Development Research (CDR), Copenhagen (1995:66-67) and they include:

(i) **Limited effects on growth-increased vulnerability:** SAP has resulted in only fragile but inadequate recovery in some countries and deterioration in others, while it has not altered the structure of the economy in terms of the balance between consumption, savings and investment, nor even the balance between agriculture and industry; despite its central position in SAPs, export values have hardly shifted especially in the light of declining terms of trade; Also, despite a widening trade gap, plugged only by aid, real imports per capita have been halved during the SAP period; Moreso, SAP has not addressed the fundamental questions: what contribution can Africa make to the future world economy and the international division of labor?, and how in the process can Africa escape from its current ensnarement in ever increasing vulnerability?

(ii) **Pushes in the wrong direction:** It is believed that SAP has pushed Africa in at least three wrong directions: Firstly, government capacity has decreased as financial cuts have not been counter-balanced by timely capacity building; secondly, aid dependency has increased for both governments and NGOs, for development and recurrent expenditure, and for imports; thirdly, African societies have become accustomed to operating under an external
 "policy command" which discourages national dialogues on societal reforms.

(iii) **Errors of design-sequencing:** At least five sequencing problems can be identified in the design and implementation of SAPs, namely: Increases in producer prices before removal of major bottlenecks in infrastructure; liberalization of financial markets and increases in real rates of interest before necessary changes in the real economy; liberalization of foreign trade without effective export promotion policies and incentives; deregulation of national markets without measures to promote an effective and competitive private sector; and introduction of cost recovery in public services before measures to ensure availability of inputs and effective services in a transitional period. These sequencing problems reflect an exaggerated faith in immediate and strong supply response from the private sector. Another related design error is the assumption that institutions can be changed overnight. Predictably, it has taken long time for both public and private institutions (and society in general) to adapt to change.

(iv) **Negative effects**: SAP has had several negative effects. For example, access to markets and to government and parastatal services has become less regular and more costly for the poorest people particularly in remote areas, and equity has almost certainly suffered as those without assets, including many women headed households, are unable to take advantage of new opportunities.

In addition to the above criticisms, a fundamental shortcoming of the Bank/Fund program and on which most other development institutions and scholars agree, is its neglect of, and inconsistency with, requirements for sustainable long-term development in Africa. Most other United Nations agencies, notably the UNICEF, UNDP, UNCTAD, and the General Assembly have expressed strong reservations about several aspects of the SAP. The core criticism is the program's neglect of "development issues." Of note is the outcome of the extra-ordinary session of the UN General Assembly, 1986, summoned to discuss Africa's acute economic crisis. The resultant strategies embodied in the United Nations Program of Action for African Economic Recovery and Development (UNPAAERD) and the follow-up program under UN-NADAF go far beyond the SAP policies to commit the international community to certain responses in order to ensure "sustainable development." Furthermore, several scholars and non-governmental organizations, notably, OXFAM, have been largely critical of the SAP framework and proffer alternative perspectives that at every level contradict the SAP policies. Even the World Bank has shifted significantly from an earlier position that SAP also promotes long-term development. In its 1994 Report (World Bank 1994:2), it admits that "Adjustment alone will not put countries on a sustained, poverty-reducing growth path. That is the challenge of long-term development, which requires better economic policies *and* more investment in human capital, infrastructure, and institution-building, along with better governance." In defence of SAP as the necessary pre-condition for development, the Report argues that "development cannot proceed when inflation is high, the exchange rate overvalued, farmers overtaxed, vital imports in short supply, prices and production heavily regulated, key public services in disrepair, and basic financial services unavailable. In such cases, fundamental restructuring of the economy is needed to make development possible." The Bank unwittingly suggests that a sequencing process which runs from SAP to development should be pursued. In essence, it implies that development concerns should be postponed until SAP succeeds. This is a fundamental source of a seemingly acrimonious controversy, and resurrects the old debate (especially between the UNECA and the Bank) as to whether Africa's problem is essentially that of development or that of "structural adjustment."

By far the most critical and articulate challenge to the SAPs in Africa was provided by the UNECA's African Alternative Framework to Structural Adjustment Program (AAF-SAP) (1989).[2] Set as an "alternative" to SAP, AAF-SAP was based on different philosophy of development and mode of analysis, as well as an entirely different perception of Africa's priorities, problems and requirements. It sets the premise that Africa's socio-economic problems are first and foremost rooted in its structural weaknesses, and therefore any meaningful analysis should start with a structural analysis of its political economy and causes of its underdevelopment. According to AAF-SAP, these features include: the predominance of subsistence and commercial activities; a narrow production base with weak inter-sectoral linkages and ill-adapted technology; neglect of the informal sector; environmental degradation; fragmentation of the African economy; openness and external dependence; and lack of institutional capability. Based on this premise, it adopts a philosophy of development which reaffirms the primacy of self-reliant and self-sustained growth as the appropriate long-term development strategy for SSA countries. Essentially, AAF-SAP was anchored on the Lagos Plan of Action which, inter alia, called for the alleviation of mass poverty, improvement in living standards, the attainment of self-sustained development and the pursuance of national and regional collective self-reliance.

With respect to the choice of policy instruments to realize the program, AAF-SAP insists on active state intervention, and prescribes policies which at every level, contradicts the SAP policies. For example, it insists on directed credit (and opposes financial liberalization); favors selective trade policies involving import controls and import management as well as export promotion; supports multiple exchange rate to be administered by government; increased domestic resource mobilization; supports capital controls; stresses the need for improving the quality of governance as to accountability, competence and regular, competitive choice of leaders; supports basic services/ human investment as a key part of transforming African economies" dynamism; seeks reduction in defense budgets to the benefits of basic services, and increase popular participation both to increase productive efficiency and to redress the bias in allocations against absolutely poor households and women; etc. Finally, it expects the international community to assist Africa with the requisite funds to finance the AAF-SAP through: debt write-offs, better terms of trade and market access, and increased financial assistance.

AAF-SAP has received mixed appraisal and acceptance by both Africans and non-Africans. While no African government is known to be following the prescriptions of AAF-SAP, none has also publicly denounced it. Among independent analysts, opinions range from those who dismiss it as full of polemics and buzz words without much rigor in analysis and consistency in recommendations, to those who see it as providing the quintessential reference point, against which to judge Africa's development performance. With the publication of AAF-SAP, and moreso the ECA's scathing criticisms of the World Bank/ UNDP impact assessment report on SAP in Africa in 1989, there ensured a war of visions between the Bank and the ECA about Africa's economic crisis. Both institutions had diametrically opposed perceptions of the problem, and the differences were seemingly irreconcilable. The Bank and its sympathizers largely dismissed the ECA and AAF-SAP as engaged in ideological but unnecessarily distracting battle. Abbott (1994: 23) could have rightly expressed the minds of many opponents of AAF-SAP when he argues that;

> AAF-SAP is not a system at all. It is simply a collection of normative statements and unproven propositions, each of which stands on its own, and bears no relationship to what has gone before, and what follows. It is not possible therefore to determine whether it will, in fact, work within the totality of the economy, or what will be the cumulative effect of any given combination of measures on the economy of the adjusting country. Without this sort of information and analysis, AAF-SAP remains a checklist of generalized comments. There is something in it for everyone, but this is unrealistic. Economics is about choice, determining priorities, and taking rational decisions. The AAF-SAP model does not provide any guidance on these points. Finally, if this model is to run, it must have a price tag, and external funding agencies have got to be convinced that, compared with conventional structural adjustment programmes, it represents better value for money. This has not been done.... It is, in effect, a distraction from the central concerns of the debate which is no longer whether SSA countries ought to adjust, but about operational details such as design, implementation, monitoring procedures and the pace and sequencing of reforms.

On other hand, some scholars have tried to rationalize a need for such an "alternative" benchmark. Adherents insist that Abbott's remarks are misconceived, and that AAF-SAP is principally a "framework" and neither a theory nor a model. It is argued that it is a broad framework, providing broad principles around which a coherent theory or model could be fashioned. For Green (1993: 68) AAF-SAP "is a step toward moulding official and academic African analysis toward an African based strategic framework and, perhaps, toward setting the foundations for more serious strategic and conceptual dialogue between the Bank and SSA countries." For some others, there is really no fundamental difference between the goals of the two programs, though (as expected and as is always the case with instrument choice in economics) there are major differences in strategies and instruments. Several analysts argue that the AAF-SAP lacks a consistent theoretical framework and that its instruments are a regurgitation of an old and failed system, but so also is the Bank's adjustment model accused of theoretical weaknesses. In particular, Mosley (1994) has demonstrated that "the recommendations attached to structural adjustment programs... lack any serious theoretical basis, and should be seen essentially as an improvisation." Given that both programs have their weaknesses, the overriding reason why the BWIs have prevailed is principally because of their financial muscle and the leverage they exercise on Africa. More fundamentally, some analysts observe that despite its weaknesses, AAF-SAP presents a better understanding of the nature of the African economic crisis than the orthodox SAP. They point to the recent admission by the World Bank that SAP is inadequate for long-run development as a vindication of AAF-SAP's stand. Moreso, the growing emphasis on "poverty eradication" as the linchpin of SAP's development agenda may be another indication that AAF-SAP may, afterall, not be wrong on the war of visions, even though the BWIs still dictate the instruments/ strategies.

Over the years since the introduction of SAPs, African states have undertaken other initiatives which point to their deep yearnings for a development paradigm and strategies that they can relate to. Several African states participate actively in the largely ineffective South-South Commission, while the strive towards Africa's "collective self-reliance" through the Abuja Treaty (establishing the African Economic Community) re-emphasizes that the so-called "old paradigm" has not completely given way. Indeed, the contractions between the BWIs and other UN-agencies in terms of the diagnosis of Africa's problems, performance evaluation, and development strategies as articulated in the UNPAAERD and UN-NADAF, and moreso, in the various reports by the ILO, UNICEF, UNCTAD, etc. underscore the fact that Africa's development debate is an unfinished business. In the mid-1990s, a mood of some consensus seems to be dominating deliberations on macroeconomic and structural reforms especially among multilateral and bilateral donors and core ministries of African governments. According to the Center for Development Research (1995:65-66), this consensus:

> takes the form of statements such as: It is time to move beyond structural adjustment towards comprehensive economic, social and political reform and development, which is owned by both government and civil society in Africa. While few would disagree with this, it is important not to forget that such statements are also admissions of a degree of failure of the whole process, economically and politically.

As the Report indicates, such a "consensus" is a tacit admission of the failure of adjustment. Moreover, the analytical framework and details of such "post-adjustment" phase are not yet being discussed in any rigorous manner as they should. But the fundamental question is whether the score card for SAP in Africa really shows a failing grade. Why is it that despite several evaluation reports by the Bank which proclaim the success of SAPs in Africa, the vast majority of other development agencies and independent analysts remain largely sceptical? Are there major differences in the methodologies for evaluation or is the debate largely ideological?

Methodologies and macroeconomic data for evaluating SAP

> There is at present little agreement in the profession either about how to estimate the macroeconomic effects of programs, or about what impact past programs of the Fund have actually had on macroeconomic variables. Despite the fact that there have been a number of studies on the subject over the past decade, one cannot say with certainty whether programs "work" or not. The question is apparently still open. (Khan 1988:1)

In the analysis of broad macroeconomic issues about SAP, the crisis of an appropriate methodology for evaluating performance generally, and in Africa in particular, is very central. It is one of the most controversial and largely unsettled aspects of the debate. Part of this crisis reflects differences in paradigms about the nature of SAP and what it should or should not do. Ostensibly, the debate about the success or failure of SAP and the choice of evaluation techniques hinge on the perceptions about the SAP philosophy, objectives, policy targets, instruments and objectives. On the other hand, the phrase "adjustment" embraces too wide a range of policies and experiences to have meaning as an independent variable. It is therefore analytically and empirically difficult to link all the policy instruments to the ultimate targets (objectives) of policy. Analysts generally focus attention on whether the SAP has been "effective" in the sense of achieving its broad objectives. Even on this, there is little agreement about the methods of evaluation, especially in respect of the benchmark period or index for comparisons, the time horizon over which the program could be said to have worked or not, the indices (variables) for measuring performance, and the separation of the impacts of SAP per se from the impacts of other factors. There is also the problem of determining whether the observed performance should be attributed to the implementation process or to the program instruments themselves.

The above considerations make any evaluation result fraught with interminable controversies about its robustness. Every analyst literally sees what one chooses to see, and chooses a convenient method of seeing it. The result is that most analysts see widely different things from what the World Bank usually sees in its reports. Hussain (1994a:152-157) has summarized a number of considerations which must be kept in mind in evaluating adjustment in Africa, and which could be sources of conflicting evidence. These include:

- First, a clear understanding of the initial conditions of the countries undergoing adjustment is absolutely critical. The protracted economic and financial crisis that preceded adjustment has made it particularly difficult to achieve a quick turn-around and improvement.

- Second, reform measures must be correctly identified for what they are and are not. If, in the first phase of reform, the emphasis was on restructuring the budget and shifting the balance in the current account through expenditure reductions, it would be incorrect to call this a structural reform. These are stabilization measures and their impact is very different from that of structural reforms. Several countries have called their stabilization efforts structural adjustment programs. The wrong labeling of adjustment, inaccurate grouping of adjusting countries, and incorrect choice of periods and duration of adjustment have been the single source of confusion in this debate.

- Third, it is important to know whether the reforms were implemented both in the spirit and in the letter. A number of countries have started adjustment programs and even borrowed from the international financial institutions, but under political pressure have made reversals and slippages during the course of their implementation. This stop-go approach has to be distinguished from a sustained implementation of planned reforms, since reversals of policies benefit some segments of the population at the expense of others.

- Fourth, exogenous shocks which were not anticipated or built into adjustment programs have a significant impact on their success. If world prices of key commodity exports decline appreciably, or rainfall is lower than expected, it would be unfair to attribute poor economic and social performance to the adjustment program alone.

- In a number of adjusting countries the external funding promised by donors or the debt relief projected under the program have not materialized. This has had a negative effect on imports and consequently on the growth of output, exports and the level of public spending. This factor should be fully taken into account when judging the success or otherwise of the adjustment program or its social consequences.

- There is always a problem of how much observed performance can be ascribed to particular policies, where so many things are all happening at once; where exogenous forces are strong, where time lags are important, and data sources are not completely reliable. This is even more so in the case of the food and agriculture sector where weather conditions are a dominant factor.

- The original premise that these programs would enable economies to revive growth and consumption in five to seven years was over-optimistic. The intensity of the distortions and of the demands made on the response capacity of African economies was underestimated.

- Finally, adjustment response has been constrained by other external factors such as losses from declining terms of trade, deepening debt, recession in the OECD countries and, more recently, by the tumultuous process of political transition which has diverted the attention of policymakers from economic management, leading in some instances to a virtual breakdown of normal economic activity.

Coming from a very senior World Bank economist (African region at the time), the above concerns can be interpreted as articulate apologies for the failures of SAP. They may also be interpreted as indicating the difficulties (or even impossibility) of correctly assessing whether or not SAP has worked, as Khan (1988), a senior Fund staffer, seems to be suggesting in the earlier quotation. However, in several evaluation reports on adjustment by the Bank staff (see Hussain 1993; Elbadawi et al. 1992; World Bank 1988a, 1988b, 1990, 1992a, 1992b, 1994; World Bank/UNDP 1989), the evidence varies from report to report, but the Bank as an institution maintains that "adjustment is working." Even though the reports have different emphasis in terms of focus, a common methodological thread spans through them. Other analysts obtain largely contradictory results and the evaluation exercises differ markedly on account of the techniques employed. Beside the more fundamental points about problem articulation, choice of variables and time frame, country aggregation, etc, both the BWI staff and independent researchers have generally applied all kinds of performance evaluation techniques, but five methods seem to stand out: the before-and-after approach; the control-group approach and the modified control-group approach; the actual—versus-target approach; the comparison-of-simulations approach; and the application of regression analysis. These methods are complementary in several ways and most scholars apply combinations of them in their studies.

The most widely used evaluation technique is the before-and-after method. Under this approach, a comparison is made between the performance of selected economic indicators during or after adjustment and a similar set of indicators prior to adjustment. This technique is used in both cross-country studies and specific country case studies. While this method is easy to implement, it is based on a strict *ceteris paribus* assumption that the underlying economic structure and the various shocks faced by the country have stayed the same in the two periods. But, "the situation prevailing before the program is not likely to be a good predictor of what would have happened in the absence of a program, given that non-program determinants can and do change from period to period" (World Bank 1990a). Evidently, the non-SAP determinants of economic performance, ranging from external factors like industrial country growth rates, terms of trade variations, and movements in international interest rates to domestic factors such as shifts in weather conditions, do change markedly from year to year (Khan 1988). This implies that the estimates derived from this approach will be: (a) biased, because it incorrectly attributes all of the changes in outcomes between the pre-SAP and post-SAP to SAP policies only; and (b) unsystematic over time, because in some cases, the non-SAP related factors might dominate SAP policies in determining economic performance for some years. For example, in Nigeria, it is likely that developments in the oil market will continue, in the foreseeable future, to be the major determinant of economic activity. Thus, if positive oil windfall coincides with SAP era, performance is bound to be salutary and vice versa.

On account of these shortcomings, the before-and-after method is a poor estimator of the "counterfactual," defined as the economic performance that would have occurred in the absence of SAP. This is a major weakness because the counterfactual is perhaps the most appealing yardstick against which to assess SAP performance and the standard most widely employed in economics to define and measure the impact of government intervention. What would have happened in the absence of SAP is by no means the only standard against which to judge the performance of SAP policies, but in many instances, it is the most realistic one. However, the crux of the problem is that the counterfactual is not directly observable and has to be estimated. The before- after method is an inadequate estimator of this counterfactual because the situation prevailing before the program is not likely to be a good predictor of what would have happened in the absence of the program, given that non-SAP determinants can and do change from time to time.

The second approach that is used mainly for cross-country comparisons is the "control-group" approach, and more recently, its variant, the "modified control-group" technique. The control-group method compares the mean changes in selected macroeconomic variables in countries with SAP policies and performance in a "control group" of non-SAP countries. This is the so-called "adjusters- versus- non-adjusters" approach. Implicitly, this method uses the "control-group" to estimate what would have happened in the countries implementing SAP in the absence of such policies. Comparisons between the "adjusters" and non-adjusters is predicated on the assumption that the two groups are similar in both the initial conditions, specific country characteristics that affect performance and the nature of external shocks affecting them. There are other problems with this technique since the defining characteristics of the two groups is "acceptance or non-acceptance of adjustment lending" from the World Bank. Grouping countries with adjustment loans together has other problems: first, the emphasis and focus of specific country programs might differ; and second, there could be tremendous diversity in the implementation of the adjustment packages. On the other hand, the method assumes that all those "without adjustment lending from the Bank" did not carry out any "adjustment" in their exchange rate policies, public sector deficits, interest rates, agricultural pricing, etc. In reality however, such assumptions are not founded. According to Mosley (1994:71),

> Given these facts, it is not clear that the policies followed by the "adjustment lending" group had enough in common to warrant their being grouped together for analytical purposes or even that the intensity of the "structural adjustment effort" which they carried out was greater than that carried out by the "non-adjustment lending" group. If we are to understand the apparently weak leverage of the structural adjustment policies on African economies, a new approach is clearly needed.

A variant of the second approach which seeks to circumvent the weaknesses of the first two approaches is the "modified control-group" approach. The strengths and weaknesses of this technique are amplified in Elbadawi et al. (1992). This takes account of the changing external environment facing each country and identifies the differences between the two groups of "adjusting" and "non-adjusting" countries. These differences are then controlled for in attempting to estimate differences in performance. This methodology thus allows for the estimation of the marginal contribution of adjustment programs in adjusting countries while: (i) explicitly taking account of the potential endogeneity of the decision to participate in SAP, since the same non-program factors that influence performance in the pre-program period are likely to influence the participation decision; (ii) controlling for other factors unrelated to programs that also affect performance; and (iii) adjusting for the counterfactual policy stance that would have prevailed in the absence of SAP (see Elbadawi et al. 1992:27). This method was used by Elbadawi (1992) and Elbadawi, et al (1992) to study the marginal contributions of SAP in relation to three groups of countries: the early intensive adjustment lending (EIAL) countries, other adjustment lending (OAL) countries, and the non-adjustment lending (NAL) countries. The results of the studies are highly revealing and raise fundamental issues about the efficacy of SAP. For example, Elbadawi (1992:21) concludes that,

> despite the similarity in terms of external shocks experienced by EIAL and NAL of SSA, the economic performance in the last group has been uniformly superior to the first. Also, despite witnessing twice as many negative external shocks compared to OAL countries, the NAL group has performed better especially in terms of domestic inflation and exports.

This method therefore exposes the "sample selection bias" and the "aggregation bias" inherent in the major evaluation techniques used by the BWIs in assessing the impacts of SAP. It points to the need to explicitly take account of non-SAP factors in the impact assessment, and questions the validity of attributing improvements or failures to SAP policies only.

The next approach often used is the actual-versus-target method. This compares the actual performance of the economy with the targets specified for these variables in the SAP package. For example, the SAP package in Nigeria envisaged a return to about five percent growth rate of the real GDP, prescribed three percent of GDP target for budget deficit, etc. Success is therefore measured by the extent to which these targets are achieved. This approach is not as widely used as the other methods, and it sheds very little light on how the country's economic performance is affected by the SAP policies. For example, if the targets are too ambitious, or if unexpected negative non-SAP factors intervene, actual outcomes could fall below targets even though the program may have produced a much better outcome than would occur without it, or under alternative set of policies. Similarly, under-ambitious targets, or positive shocks, would lead to meeting or even exceeding of targets, even if the policies produced weaker effects than would occur in its absence.

Another technique of performance evaluation is the "comparison-of-simulations approach." Unlike the other approaches, this one does not infer SAP effects from actual outcomes in the countries concerned. Its principal use is in evaluating the design and effectiveness of policies in general, as well as in examining the likely effects of alternative policy packages. This technique employs structural macroeconomic models to evaluate the counterfactual policy scenarios. While some attempts have been made to construct macroeconomic models of some developing countries for such purposes, it is very difficult to build a model that covers the whole range of policy measures contained in a typical SAP package. Moreover, since models are often very aggregative and rely on specific assumptions, the results of simulations might differ under different models with different sets of assumptions. Also, it is known that the essence of policy reforms is to change the underlying structural parameters of the economy, which the models assume to be constant. In effect, the results of policy simulations might be misleading since models often assume the parameter estimates to remain constant across policy regimes.

The last but popularly used method of impact assessment is the regression analysis. This is mostly used in estimating the impacts of different policy changes under SAP on economic growth. More broadly, the regression analysis is often used to gauge the differential impacts of domestic policies and external shocks in explaining performance. In the literature, some of the variables that often enter as arguments in most regression analysis include: index of real exchange rate misalignment, fiscal deficits, inflation, population, measures of human capital formation, terms of trade, net capital flows, and the growth of world demand. While this technique has several appealing features, its major shortcomings relate to the ad-hoc procedure for the selection of regressors, the potential multicollinearity problems among the regressors (especially the policy variables), and often exclusion of non-SAP factors which could, in fact, have greater explanatory powers than the SAP policy variables.

In various impact assessment reports of economic performance under SAP in Africa, the World Bank in particular has used a combination of these approaches and battled with the overarching burden of "proving that adjustment is working." Appendix table 2.1 summarizes the results of 1988b, 1990b, and 1992b reports of the World Bank. The reports differ in terms of emphasis (with greater focus on poverty in the 1990 report and emphasis on private investment in the 1992 report). A common methodological thread in the three reports was the comparison of adjustment lending with the non-adjustment lending countries. Aside from investment ratio among the "selected" variables, it is evident from table 2.1 that the overall performance of adjustment-lending countries was better than that of non-lending countries. However, for the low-income and sub-Saharan African countries, the growth and export performance of non-lending countries was better than that of the countries that received adjustment loans. This point was corroborated by Elbadawi et al. (1992:5) who observed that for sub-Saharan Africa, the "World Bank adjustment lending has not significantly affected economic growth and has contributed to a statistically significant drop in the investment ratio."

Beside the worrisome implication of the results obtained even by the World Bank staff, the methodology of comparing "adjusting" with "non-adjusting" countries is questionable. Mosley (1994:71) notes that "in the first instance, there were enormous variations in the packages prescribed: some centered on trade liberalization, some on agricultural price reform, some on public enterprise reform, some on all of these. In the second place, there were for political reasons tremendous diversity in the implementation of the conditions prescribed among those countries which did accept adjustment loans."

Prior to the 1994 Report, the World Bank/UNDP report of 1989 was the one that received the most scathing criticisms from scholars especially the UNECA. The report makes an attempt to distinguish countries not just in terms of whether or not they accepted adjustment loans, but rather in terms of the intensity of their adjustment effort or none at all (see tables 2.2 and 2.3). Just as in other reports, it also concludes that the "strong adjusters" perform better than weak and non-adjusters. The criticisms leveled against this report can in some sense, be generalized for many of the other reports. A careful study of the tables 2.1 to 2.3 shows the different reference periods for analysis, with the beginning period differing from report to report. Comparison of results across various reports is therefore difficult. Furthermore, there is the problem of "sample selection bias" in the sense that the countries which are often grouped as "non-adjusting" countries are mostly ones ravaged by wars or ones that do not have any discernible economic programs. Helleiner (1992) argues that of the 14 "strong" adjusters, "fully half of them were in countries that had already been realizing quite respectable rates of growth over the previous two decades and had been characterized... as "fast growers" over the 1960-82 period." The result one gets therefore depends on the "selection" of countries, time period, and variables.

The UNECA ferociously criticized the 1989 report on two broad fronts. A set of the criticisms come under the accusation of "one dimensionality and selectivity of data." By this, the Bank is accused of (a) taking a pre-determined position and selecting the appropriate assumptions, data and level of analysis to support it; (b) choosing a limited and convenient grouping of countries to show that the adjusters did better than the non-adjusters; and (c) introducing a new category of important and not-so-important country groups without specifying the criteria for evaluating their economic performance, or including a data set and sufficient documentation to authenticate its findings. Second, the Bank is also accused of "a simplified approach to Africa's crisis" to the extent of (i) minimizing the nature of the crisis as well as Africa's vulnerability to external factors; (ii) addressing solely economic issues without considering the social impact of structural adjustment programs; and (iii) not substantiating its claim that adjustment raises the living standards of the poor. Two of the statistical tables produced by the UNECA to debunk the Bank's findings address two important issues: first, it demonstrates how the Bank is guilty of "fiddling the figures" by changing the base years and periods to prove a predetermined position; and second, it shows that alternative grouping of countries and base years, and re-calculations can give results opposite to the ones obtained by the Bank (see tables 2.4 and 2.5). Even though the UNECA does not stress that it is better not to adjust (as table 5 clearly shows), it simply drives home the point that "any attempt to establish a one-to-one relationship between growth trends and the adoption or non-adoption of SAPs is prone to over-simplification and fallacies" (see UNECA 1989b). This observation still haunts the entire evaluation process and continues to raise questions about causality relationships between SAP measures and economic performance.

From the foregoing, it is evident that issues of choice and definition of variables, base year period, and methods for quantification of results continue to bug the empirical evaluation process. Part of the controversy about the choice of variables relates to the debate about what should be the focus of SAP in the first instance. Some argue that SAP should be evaluated on the basis of its stated objectives for individual countries. The problem however is that in several cases, the objectives are stated in a manner that is not easily quantifiable. In Nigeria for instance, the objectives of SAP include:

(i) to restructure and diversify the productive base of the economy in order to lessen the dependence on the oil sector and on imports;
(ii) achieve fiscal and balance of payments viability over time;
(iii) to lay the basis for sustainable, non-inflationary or minimum inflationary growth; and
(iv) lessen the dominance of unproductive investments in the public sector, improve the sector's efficiency and intensify the growth potential of the private sector.

But how do we know if *significant* "diversification of the productive base of the economy" has occurred, or if we have "lessened the dominance of unproductive investments in the public sector," or whether "basis for sustainable non-inflationary growth" has been laid. The core issue here is a problem of measurement and thus the validity of statistical inference in assessing SAP based on its defining objectives in each country. Besides, comparison across countries might be difficult if the specific country objectives differ.

Contrary to the above observation, the World Bank believes that economic growth is the overriding objective and therefore the most appropriate summary indicator of performance (irrespective of variations in country-specific objectives). According to the Bank (1994:132) "the most telling indicator of the success of adjustment efforts is the change in GDP per capita growth." Little doubt therefore that the major focus of the BWIs impact assessments has been on GDP per capita, the ratios of exports, investments, savings, and imports to GDP. Success is said to have occurred if changes in these variables have been positive or exceeded levels prior to adjustment. In the 1994 report, the Bank also argues strongly that countries which have improved their policies along the lines of SAP have experienced strong turnaround. But to what extent are the criticisms of the UNECA against the 1989 report vitiated by the results of the 1994 study? How robust are the methodologies and results, and to what extent can we attribute performance to SAP? A few examples from the 1994 report could illuminate these questions.

The major technique employed is to determine countries' "adjustment effort" and to match it with "adjustment outcomes". The emphasis is therefore on the effectiveness of implementation, and comparison is made between "strong adjusters" and "weak adjusters." A set of indicators for the overall macroeconomic stance of countries is developed, and countries are classified depending on whether their policies are judged to be: Adequate, Fair, Poor, or Very Poor; or in other instances, countries are judged to have had large improvement, small improvement, or deterioration in macroeconomic policies during the adjustment period. After comparing the performance of countries vis-a-vis the macroeconomic stance, the report concludes that "Of the twenty-nine countries studied in this report, the six with the most improvement in macroeconomic policies...enjoyed the strongest resurgence in economic performance" (p.1). This claim is, unfortunately, not supported by the evidence presented by the Bank.

A careful study of appendix tables 2.6, 2.7, and 2.8 reveals a lot of the absurdities in the results. The first point that strikes a reader while comparing tables 2.6 and 2.8 is that countries can enter or drop out of the "good" performance list depending on the period under consideration. Over the 1987-91 period, 6 countries are adjudged to have "large improvement in macroeconomic policies." When the same analysis is carried out for only 1990-91, only Ghana's policies are judged "Adequate," while 13 others are ranked "Fair." Moreover, countries said to have a "deterioration" in their policies in table 2.6 (Togo and Gabon) turn out with "Fair" policies in table 8, while Tanzania and Zimbabwe which had "large improvement" in table 2.6 get demoted to ones with "poor" policies in table 2.8. In both tables, Ghana remains the four-star general of SAP in terms of its policy stance, but it does not have the best performance record.

Table 2.7 provides an interesting summary of the mapping of performance and policy stance. Surprisingly, the Bank's results show that Mozambique (which was at war for most of the period, and which is judged to have had very poor policy stance) is the best performer in terms of GDP growth rate (which is the Bank's defining performance criterion). Only 14 out of the 29 countries studied (less than 50%) had positive growth rates. Out of this 14, only four have large improvement in policies; five had small improvement while the other five are said to have had deterioration in their policies. Curiously, two of the six countries with "large improvement" in policies (33%) actually recorded negative growth rates, whereas five out of the eleven with "deterioration" in policies recorded positive growth rates. The veracity of the claim that the six countries with large improvements experienced the most resurgence in their growth is therefore questionable.

The basis for the conclusion seems to lie in the computation of "mean" performance among different groups. But, it all depends on whether a simple or weighted average is taken and the selection of the sample size in each group. If for example, one selects the best four performers in each of the three categories of large and small improvements, and deterioration in policies, it would be difficult to prove that the mean performance of the "large improvement" group differs in any statistically significant manner from the other groups. Another example of different base years and contradictory results is shown in table 2.10. In computing change in real protection coefficient, the pre-SAP is chosen as (1981-83), and the SAP period as 1989-91 whereas changes in agricultural growth rate was calculated for the periods

1981-86 and 1987-91. Even this table demonstrates the absurdity of attributing performance to policy changes alone. For example, Côte d' Ivoire and Chad are reported as having the highest growth rates of 8.2% and 6.9% whereas they had increased the overall taxation of export crop producers to the levels of 23.2 and 27.4 percents respectively. Ghana had the largest decrease in the overall taxation of agricultural export producers but the change in agricultural growth rate was a mere 2.2%. Again, if a weighted mean performance is calculated, it is possible that the group with small decrease in overall taxation (which includes Nigeria' s growth of 2.7%) might outperform the group with largest decrease in taxation.

It is difficult to know which of the two tables (2.6 or 2.8) should be taken as basis for evaluating performance vis-a-vis policy stance. Matching the ranking by overall GDP growth rate (table 2.7) with the classification in table 2.8 produces a mixed grill. It is evident that aside from Ghana, the other five countries with "best performance" records—Mozambique, Nigeria, Uganda, Tanzania and Sierra Leone are ones with relatively poor records in their macroframework as at 1990-91. Nigeria and Uganda are judged to have fair record, but others are seen to be poor (for Tanzania) and very poor for (Sierra Leone and Mozambique). Throughout the Report, Mozambique scores very poorly on all measures of economic policy stance, and it notes on page 135 that Mozambique' s performance is not attributable to favourable external environment. The question then is what explains the growth performance of countries such as Mozambique?

Even on the account of the Bank' s own evidence, there is no discernible systematic pattern about the relationship between turnaround in growth and macroeconomic stance of the country. Since the study has ruled out any significant impact of the external environment, the job of explaining the determinants of most countries" growth performance remains largely unaccomplished.

As evidenced from tables 2.6 and 2.8, the choice of the study period can significantly alter the results. The 1981-86 period is designated the pre-adjustment period, but the fact is that several countries started their reform programs as far back as 1982. Also, the impacts of exogenous factors on the GDP growth during that period was not controlled for in the analysis. For example, the period coincided with two world recessions, 1979-81 and 1982-83, and also the great African drought of 1983-85. Many countries were adversely affected by these events during the period and it would not surprise anyone if their average growth rate was negative. Again, even though the computations do not validate the Bank' s arguments, it is important to note the inelegance of the mathematics employed. Two examples—Mozambique and Chad—demonstrate the point. As indicated in table 2.6 Mozambique had an average negative growth rate of minus 5.9% during the 1981-86 while Chad had positive rate 4.5%. In the 1987-91 period, Mozambique had an average rate of 1.7% while Chad had 2.6% growth. The mathematician comes on board and calculates the "turnaround" in the two countries by taking the differences in the two periods to obtain 7.6% for Mozambique and minus 1.9% for Chad. The report concludes that Mozambique has "performed better" than Chad even though Chad's growth rates during the two periods were positive and much higher than Mozambique's. This is a classic case of convenient use of statistics but even this "cleverness" has done much to confound, and very little to clarify, the issues.

We draw an example from a country case study (Nigeria) conducted by the Western Africa Department of the World Bank (1994) to caution about generalizations on the causes of recovery. The GDP growth rate is used as the major performance indicator. According to the Report, all the SAP years, except 1987, recorded significant economic growth. After a meagre 2.2 percent growth in 1986, and a decline of 0.3 percent in 1987, the GDP (at 1984 factor cost) grew by about 7 percent in 1988, and thereafter recorded an average annual growth rate of about 6.1 percent. In effect, the economy grew at an average annual growth rate of 5 percent throughout the six year SAP era (1987-92) compared to a decline of about 2 percent per annum in the preceding six years (1981-86). While attributing the strong recovery to the gamut of supply incentives in the SAP policies, the Western Africa Department (1994:53) cautions that,

> ...policy change was not the whole story. The post-SAP period coincided with the return of normal rainfalls after an extended and devastating drought. In 1988, two years after the adoption of the SAP, oil prices stabilized, although at relatively low levels. Furthermore, the upgrading and rehabilitation of rural infrastructure begun before 1986 was taking effect.

The critical issue really is how much of the growth is explained by the policies and how much by good luck with weather and oil prices. The statistics show that much of the trend in GDP growth rates is tracked by the trend in the oil sector especially since 1985. One could therefore argue that good weather and the oil sector dominate macroeconomic policies as predictors of economic performance. This is an important empirical issue, and a growth accounting model to isolate the sources of economic growth in Nigeria and other African countries would be required to resolve the debate.

The turnaround in the SAP era in comparison with pre-SAP period seems to be exaggerated in some sense. This is another demonstration of the impact of choice of base year on evaluation outcomes. For Nigeria, if the "before-after" comparison is done using 1985 rather than the average of previous years, it could be inferred that the economy performed better in 1985 (with

5.9 growth rate) than the average of 5.1 percent during the SAP period. For some analysts therefore, the SAP might have, in fact, slowed down a recovery process that was already in place. These critics of the "SAP success" argue that the policies pursued in 1984 and 1985 under Buhari's government could have turned the economy around. To them, the year 1985 should be the relevant reference period rather than an average of past six years, since recovery was already underway in 1985. Furthermore, it is an important counterfactual question whether the economy would not have performed better under the Buhari-type austerity measures if the World Bank and IMF had agreed to grant the type of external debt relief measures as they did under SAP.

A common weakness of the earlier studies is the unfounded attribution of economic recovery to "improved macroeconomic policies." The inability of the evaluation criteria to control for the impacts of non-SAP factors and initial conditions on GDP, the changing base-period for analysis, questionable calculation of index of reform effort, and the procedure for manipulating the statistics, could be among the reasons why countries enter and drop-out of "good performers" list. As a further example, consider the regression analysis reported on page 140 of the 1994 Report. The regressions attempt to demonstrate that domestic policies rather than external variables explain the economic performance of African countries (see table 2.9). The arbitrary manner of adding and dropping variables from one regression to the other, the conflicting signs of the coefficients, the choice and definition of variables, etc are some of the methodological flaws that raise questions about their robustness (see Soludo 1996c for detailed criticisms of the regression results). In essence, both the individual country analysis and the cross-country regression results have not proved that observed performance can be attributed to SAP policies.

Another aspect of the UNECA's criticisms of the 1989 report which has not been adequately addressed in the 1994 report relates to the consistency of SAP with poverty reduction and development. A development-oriented assessment of SAP would have placed greater weights on some indicators as employment, poverty level, income distribution, access to educational facilities, extent of structural transformation (e.g. increasing shift from agriculture to industry), etc in assessing SAP performance. The discordant views of the BWIs about the compatibility of SAP to long-term development makes it difficult to judge whether or not SAP is designed to achieve improvements in these social indicators in the first instance. On the one hand, there is a large amount of work by the BWIs on "social dimensions of adjustment," "long-term perspective study on Africa," etc, but on the other hand, several reports of the same institutions dichotomize between the goals of SAP and the requirements of long-term development (see, for example, World Bank 1994). Tables 2.11 and 2.12 show changes in some social spending of selected countries (notably education and health) which are critical for sustaining long-term development. Table 2.11 shows that aside from Zimbabwe, the countries with highest expenditures on these critical areas are mainly countries judged to have either small improvement or deterioration in policies. Table 2.12 is even more revealing because it demonstrates that high performers with most improvements in policies (such as Nigeria, Sierra Leone, The Gambia, etc) have also experienced alarming reductions in health and educational spending. Where then lies these countries" capacity to sustain the perceived "turnaround" in economic growth?

One serious methodological problem that is often glossed over in much of the discussions is the availability and quality of African macroeconomic data. The seriousness of this issue, and more so, its implications for macroeconomic research and inference have led to the commissioning of several studies by the African Economic Research Consortium, Nairobi, to examine the problem. As Helleiner (1992:54) summarizes,

> It is ... well known that African data leave a great deal to be desired. Even in the series for such usually fairly reliable variables as export and import values there are enormous gaps and uncertainties. (In Tanzania, for example, informed local observers suggest that because of smuggling and underinvoicing, recorded exports today may only be about half of their actual value.) Informal economic activities have vastly increased in their importance in all of the poorly performing African economies, and what transpires in them—production and trade, incomes earned, prices, etc.—is very imperfectly recorded if it is estimated at all.

Generally, GDP figures in Africa are heavily dependent on the agricultural sector in most African countries, and this depends on the weather. In reality, "many of the figures for agricultural production are little better than informed guesswork and by no means capable of bearing an analysis of fractions of a percent growth which often hands on them" (Center for Development Research 1995:8). Beside agricultural data, it is widely reported that the bulk of economic activities has been driven "underground" into the informal sector whose activities are hardly captured in formal statistics. In most African countries, census figures are highly politicized and grossly unreliable. What figures exist are usually products of guesswork and "projections" based on certain assumptions. Take Nigeria as an example. The BWIs still project Nigeria's population to be anything over 110 million whereas national census figures put the population at around 90 million. With more than 20 million people in dispute, the reliability of what emerges as Nigeria's per capita income is anybody's guess.

In many countries also, it is the case that significant discrepancies exist between various sources of macroeconomic data. What you get therefore depends on the source of your data. In Nigeria, for instance, even from the same source say, the Federal Office of Statistics, one often also encounters large discrepancies depending on the particular edition of the publication used. In situations where several data are neither available on continuous basis nor reliable, the robustness of fanciful econometric models, etc is certainly called into question. In recent times, a more disquieting issue has also emerged regarding the politicization of certain macroeconomic data (such as GDP, budget deficits, and inflation) by governments so as to qualify for certain concessions from the BWIs. These data constraints raise the need for some restraints in making sweeping generalizations on the basis of the "partial" or "invented" evidence.

In conclusion, the various World Bank studies cited above embody the weaknesses of the "before-and-after" methodology and more so, demonstrate the shortcomings of using a single, summary indicator for evaluating performance. The narrow focus on GDP per capita is clearly inadequate, and the example cited in the case of the 1994 World Bank report draws attention that such a measure could in fact produce embarrassing results. In a situation where three out of five "best performers" have "bad" or "very bad" policies, the reliance on such index as a summary measure of performance could be anything but convincing. Moreover, such an emphasis on GDP, and the contradiction in results obtained, go further to demonstrate the contradictions and inadequacies of the paradigm underlying SAP in terms of emphasis on growth rather than on development, and the obdurate insistence that only a particular kind of policies is consistent with growth. The fact that those with the "best" policies according to the dictates of this paradigm are not necessarily the ones with "best" performance record according to its evaluation criterion, speaks volumes about the inherent inconsistencies of the framework. Also, since the results of most evaluation studies differ, both across and within time and evaluation criteria employed, there is evidence of potential "instability" of the measures of performance. This raises the urgency for alternative performance indicators that might be consistent with a broader, time-consistent analytical framework.

Towards Alternative Frameworks

The foregoing analysis demonstrates the crisis of analytical and methodological frameworks in defining and evaluating Africa's economic agenda. Defining a consensus economic development model and policy instruments for Africa has been the source of acrimonious debate over the years. Such a debate is not likely to abate in the foreseeable future, especially in the light of the crisis of theoretical paradigm among African social scientists. Most of these scientists were better schooled in the neo-Marxist/structuralist tradition.

With the collapse of east European "true socialism" and the ascendancy of liberalism as the dominant mode of analysis and policy, there has developed a crisis and near vacuum. There is a half-hearted reluctance to abandon the "failed" neo-marxist/structuralist paradigm, and also a contempt for the liberal orthodoxy, but without an alternative coherent framework. One consequence of this is that African scholars become reactive rather than proactive in charting Africa's development agenda: merely indulge in highly polemical criticisms of the liberal agenda as inappropriate, but proceed to make prescriptions that are internally inconsistent at best, and which derive from smudged eclectic of neo-Marxist/structuralist paradigm and "pragmatic market reforms" (see Himmelstrand). On the other hand, the liberal agenda continues to be defined by the BWIs, and African scholars respond with criticisms. This ding-dong affair has certainly had its positive effects in "forcing" the BWIs to defend or rethink some of their positions. It has however, done little to advance the African framework.

A critical question at the heart of the crisis is whether there can be an African theory/ strategy of development that is fundamentally different from other developing countries: one that is essentially African in origin and character and which can be inconsistent with the trends in the rest of the world. In other words, is there a need for an African development paradigm or should Africa strive to learn from the accumulated experiences of other countries which have managed to escape the throes of underdevelopment—in which case the framework should continue to be ad-hoc? There could be potentially many answers to these questions. Suffice it to note that one factor which could make the evolution of such a "consensus African theoretical framework" difficult is that African countries are as diverse as they are similar. While there are several characteristics that largely define sub-Saharan Africa, there are also much in individual country features that might make them comparable to countries in other regions. Thus, the challenge of such attempt at "African unique theorizing" is to cautiously distinguish between those aspects of the framework that relate to "Africa specific" issues and those that relate to the "general underdevelopment" issues. Tangential to this and deriving from the apparent differences among African economies themselves, the relevant framework may not be one that provides "one general case" that applies equally to all African countries. Care might be taken to map out a few typologies of the African economies (defined by certain characteristics — economic structure, export composition, initial conditions, external debt burden, etc.) and delineate the variegated nature of the analytical framework. If one is not careful, such attempt might end up applying to only "special cases" of the African experience.

The trend towards a consensus broad framework for Africa's development relies mainly on "accumulated experiences in Africa and elsewhere." This emerging consensus as aptly summarized by the World Bank (1994:2) is that "Adjustment alone will not put countries on a sustained, poverty-reducing growth path. That is the challenge of long-term development, which requires better economic policies and more investment in human capital, infrastructure, and institution-building, along with better governance." What has been lacking however, is a broad policy frontier that integrates the SAP-cum-development requirements into a holistic framework. The World Bank implies that there should be a sequencing process that runs from adjustment to development, which implies that development concerns cannot be pursued unless and until SAP succeeds (see World Bank 1994:2). We take exception to this and hypothesize instead that both processes are so intertwined that one cannot succeed on a sustainable basis without the other. As Helleiner (1992:57) argues:

> There is a growing awareness that the most important element in a new strategy of economic development for SSA is the adoption of a much longer time horizon than is implicit in current structural adjustment programs and many current evaluations. There are no quick "fix-its" for Africa's problems. Many of the most important investments and policy changes have very long gestation periods. It is not helpful to attempt repeated short-term evaluations of the adjusting countries' progress according to conventional growth accounting yardsticks, or probably according to most others, still less by one agency after another. Such evaluation can be enormously demanding of scarce local decision-makers' time and skills, disruptive of local planning and operations, and ultimately unproductive. Continuous "plucking up by the roots" of a fragile plant to see how it is doing has never been a good husbandry. Rather, what is required is a sustained and internally consistent set of government policies backed by a predictable flow of key inputs, most notably of the necessary external resources. Predictability and sustainability are almost certainly more important than attaining near perfection on the finer points of policy dispute upon which the African governments are at present receiving unending, continually changing (and largely unsolicited) advice.

The key words here are long-term vision, sustainability of policy regimes, and development-orientation to SAP measures. The emerging consensus is a framework where development is not seen as the "add-on" but the fundamental essence of adjustment. Much of the discussions of the finer details of African development have focused on strategies to make agriculture the rubric of such agenda (see Stewart et al. 1992; Olofin 1995; etc.). Here we simply raise issues about the broad framework, and more so, focus on the implications of this SAP-cum-development paradigm for a more embracing evaluation criteria. Our premise is that Africa is a part of the global system, which is currently characterized by increasing competition, regionalism, and with the rules of the game set by GATT and the recently inaugurated World Trade Organization (WTO)—to which most African countries are signatories. Since Africa cannot live in isolation, it inevitably has to abide by the rules of international trade and competition. To do so, it has to build the capacity to manage shocks and compete in a world characterized by asymmetrical power and economic relations. This requires that the pre-eminent emphasis of economic policies should be capacity-building and this demands a fundamental re-orientation of the SAP instruments to accomplish this. There are several ways to achieve aims, and rather than the apparently ineffectual focus on short-term demand management and obdurate faith in the markets, a case can be made for a framework that focuses on activist supply side policies with readjusted (but not necessarily diminished) aggregate demand (see Soludo 1996b for details on the theoretical consistency of such a framework). This framework would necessarily require the guidance of an activist, developmental state which is capable of providing the market-compatible incentive and institutional structures, as well as actively wheeling the economic development agenda towards a clearly defined long-term vision.

The development-oriented framework demands a new evaluation criteria for both policy stance and outcomes. The current evaluation criteria based solely on a few macroeconomic variables—budget deficits, inflation, and exchange rate—for policy stance, and a single indicator (GNP per capita) for evaluation of outcomes are grossly inadequate and potentially misleading. Experiences of past evaluation exercises have demonstrated that proving that countries with best policies (as measured by those few variables) have necessarily had the best growth (as measured by GNP per capita) has remained a very tall task.

We propose a broader criteria which, for want of a better nomenclature, we describe as Composite Economic Development Index (CEDI). This index is akin to the Human Development Index (HDI) but focuses more on economic development which is the domain of our analysis. Under the CEDI, the current SAP/stabilization measures on the one hand, and the broader development indicators on the other, are adequately weighted in the computation of the index. The realization that development cannot proceed on a sustainable basis with triple digit inflation or unsustainable balance of payments dictates that a realistic composite measure should not ignore stabilization issues.

Two broad measures—one for policy stance and the other for evaluation of outcomes could be computed. Let us consider the measure of policy stance. Under the CEDI, the usual SAP stabilization indicators—budget deficits, inflation, and exchange rate misalignment could continue to weigh heavily in the computation of the SAP aspect of the policy stance. In addition to this, development-oriented policy stance would also be computed. This could be set in terms of percentage of government expenditures devoted to certain critical sectors—education, infrastructural development, health, and environmental protection. Alternatively, such minimum government spending on these sectors could be set as percentages of previous year's GDP. This measure has enormous implications for gauging resource-gaps that should be met by external resource inflows. If for example, different percentages of GDP are assigned as minimum expenditures on those four sectors (with government consumption also pegged at some level, and other sectors—military, etc treated as residuals), it would immediately highlight the narrow revenue base relative to the basic minimum spending for sustainable development. By the time we calculate the required spending vis-a-vis the total revenues (less debt service payments) the nature of the resource-gap would be obvious. This orientation would force attention to the nature of the debt service payments in these economies relative to their requirements for long-term development. For years now, attention has been on demand contraction and increased debt service payments or increased external lending for balance of payments purposes. Unfortunately, much of the spending cuts have severely hurt the very investments which are considered the fulcrum of long-term development. If we care about long-term development of African economies, the issues of investments in infrastructure, education, health, etc can no longer be issues for the appendix in policy designs. A major way to bring them to the center-page is by assigning minimum levels of investments in them. Once this is done, the resource gap that emerges would also have implications for considerations of debt relief/ forgiveness measures.

In the computation of a country's policy stance, the deviations of its spending on the four sectors relative to the minimum standards would be computed and weighted into a composite measure. Such a measure would then be combined with the measure of macroeconomic policy stance (à la current SAP measures). In this way, an overall index of policy stance would be derived—which combines both SAP and development-oriented measures.

A CEDI evaluation index would also follow similar considerations as in the computation of policy stance. First, the SAP evaluation criterion should be broadened to focus, not only on GNP per capita, but also on savings mobilization, and structural realignment—shift from nontradeables to tradeables, switch from agriculture to manufacturing sector, and diversification of the export base. Indicators of performance along these fronts should figure into the computation of SAP performance index which should be appropriately weighted in the total CEDI evaluation index. Development-oriented indicators should also account for a significant proportion of the index. Such development indicators would include school enrolment, changes in infrastructural development, social indicators (access to health care and life expectancy, income distribution, etc.), technological changes and factor productivity. A composite index with all these as elements would be a more realistic measure, and this is the challenge to statisticians. When this development aspect is integrated with the SAP component, a more general CEDI evaluation index would emerge. It would be a broader measure than the HDI, since it not only enables us to gauge the extent of policy misalignment, but also indicates the extent of *capacity* in the economy and thus its long-term "competitiveness" in the global context.

Thus, when analysts infer that growth has occurred but that social/development indicators have deteriorated, some indication could be given as to whether the rise in income has more than offset the fall in social indicators to produce an overall improvement or vice versa. However, assigning weights to the separate indices to produce the composite index will be highly subjective since the weights attached would depend on individual preferences about the importance of particular indices, etc. For example, within the social/development group, an analyst might assign equal weights to education, infrastructural development, etc, while another might assign more weight to education than health, etc. Whatever the demerits of this CEDI measure, it will still be a superior index to the over-arching emphasis on GNP per capita.

Even without going into the details of computing such indices, a casual examination of the results of the various World Bank Reports reveals that when the social cum development indicators of performance are taken into account, "the best performers" on macroeconomic policy stance or GNP per capita growth could turn out with poor performances. Indeed, depending on the period of analysis, some of the seemingly poor performers on account of macroeconomic stance could turn out as star generals when other indices are considered. Beside providing a more comprehensive indicator of policy stance and performance, the emphasis on CEDI could also force a new orientation and strategy to "bring development issues back" in policy analysis and policy making. Of course, the computation of the CEDI would require enormous data input. As we noted earlier, the availability and reliability of such data in many African countries are questionable. Thus, one offshoot of this framework could be to constrain governments and development agencies to invest resources in capacity-building in the area of collecting and building reliable national databanks. A lot of capacity would be needed to harmonize and strengthen existing agencies for data collection, and this would require understanding and partnerships with bilateral and multilateral donor agencies.

In addition to the computation of the CEDIs, future performance evaluations should have at least two other features: first, it should seek to circumvent the weaknesses of the most popular techniques by taking account of the refinements in the "modified control-group approach." This controls for non-policy factors in explaining performance; and second, any evaluation technique should emphasize the construction of counterfactuals in terms of what would have happened without the policies.

The emphasis on counterfactuals is needed to shed more light on whether or not alternative policies could have performed better if exogenous factors such as debt relief and resource flows were extended to a country as under SAP. It is common knowledge that countries that implemented SAP policies also received greater financial assistance. An evaluation technique that disentangles the effects of such factors and evaluates potential performance under alternative policies would shed light on the implications of combinations of policies for performance. This would provide some insights as to which kinds of policies produce superior outcomes. Counterfactual analysis is not easy, and at times, requires the development of economy-wide models to simulate such counterfactual scenarios. This is not an easy task either, and the data requirements could be daunting.

One conclusion that emerges from the discussions in this chapter is that both the existing policy framework and the performance evaluation criteria are grossly inadequate. Charting alternative visions and policy instruments is obviously not an easy task. But African development cannot wait until consensus is reached on the finer points of the debate. We propose that policy frontiers and evaluation techniques must derive from the weaknesses of previous attempts and learn from the accumulated experiences of other regions. The fundamental lesson of such experiences is that "development issues have to be brought back" into the center-stage of policy dialogue. Reconciling such development-orientation with the content and instruments of the ongoing SAP is the challenge of policy making in the next decade. In other words, the challenge should be on how to design development-oriented stabilization/ adjustment measures.

Notes

1. The other schools, namely, the neo-Marxists, Keynesian, and dependency school, are included within the structuralist paradigm. See Elbadawi, et al. 1992; Little, et al. 1993; Mkandawire 1989; and Toye 1994 for discussions on the intellectual climate that informed SAP policies.
2. See Abbott (1994) for an excellent review of the AAF-SAP and comparisons with the World Bank sponsored SAPs.

Works Cited

Abbott, G.C. "Two Perspectives on Adjustment in Africa." *African Review of Money Finance and Banking*, 1, 2 (1994).

Ajayi, S.I. "The State of Research on the Macroeconomic Effectiveness of Structural Adjustment Programs in Sub-Saharan Africa." In: R. Van Der Hoeven and F. Van Der Kraaij, eds., *Structural Adjustment and Beyond in Sub-Saharan Africa*, The Hague, Netherlands, Ministry of Foreign Affairs (1994).

Bates, R. "The Reality of Structural Adjustment: A Sceptical Appraisal." In: S. Commander, ed., *Structural Adjustment and Agriculture: Theory and Practice in Africa and Latin America*, ODI, London (1989).

Befekadu, D. "An African perspective on Long-term Development in Sub-Saharan Africa." In: G. Cornia and G.K. Helleiner, eds., *From Adjustment to Development in Africa*, Martins Press, New York (1994).

Browne, R.S and R.J. Cummings *The Lagos Plan of Action Vs. The Berg Report: Contemporary Issues In African Economic Development*. 2nd ed., Virginia, Brunswick Publishing Company (1985).

Camdessus, M. "Opening Remarks." In: V. Corbo, M. Goldstein, M. Khan, eds., *Growth-Oriented Adjustment Programs*. International Monetary Fund/The World Bank, Washington, DC (1987).

Cassen, R. "Structural Adjustment in Sub-Saharan Africa." In: W.V.D Geest, ed., *Negotiating Structural Adjustment in Africa*, New York, UNDP (1994).

Central Bank of Nigeria, CBN, *Perspectives of Economic Policy Reforms in Nigeria*. Lagos, Central Bank of Nigeria (1993).

Centre for Development Research, "Structural Adjustment in Africa: A Survey of the Experience." Copenhagen, March (1995).

Cornia, G.A, R. van der Hoeven, and T. Mkandawire, eds., *Africa's Recovery in the 1990s: From Stagnation and Ajustment to Human Development*. A UNICEF Study, London: Macmillan (1992).

Davies, J.M., ed. *Macroeconomic Adjustment: Policy Instruments and Issues*. Washington,DC, IMF Institute, International Monetary Fund (1992).

Demery, L. "Structural Adjustment: Its Origins, Rationale and Achievements." In: G.A. Cornia and G.K. Helleiner, eds., *From Adjustment to Development in Africa: Conflict, Controversy, Convergence, Consensus?*, St. Martins Press, New York (1994).

Dornbusch, R. and L.T. Kuenzler "Exchange Rate Policy: Options and Issues." In: R. Dornbusch, ed., *Policymaking in the Open Economy: Concepts and Case Studies in Economic Performance*, Oxford, Oxford University Press (1993).

Elbadawi, I. *World Bank Adjustment Lending and Economic Performance in Sub-Saharan Africa*. Policy Research Working Paper, No.1001, Washington, DC, The World Bank (1992).

Elbadawi, I., D. Ghura, and G. Uwujaren "Why Structural Adjustment Has Not Succeeded in Sub-Saharan Africa." Policy Research Working Papers, No. 1000, Washington, DC, The World Bank (1992).

Federal Government of Nigeria *Structural Adjustment Program for Nigeria 1986-1988*. Lagos: Government Printers (1986).

Ghai, D. ed., *The IMF and the South: The Social Impact of Crisis and Adjustment*. London, Zed Books Ltd. (1991).

Green, R. "The IMF and the World Bank in Africa: How Much Learning?" In: T.M. Callaghy and J. Ravenhill, eds., Hemmed. In: Responses to Africa's Economic Decline, Columbia University Press, New York (1993).

Haggard, S. and Kaufman, R.F. "Institutions and Economic Adjustment." In: S. Haggard and R.F. Kaufman, eds., *The Politics of Economic Adjustment*, Princeton: Princeton University Press (1992).

Helleiner, G.K. "Structural Adjustment and Long-Term Development in Sub-Saharan Africa." In: Stewart, F., S. Lall, and S. Wangwe, eds., *Alternative Development Strategies in Sub-Saharan Africa*, St. Martins Press, New York (1992).

Himmelstrand, U.; K. Kinyanjui and E.K. Mburugu, "In Search of New Paradigms." In: U. Himmelstrand, et al., eds., *African Perspectives on Development: Controversies, Dilemmas and Openings*, E.A.E.P, Nairobi (1994).

Hirshman, A. *The Strategy of Economic Development*. Yale University Press, New Haven (1958).

Hussain, I. "Structural Adjustment in Sub-Saharan Africa: A Preliminary Evaluation." Paper presented at EDI/NY Office of the World Bank Seminar, New York, 18 May (1993).

Hussain, I. "Structural Adjustment and the Long-Term Development of Sub-Saharan Africa." In: R. Van Der Hoeven, and F. Van Der Kraaij, eds., *Structural Adjustment and Beyond in Sub-Saharan Africa*, The Hague, Netherlands, Ministry of Foreign Affairs (1994a).

Hussain, I. "Why Do Some Economies Adjust More Successfully than Others?" Lessons from Seven African Countries." Policy Research Working Paper, No.1364, Washington, DC, The World Bank (1994b).

Hussain, I. "The Macroeconomics of Adjustment in Sub-Saharan African Countries: Results and Lessons." Policy Research Working Paper, No.1365, Washington, DC, The World Bank (1994c).

Khan, M.S. "The Macroeconomic Effects of Fund-Supported Adjustment Programs: An Empirical Assessment." IMF Working Paper, WP/88/113, International Monetary Fund (1988).

Lewis, W.A. "Economic Development with Unlimited Supplies of Labor." Manchester School of Economics and Social Studies, 22, 2 (1954).

Lipumba, N.H.I. *Africa Beyond Adjustment*. Overseas Development Council, Washington, DC (1994).

Little, I.M.D, et al., *Boom, Crisis, and Adjustment: The Macroeconomic Experience of Developing Countries*. New York, Oxford University Press (1993).

Mkandawire, T. *Structural Adjustment and Agrarian Crisis in Africa: A Research Agenda*.CODESRIA Working Paper 2/89, Dakar, Senegal (1989).

Mongula, B.S. "Development Theory and Changing Trends in Sub-Saharan African Economies 1960-89." In: U. Himmelstrand, et al., eds., African Perspectives on Development: Controversies, Dilemmas and Openings, E.A.E.P, Nairobi (1994).

Mosley, P. "Decomposing the Effects of Structural Adjustment: The Case of Sub-Saharan Africa." In: R. Van Der Hoeven, and F. Van Der Kraaij, eds., *Structural Adjustment and Beyond in Sub-Saharan Africa,* Haque, Netherlands, Ministry of Foreign Affairs (1994).

Nashishibi, K. and S. Bazzoni "Alternative Exchange Rate Strategies and Fiscal Performance in Sub-Saharan Africa." *International Monetary Fund Staff Papers* 41(1) (1994).

Olofin, S. "Economic Development in Sub-Saharan Africa." Paper presented at Project LINK Conference held at the University of Pretoria, South Africa; September 25-29 (1995).

Prebisch, R. "Commercial Policies in the Underdeveloped Countries." *American Economic Review,* Papers and Proceedings (1959).

Prebisch, R. *Towards a New Trade Policy for Development.* Report by the Secretary-General, UNCTAD, New York: UNCTAD (1964).

Ranis, G. "The Political Economy of Development Policy." In: G.M. Meier, ed., *Politics and Policymaking in Developing Countries,* ICS Press San Francisco (1991).

Sahn, D.E. "Public Expenditures in Sub-Saharan Africa during a Period of Economic Reforms." World Development, 20 (5) (1992): 673-93.

Soludo, C.C. *North-South Macroeconomic Interactions: Comparative Analysis using MULTIMOD and INTERMOD Global Models.* Brookings Discussion Papers in International Economics, No. 95, July 1992, The Brookings Institution, Washington, DC (1992).

Soludo, C.C. *Growth Performance in Africa: Further Evidence on the External Shocks Versus Domestic Policy Debate.* Development Research Paper Series No. 6, United Nations Economic Commission for Africa, Addis Ababa (1993a).

Soludo, C.C. *Implications of Alternative Macroeconomic Policy Responses to External Shocks in Africa.* Development Research Paper Series DRPS No.8, United Nations Economic Commission for Africa, Addis Ababa (1993b).

Soludo, C.C. "An Evaluation of Structural Adjustment Programs in Africa: Issues, Methods, and Consequences for Nigeria." In: Onuoha, J, and Ozioko, eds. *Contemporary Issues in Nigerian Social Sciences,* Enugu, Acena Publishers Ltd. (1995a).

Soludo, C.C. "A Framework for Macroeconomic Policy Coordination Among African Countries: The Case of the ECOWAS Sub-Region." *African Economic and Social Review.* 1, 1 and 2 (1995b).

Soludo, C.C. *Macroeconomic Adjustment, Trade and Growth: Policy Analysis Using a Macroeconomic Model of Nigeria.* The AERC Research Paper No. 32, Nairobi (1995c).

Soludo, C.C. *Choice of Optimal Mix of Fiscal and Monetary Policy Rules: Empirical Evidence from a Model of Nigeria.* Forthcoming under the AERC Research Paper Series (1995d).

Soludo, C. C. "Intertemporal Fiscal Policy Rules, Deficit Reduction Plans and Macroeconomic Performance in Nigeria." Paper presented at the Senior National Policy Seminar organized by the Nigerian Economic Society (March 6-8), Lagos (1996a).

Soludo, C.C. "Theoretical Frameworks of SAPs in Africa: Some Better Ideas from Dead Economists." In: *Beyond Adjustment: The Management of the Nigerian Economy,* Nigerian Economic Society Conference proceedings (1996b).

Stewart, F., S. Lall, and S. Wangwe "Alternative Development Strategies: An Overview."

In: Stewart, F., S. Lall, and S. Wangwe, eds., *Alternative Development Strategies in Sub-Saharan Africa*, St. Martins Press, New York (1992).

Toye, J. "Structural Adjustment: Context, Assumptions, Origin and Diversity." In: R.V.D Hoeven, and F.V.D Kraaij, eds., *Structural Adjustment and Beyond in Sub-Saharan Africa*. Haque, Netherlands, Ministry of Foreign Affairs (1994).

United Nations Economic Commission for Africa (UNECA) *African Alternative Framework to Structural Adjustment Programs for Socio-Economic Recovery and Transformation*. Addis Ababa, UNECA (1989a?).

UNECA *Statistics and Policies: ECA's Preliminary Observations on the World Bank Report: Africa's Adjustment and Growth in the 1980s*. Addis Ababa, April (1989b).

Western Africa Department, The World Bank *Nigeria: Structural Adjustment Program- Policies, Implementation, and Impact*. Washington, DC, The World Bank (1994).

World Bank *Accelerated Development in Sub-Saharan Africa: An Agenda for Action*. Washington, DC, The World Bank (1981).

World Bank *Adjustment Lending: An Evaluation of Ten Years of Experience*. Policy and Research Series No.1, Washington, DC, The World Bank (1988a).

World Bank) *Report on Adjustment Lending*. Washington, DC, The World Bank (1988b.

World Bank *Sub-Saharan Africa: From Crisis to Sustainable Growth, A Long-Term Perspective Study*. Washington, DC, The World Bank (1989).

World Bank *Adjustment Lending Policies for Sustainable Growth*. Policy and Research Series, No.14, Washington, DC, The World Bank (1990a).

World Bank *Report on Adjustment Lending II: Policies for the Recovery of Growth*. Washington, DC, The World Bank (1990b).

World Bank *Making Adjustment Work for the Poor: A framework for Policy Reform in Africa*. Washington, DC, The World Bank (1990c).

World Bank *Structural and Sectoral Adjustment Operations, The Second OED Review*. Washington, DC, The World Bank (1992a).

World Bank *Adjustment Lending and Mobilization of Private and Public Resources for Growth*. Policy and Research Series No.22, Washington, DC, The World Bank (1992b).

World Bank *The East Asian Miracle: Economic Growth and Public Policy*. New York, Oxford University Press (1993).

World Bank *Adjustment in Africa: Reforms, Results, and the Road Ahead*. New York, Oxford University Press (1994).

World Bank/ UNDP *Africa's Adjustment and Growth*. Washington, DC, The World Bank/UNDP (1989).

Table 2.1: World Bank Assessment of Impact of Structural Adjustment

Study	Country Group	Real growth rate of GDP			Investment/GDP		Exports/GDP			Growth of real per capita private consumption
		1982-7	1985-8	1986-90	1985-8	1986-90	1982-7	1985-8	1986-90	1985-8
World Bank 1988	Intensive adjustment lending	4.7					8.5			
	Non-adjustment lending	4.8					7.1			
World Bank 1990	Intensive adjustment lending									
	All countries		4.2	18.6				28.1		1.4
	Low-income		3.9	16.6				29.6		1.0
	Non-adjustment lending									
	All countries		2.7	20.0				24.6		-0.2
	Low-income		2.7	15.5				15.5		0.2
World Bank 1992	Intensive adjustment lending									
	All countries			4.2		17.9			28.4	
	Sub-Saharan Africa			3.5		16.3			28.0	
	Non-adjustment lending									
	All countries			2.4		18.4			28.4	
	Sub-Saharan Africa			3.9		15.6			31.7	

Source: World Bank (1988), Table 3.2; (1990a), Table 2.6; (1992), Table A1, p. 27

Table 2.2: Summary of Policy Indicators

Indicator	Period	Countries with strong reform programs	Countries with weak or no programs
Government expenditure (percentage of GDP)	1980-83	31.9	28.9
	1987-87	29.9	30.3
Government revenues (percentage of GDP) [a]	1980-83	23.2 (19.5)	20.9 (17.7)
	1986-87	23.9 (18.9)	215 (17.6)
Fiscal balance (percentage of GDP) [b]	1980-83	-87 (-12.4)	-8.0 (-11.2)
	1986-87	-6.0 (-11.0)	-8.9 (-12.7)
Real central bank discount rate (percent)	1980-82	-7.0	-7.6
	1986	-1.0	-12.2
Consumer price changes (percent per year)	1980-85	18	23
	1986-7	16	35
Nominal exchange rate (SDR per local currency) (index, 1980-82 = 100)	1986-87	48	55
Agricultural incentives [c]	1986	146	108
Real export crop prices (index, 1980-82 = 100)	1986	115	90
Nominal protection coefficient for major export crops [d]	1980-84	0.9	0.8
	1986	1.7	1.1

Note: Country coverage varies by indicatory depending on available data over the entire period covered. Averages are unweighted.
a. Figures in parentheses exclude grants.
b. Figures in parenthesis exclude countries recently affected by strong external shocks.
c. Crop years: for example, 1986 = 1986/87. Country groups for agricultural indicators differs reflecting specific reforms in the pricing and marketing of export and food crops (see Appendix A).
d. The implied subsidy when the NPC exceeds one reflects to some degree recent currency appreciation in CFA franc zone countries rather than increases in nominal producer prices.

Source: World Bank/UNDP Africa's Adjustment and Growth in the 1980s, Washington DC, March 1989, Table 19.

Table 2.3: Summary of Economic Performance Indicators
(average annual percentage change unless indicated otherwise)

Indicators	Period	All countries		Countries not affected by strong shocks	
		Strong reform programs	Weak or no reform programs	Strong reform programs	Weak or no reform programs
Growth of GDP (constant 1980 prices)	1980-84 1986-87	1.4 2.8	1.5 2.7	1.2 3.8	0.7 1.5
Agricultural production	1980-84 1985-87	1.1 2.6	1.3 1.5	1.4 3.4	1.8 2.6
Growth of export volume	1980-84 1985-87	-1.3 (-11.0) 4.2 (-2.0)	-3.1 (-0.9) 0.2 (-2.5)	-07 (-4.7) 4.9 (3.5)	-5.7 (-2.1) -3.3 (-6.0)
Growth of import volume (excluding oil exporters)	1985-87	1.7 (-7.7) 4.8 (6.8)	-2.7 (-3.0)	6.1 (7.4)	-4.0 (-2.2)
Growth of real domestic investment	1980-84 1985-87	-8.1 -0.9	-3.7 -7.0	-3.5 1.9	-7.0 -4.8
Gross domestic savings (percentage of GDP)	1982-84 1986-87	9.9 10.7	2.3 6.0	7.8 10.7	0.9 5.6
Growth of per capita consumption (real)	1980-84 1985-87	-2.3 -0.4	-1.1 -0.5	-2.4 0.7	-1.5 -0.9

Note: Country coverage varies by indicatory depending on available data over the entire period covered. Averages are unweighted except as noted. Growth rates are computed using least squares. Periods are inclusive. Figures in parenthesis are weighted averages of country growth rates based on total values summed across countries.

Source: Africa's Adjustment and Growth in the 1980s, op cit, Table 20.

Table 2.4: Variety of Base Years and Periods Used

Topic	Base year Period = 100
Terms of Trade	1970-1973
Export Prices	1970-1972
Export Volumes	1970
Real Effective Exchange Rate	1971-1972
Inflation and Nominal Exchange Rate	1971-1972
Real Agricultural Producer Prices	1980
Agricultural Production	1979-1981
GDP	1970

Source: *Statistics and Polices: ECA Preliminary Observations on the World Bank Report: "Africa's Adjustment and Growth in the 1980s."* UN Economic Commission for Africa, April 1986, p. 6.

Table 2.5: Growth of Gross Domestic Production in Africa, 1980-87: Stant 1980 US Dollars, market prices (in percent)

Country Groups	1980-81	1981-82	1982-83	1983-84	1984-85	1985-86	1986-87	1980-87 Ave.
Strong Adjusting	-3.01	0.33	-3.85	-4.31	6.33	2.82	-1.97	-0.53
Weak Adjusting	5.44	3.46	0.66	1.29	0.13	4.01	1.88	2.00
Non-Adjusting	3.92	3.35	3.53	3.68	6.40	3.62	-2.51	3.50
Sub-Saharan Africa	-1.50	1.01	-2.37	-2.94	5.44	3.09	-1.48	0.24
North Africa	-2.27	3.12	3.63	2.78	1.90	0.19	1.29	1.50
Africa (All)	-1.52	1.81	-0.06	-0.66	3.98	1.92	-0.38	0.73

Source: World Bank data files; country coverage and classification of strong adjusting, weak adjusting and non-adjusting countries according to World Bank: Average annual growth rates were calculated as arithmetic average (preliminary), published in *Statistics and Policies*. Source: UNECA (1989).

Table 2.6: GDP Per Capita Growth

Country	Average annual growth rate (Percent)		Differences between the two periods (Percentage point)
	1981-86	1987-91	
Large improvement in macroeconomic policies			
Ghana	-2.4	1.3	3.7
Tanzania	-1.7	1.3	2.9
The Gambia	1.2	0.3	-0.8
Burkina Faso	2.2	0.4	-1.7
Nigeria	-4.6	2.4	7.0
Zimbabwe	0.3	1.0	0.7
Mean	**-0.8**	**1.1**	**2.0**
Median	**-0.7**	**1.1**	**1.8**
Small improvement			
Madagascar	-3.7	-2.1	1.6
Malawi	-1.4	0.7	2.2
Burundi	2.1	1.2	-0.9
Kenya	-0.5	0.9	1.5
Mali	0.4	-1.2	-1.6
Mauritania	-0.9	-1.0	-0.1
Senegal	0.4	-0.2	-0.6
Niger	-4.9	-2.4	2.5
Uganda	-1.5	2.8	4.3
Mean	**-1.1**	**-0.1**	**1.0**
Median	**-0.9**	**-0.2**	**1.5**
Deterioration			
Benin	1.1	-2.0	-3.1
Central African Republic	-0.1	-2.8	-2.6
Rwanda	0.4	-5.0	-5.5
Sierra Leone	-2.1	0.8	2.6
Togo	-2.8	-1.4	1.4
Zambia	-3.2	-2.3	0.9
Mozambique	-5.9	1.7	7.6
Congo	4.1	-0.7	-4.9
Cote d'Ivoire	-4.2	-6.8	-2.6
Cameroon	4.6	-7.9	-12.5
Gabon	-2.7	-1.9	0.9
Mean	**-1.0**	**-2.6**	**-1.6**
Median	**-2.1**	**-2.0**	**-2.6**

Table 2.6 cont..

Country	Average annual growth rate (Percent)		Differences between the two periods (Percentage point)
	1981-86	1987-91	
Unclassified			
Chad	4.5	2.6	-1.9
Guinea	-	-	-
Guinea-Bissau	2.9	1.5	-1.4
Medians			
All countries	-0.7	0.1	0.3
Low-income countries	-1.2	0.6	1.1
Middle-income countries	0.3	-1.3	-1.6
Countries with fixed exchange rates	0.4	-1.7	-1.8
Countries with flexible exchange rates	-1.5	0.9	1.5
Oil-exporting countries	0.7	-2.0	-2.0

- Not available
Source: World Bank (1994:138)

Table 2.7: Ranking of Overall Performance vis-à-vis Policy Stance 1981-91

Rank	Country	Overall GDP Growth Rate	Policy Stance
1.	Mozambique	7.6	**Deterioration**
2.	Nigeria	7.0	Large improvement
3.	Uganda	4.3	Small improvement
4.	Ghana	3.7	Large improvement
5.	Tanzania	2.9	Large improvement
6.	Sierra Leone	2.9	**Deterioration**
7.	Niger	2.5	Small improvement
8.	Malawi	2.2	Small improvement
9.	Madagascar	1.6	Small improvement
10.	Kenya	1.5	Small improvement
11.	Togo	1.4	**Deterioration**
12.	Gabon	0.9	**Deterioration**
13.	Zambia	0.9	**Deterioration**
14.	Zimbabwe	0.7	Large improvement
15.	Mauritania	-0.1	Small improvement
16.	Senegal	-0.6	Small improvement
17.	The Gambia	-0.8	Large improvement
18.	Burundi	-0.9	Small improvement
19.	Guinea - Bissau	-1.4	Unclassified
20.	Mali	-1.6	Small improvement
21.	Burkina Faso	-0.1.7	Large improvement
22.	Chad	-1.9	Unclassified
23.	Cote d'Ivoire	-2.6	Deterioration
24.	Central Africa Rep.	-2.6	Deterioration
25.	Benin	-3.1	Deterioration
26.	Congo	-4.9	Deterioration
27.	Rwanda	-5.5	Deterioration
28.	Cameroon	-12.5	Deterioration

Source: Derived from Table 2.6.

Table 2.8: Ranked by Overall Macroeconomic Policy

(a)	**ADEQUATE**			
	1.	Ghana		* * * *
(b)	**FAIR**			
	2.	Burundi	* * *	
	3.	The Gambia		* * *
	4.	Madagascar		"
	5.	Malawi	"	
	6.	Burkina Faso	" "	
	7.	Kenya		"
	8.	Gabon	"	
	9.	Mauritania		"
	10.	Nigeria	"	
	11.	Senegal	"	
	12.	Togo		"
	13.	Mali		"
	14.	Uganda	"	
(c)	**POOR**			
	15.	Central African Republic	* *	
	16.	Niger		* *
	17.	Benin		"
	18.	Rwanda		"
	19.	Tanzania		"
	20.	Zimbabwe		"
	21.	Côte d'Ivoire		"
	22.	Cameroon		"
	23.	Congo		"
	24.	Mozambique		"
	25.	Sierra Leone		"
	26.	Zambia		"

Source: World Bank (1994 : 58).

Table 2.9: Regressions for Analyzing the Changes in Growth (dependent variable = change in GDP per capita growth, 1981 – 86 to 1987 – 91

Regression	Constant	Overall macro-economic policies	Exchange rate policy	Inflation	Overall fiscal deficit	Net external transfers	Income from changes in the terms of trade	GDP per capita growth 1981-86	Adjusted R^2
I	-1.84 (-3.50)	2.11 (4.4)						-1.04 (-6.3)	0.75
II	-1.85 (-4.4)		0.59 (2.6)	-0.43 (-1.8)	-1.04 (-4.5)	0.20 (0.8)	-1.08 (2.9)	-0.98 (-7.2)	0.85
III	-1.13 (-2.2)		1.09 (3.8)			-0.06 (-0.2)	-0.05 (-0.1)	-0.84 (-5.1)	0.70
IV	-0.62 (-1.0)			-0.47 (1.1)		0.22 (0.6)	0.04 (0.1)	-0.92 (-4.4)	0.53
V	-2.00 (-4.0)				-1.28 (-5.0)	0.58 (2.2)	-1.06 (-2.5)	-1.16 (-7.5)	0.79

Note: Numbers in parentheses are t-statistics.
Source: World Bank (1994: 140)

Table 2.10: Change in Agricultural Taxation and Agricultural Growth

Country	Change in the real protection coefficient, 1981-83 to 1989-91 (Percent)	Difference in average annual agricultural growth rate between 1981-86 and 1987-91 (percentage points)
Large decrease in overall taxation of export crop producers		
Ghana	341.0	2.2
Guinea	325.8	-
Madagascar	117.1	0.9
Malawi	78.3	2.3
Uganda	33.9	6.4
Central African Republic	31.5	-2.9
Tanzania	30.6	1.9
Mean	136.9	1.8
Median	78.3	2.0
Small decrease in overall taxation		
Burkina Faso	17.9	-2.3
Rwanda	15.2	-0.1
Burundi	15.0	-1.6
Togo	10.9	-1.8
Gabon	10.7	1.5
Mali	9.4	2.5
Kenya	8.9	-0.3
Congo	4.5	1.9
Nigeria	1.3	2.7
Niger	1.1	-
Mean	9.5	0.3
Median	10.0	-0.1
Increase in overall taxation		
Mozambique	-2.0	5.2
Zimbabwe	-3.1	-4.7
The Gambia	-10.3	-10.0
Côte d'Ivoire	-23.2	8.2
Chad	-27.4	6.9
Benin	-27.6	-0.1
Senegal	-28.3	-2.2
Sierra Leone	-33.4	1.4
Cameroon	-34.7	-4.7
Guinea-Bissau	-70.3	-5.0
Zambia	-76.0	-1.6
Mean	-30.6	-0.6
Median	-27.6	-1.6

-Not available.
Note: Mauritania is excluded because it has no major export crop.
a. An increase in the real protection coefficient constitutes a decrease in agricultural taxation.
Source: World Bank (1994: 245)

Table 2.11: Social Spending in Selected Countries

Country	Health expenditures as a percentage of GDP		Education expenditures as a percentage of GDP	
	1981-86	1987-90	1981-86	1987-90
Large improvement in macroeconomic policies				
Ghana	0.8	1.3	2.2	3.4
Tanzania	1.3	0.6	-	-
The Gambia	2.3	1.5	4.6	3.3
Burkina Faso	0.7	0.6	1.9	1.6
Zimbabwe	2.3	2.9	7.5	8.7
Mean	**1.5**	**1.4**	**3.2**	**3.4**
Median	1.3	1.3	3.4	3.3
Small improvement				
Madagascar	1.0	1.2	3.6	3.0
Malawi	2.0	2.0	3.7	3.0
Kenya	1.8	1.6	5.3	5.8
Mali	0.8	0.7	3.0	2.7
Niger	0.8	1.3	2.5	3.1
Uganda	0.4	0.4	1.3	1.3
Mean	**1.1**	**1.2**	**3.3**	**3.1**
Median	0.9	1.2	3.3	3.0
Deterioration				
Sierra Leone	1.1	0.4	2.4	1.0
Togo	1.7	1.3	5.6	4.4
Zambia	2.3	1.7	4.5	2.2
Cameroon	0.9	0.8	2.5	3.1
Mean	**1.5**	**1.1**	**3.7**	**2.7**
Median	1.4	1.1	3.5	2.7
All countries				
Mean	**1.3**	**1.2**	**3.6**	**3.3**
Median	1.1	1.3	3.3	3.0

- *Not available.*
Source: Nashishibi and Bazzoni (1994).

Table 2.12: Change in Real Health and Education Expenditures in Selected Countries, 1980-89 (percent)

Country	Health	Education
Burkina Faso	11.7	13.4
Cameroon	18.5	60.4
The Gambia	-	-64.0
Ghana	-	136.4
Kenya	1.5	33.8
Madagascar	6.2	-12.8
Malawi	3.4	-15.8
Niger	36.7	-2.1
Nigeria	-50.5	-70.3
Sierra Leone	-	-82.8
Togo	-	-4.7
Uganda	-24.1	-4.5
Zambia	-3.4	-18.3
Zimbabwe	37.6	30.9
Mean	**3.8**	**0.0**
Median	**4.8**	**-4.6**

- *Not available:*
Source: Sahn (1992)

Chapter 3

Trade Liberalization, Regional Integration, and African Development in the Context of Structural Adjustment

T. Ademola Oyejide

Introduction

The typical African economy has a substantial exposure to the rest of the world through external trade. On average imports and exports accounted for a sizeable part of each African country's gross domestic product (GDP) in the early 1990s (Oyejide 1995). This level of external exposure has significant implications for successful stabilization and structural adjustment in African countries, the continent's longer-term development prospects and its choice of development strategies and policies.

Many of the African countries have, since the early 1980s, been implementing, in varying degrees, fairly comprehensive packages of policy reforms under the general umbrella of structural adjustment programs (SAPs) that were initiated, in most cases, and actively supported and funded by the World Bank and the International Monetary Fund (IMF). Trade policy is a prominent component of African SAPs. This reflects partly the substantial exposure of African economies to world market developments noted above; but it also reflects the fact that the original diagnostic study (World Bank 1981), which prepared the way for the SAP initiative attributed much of the fault for Africa's poor economic performance in the 1970s to misguided trade policies. These considerations probably explain why trade policy reform "carried great weight in many of the reform packages proposed to (and increasingly implemented by) African governments" (Rodrik 1988:2). The trend which started in the late 1970s has continued into the mid-1990s such that in many African countries, "trade policy recommendations are prominent in normative analyses of current development problems" (Helleiner 1995:1).

This paper examines the relationship(s) between trade liberalization, regional integration and African development as moderated by the over-riding stabilization and structural adjustment policy framework. It begins, therefore, with a review of the key elements of stabilization and structural adjustment policy reform paradigm as applied in Africa, paying particular attention to the role of trade and trade policy in this paradigm, their trade-theoretic underpinnings and the postulated trade-development linkages that are often used to justify trade policy recommendations under SAP. Secondly, the chapter focuses on Africa's long-running romance with regional integration and examines its trade liberalization experience both within and outside specific regional integration arrangements. In this context, questions are raised regarding the consistency between national-level trade policies induced by SAP and corresponding trade policy measures that were implicitly mandated by explicit commitments entered into in the context of regional arrangements.

Thirdly, available evidence is marshalled to tease out the impact of trade policy reform in Africa between the early 1980s and the mid-1990s. Finally, the chpater concludes by pulling together the key implications of Africa's SAP-induced experience with regional integration and trade liberalization for its longer-term development strategy.

Trade Policy in the Context of Stabilization and Structural Adjustment

Since the early 1980s, SAPs have provided the over-riding framework for the articulation and implementation of key economic policy reforms in most African countries. This general trend applies with particular reference to trade policy. An analysis of African trade policy reform experience since the mid-1980s would, therefore, be in-complete without an understanding of the role and significance assigned to trade policy in the context of SAPs.

Elements of Stabilization and Structural Adjustment Programs

African SAPs have several dimensions; some of these have changed and others have been added on in an evolutionary learning process that has characterized both the development and application of the conceptual underpinnings of the SAPs. In the words of Ndulu (1995:5):

> A combination of macroeconomic stabilization measures and structural reforms was adopted, albeit with a significant variation in the relative emphasis on the two features over time. The evolution of the adjustment process and

features constituted a "costly policy learning curve" drawing from actual experience with implementation and prompted by the criticism of "adjustment orthodoxy" along the way.

The evolutionary process has gone through at least three broad phases. These can be identified as the "stabilization" phase which spanned the early 1980s to the mid-1980s; the "adjustment with growth" phase which ran over the second half of the 1980s; and the "adjustment with poverty alleviation" phase which began in the early 1990s.

The first phase focussed primarily on restoring macroeconomic balance and stability by eliminating balance-of-payments and fiscal deficits and thus closing the existing external and domestic resource gaps. This phase employed standard demand management and restraint instruments which, not unexpectedly, generated drastic import compression and public investment decline. It soon became clear that stabilization measures could not, on their own, restore external and internal equilibrium and thus macroeconomic stability on a sustainable basis; and that sustained growth is required to achieve this objective in the medium to long term. Hence, the second phase added "supply-side" measures to demand management. Supply-side measures aimed at re-aligning incentive structures in favor of tradeables were expected to encourage output growth. In addition, increased external resource flows to adjusting countries became a key element of the reform package as a means of bridging the resource gaps (which stabilization measures could not immediately close without permanently damaging longer-term growth prospects) given that economic restructuring could not be accomplished quickly.

Concerns about the cost of adjustment and the unequal sharing of this burden especially since adjustment was turning out to be a long drawn-out process generated part of the impetus for the emergence of the third phase. This phase, therefore, includes special strategies for poverty alleviation, including provision of safety nets and targeted poverty programs. In addition, there is greater concern for capacity-building, human capital development, and for improving governance.

Trade and trade policy issues have been important elements of the three phases briefly described above. The restoration and subsequent maintenance of macroeconomic balance and stability could be assisted by appropriate trade policy measures. Similarly, re-alignment of incentives in favor of tradeables and the enhancement of efficiency with which resources are allocated and used are key fundamentals for promoting sustainable longer-term growth. Both also imply a significant role for trade policy. In the view of the World Bank (1994), poor or misguided trade policies accounted for Africa's economic stagnation of the 1970s and their reform would be necessary to restore both macroeconomic equilibrium and growth. More specifically, it is claimed (World Bank 1994:20) that:

> The main factors behind the stagnation and decline were poor policies—both macroeconomic and sectoral—emanating from a development paradigm that gave the state a prominent role in production and in regulating economic activity.··· Protectionist trade policies and government monopolies reduced the competition so vital for increasing productivity···. More important, the development strategy had a clear bias against exports, heavily taxing agricultural exports, one of the largest suppliers of foreign exchange.

In a further elaboration of the role of trade and trade policy in SAPs and its associated development perspective, the same source (pp. 23-24) declares that:

> Most (African) economies followed an inward-oriented, import-substitution strategy, supplemented by widespread use of tariff and nontariff barriers to reduce external competition, mainly in manufacturing···.This protectionism was another unfortunate policy choice, because competition increases productivity while trade restrictions increase input prices and the cost of capital, choking growth.

By implication, the development strategy favored by SAP framework would prominently feature an outward-oriented trade regime in which tariff and nontariff barriers against imports are liberalized and anti-export bias eliminated.

Trade Theory

The conventional and, in many ways, still dominant construct in international trade theory focuses on the growth-transmitting effects of international trade emanating from both the static and dynamic gains from trade. This focus derives directly from the early theoretical insights regarding the importance of trade as a means of widening markets, enhancing division of labor and, hence, raising the level of factor productivity.

These insights have been codified in the principle of comparative advantage which, in turn, leads to the neo-classical prescription that countries should "specialize in the production and··· export of those commodities in whose production (they) enjoy a comparative advantage" (Viner 1937:348). The principle also serves as the intellectual backbone of the more recent neo-liberal trade policy advocacy which argues that liberalization of markets is the key to policy reform. This rests, in effect, on traditional trade theory's focus on the allocative efficiency that could be achieved when markets are "free" and promote the actualization of comparative advantage and thus the maximization of the gains from international trade. In addition to the trade gains associated with static allocative efficiency, neoclassical trade theory also suggests that dynamic gains can accrue from increased access to technology and the exploitation of economies of state which are absent in small domestic markets.

Based on this conventional theory, the neo-liberal policy stance holds that economies that are shielded from international trade competition can be expected to have firms that operates at inefficiently small scales in the protected domestic market. As such economies are divorced from the international price structure, they could fail or be extremely sluggish in responding to international relative price changes. Insulated economies thus share various characteristics that hurt their international/competitiveness which can be expected, in turn, to adversely affect their growth prospects.

The neoclassical theory of international trade and the associated neoliberal trade policy stance have, of course, been confronted and challenged by alternative perspectives on the trade-development nexus. One such alternative view focuses on the unequal distribution of the gains from trade (Evans 1989). A major plank of this view rests on the idea that the structure of trade of the developing countries and the underlying dynamics of the world economy imply that the gains from active participation in international trade are biased against the developing countries because their patterns of specialization fail to bring about significant dynamic and growth-inducing effects and they tend to specialize in those goods and sectors in which linkages with the rest of the economy are tenuous and weak. In the extreme case, the developing countries such as those in Africa, that specialize in primary products tend to suffer from endemic long term deterioration of their terms of trade, emanating partly from the low income elasticity of demand of these commodities in the world market.

A more theoretically rigorous and less ideological challenge to neoclassical trade theory orthodoxy emerged in the 1980s. Its main point of attack is the perfectly competitive model on which so much of neoclassical trade theory rests. This "new" trade theory suggests that models incorporating imperfect competition, economies of scale, and learning effects constitute a more realistic framework for understanding international trade (Caves 1985; Helpman and Krugman 1985; Helleiner 1987; Rodrik 1987; Srinivasan 1989).

This theory views economies of scale as an important explanation of trade and specialization and a major factor of intra-industry trade. More importantly, it demonstrates that the effects of trade policy can be quite different from those postulated in conventional neoclassical models depending on how trade policy changes affect domestic market structures. Its recognition of the role of policy in "creating" international competitiveness and modifying dynamic comparative advantages goes some way in rehabilitating the significance of policy intervention, particularly of the "strategic" trade policy variety.

Trade, Trade Policy, and Development

The debate on the role of international trade in economic development has involved not only theoretical arguments but also empirical analysis. Conventional trade theory based on the principle of comparative advantage postulates that the expansion of trade is beneficial to all trading partners, that when commercial policies approximate free trade conditions the gains from trade are likely to be maximized and, by implication, overall economic growth is maximized.

This postulated link between trade and growth has been consistently challenged. Lewis (1978) deflates this relationship somewhat by suggesting that "the engine of growth should be technological change, with international trade serving as a lubricating oil and not as fuel." Diaz-Alejandro (1980) goes further by arguing that the purported link between trade and growth has not been rigorously and convincingly established. More specifically, he claims that "when all is said and done, we remain unsure as to whether and when trade is the engine, the handmaiden, the brake, or the offspring of growth (300)." Finally, Helleiner (1985:6) offers "good reasons for questioning whether there exist any such links" between trade and development and notes that
"the purported link between proper static allocation and growth remains… a "black box" about which it is best (and most honest for us to be cautious."

A variant of the trade-development nexus suggests that export expansion is positively related to output growth. Although this idea had been discussed and even empirically tested earlier, the 1987 *World Development Report* re-energized the debate on it by claiming that abundant evidence based on the experience of the previous thirty years provides strong support of the idea that growth in developing countries is closely related to their export growth. The key point here is that this evidence could provide a basis for the neo-liberal trade policy advocacy in favor of outward-oriented trade regimes and focus on export expansion which are also prominently reflected in the design of many African SAPs.

The postulated relationship between export performance and economic growth has been tested for many developing countries as well as for Africa. In her review of some of the earlier evidence, Krueger (1980:289) affirms that
"the relationship between export performance and growth is sufficiently strong that it seems to bear up under many different specifications of the

relationship." Similar studies focusing specifically on Africa (e.g., Oyejide 1975 and Fosu 1990) have reached conclusions that are broadly in line with a confirmation of the postulated positive relationship between exports and growth. But there are also studies that have failed to support this position. For instance, Taylor (1988:8) finds that "there is no obvious relationship between performance and overall openness to trade." In the same way, Helleiner (1986:146) concluded in his study that there is "no statistically significant link between the change in export share of GDP and growth; indeed, the sign on this relationship is consistently negative."

In spite of the mixed results obtained so far, two interesting elements of the basic idea appear to stand out and could be used with appropriate caution, as broad guides to policy. First, both Michaely (1977) and Helleiner (1986) find that export expansion in the poorer economies is not necessarily accompanied by increased overall economic growth. They infer that the validation of the export-led growth hypothesis appears to hold, if at all, only with respect to those developing countries whose level of development places them above a certain "threshold" level of per capita income. In the words of Michaely (1977): "this seems to indicate that growth is affected by export performance only once the countries achieve some minimum level of development (52)." Second, one of the findings of Balassa (1985) is that the greater the share of manufactured exports (in overall exports) the greater tends to be the contribution of exports to growth.

Taken together, both of these elements might easily feed into Helleiner's (1988) admonitions regarding the developmental implications of various categories of exports. Apparently, the lack of statistically significant relationship between export expansion and economic growth in low-income developing countries derives largely from the virtual absence of manufactures from their exports. Since the developmental implications of various categories of exports are different, broad selectivity should be exercised in promoting exports to ensure that only those export sectors which are particularly "developmentally nutritious" are given high priority.

Regional Integration and Trade Liberalization in Africa

Given the prominence of trade policy and its implications in African SAPs, an analysis of the African experience with regional integration and trade liberalization in the context of structural adjustment should directly confront several pertinent questions. These include the following: How and to what extent have the structural adjustment programs affected African regional integration schemes (RIs)? Have the objectives and operational modalities of the RIs been consistent and complementary with those of the SAPs? In cases of conflicts, which policy framework (RIs or SAPs) has been sustained and why? These questions will guide the analysis presented in this section.

Experience with Regional Integration

The African continent reputedly contains the largest number of regional and sub-regional groupings in the world (Torre and Kelly 1992). This implies that over the last three to four decades Africa has apparently chosen to organize and manage certain elements of its affairs through these regional and sub-regional bodies. From the economic perspective, the small size of the typical African economy and the perceived disadvantages associated with smallness would appear to be the main reason for the establishment of regional integration schemes.

One important area in which the disadvantages of smallness are particularly pronounced relates to the predominant import-substitution industrialization (ISI) strategy which was more or less universally adopted across Africa during the 1960s and 1970s. In the context of this development strategy, the small population of each African country combined with low per capital income placed sharp limitations on development and growth prospects within the limited market of each country. The disadvantages imposed by the narrow confines of each national market suggested regional economic integration as a means of widening the size of the market and thus allowing the benefits of greater specialization and economies of scale to be realized. In addition, import substitution in the context of larger regional markets could also be more efficient to the extent that the integration of national markets generates greater competition within the region which, in turn, promotes higher levels of productivity overall. The basic economic rationale for African RIs is succinctly articulated by McCarthy (1996:6) as follows:

> ··· the small size of most developing countries, notably those in Africa, restricts the ability of these countries to benefit from lower unit costs in a process of inward-looking development. Extending the market through regional integration suggests itself as the logical way to enable manufacturers to produce at lower unit costs for a larger protected market. The latter is regarded as a more viable source of the benefits of economies of scale than the world market because of the anticipated problems of access and the higher transaction costs of producing for the world market.

Deriving broadly from these considerations, African RIs have had three main types of objectives (Oyejide 1993). These include promotion of the growth of intra-regional trade by removing tariff and non-tariff barriers; promotion of regional development through sector planning and coordinating and mobilizing of funds for executing regional infrastructural and large scale manufacturing projects; and enhancing the region's market power in the world while reducing its external dependence. The fundamental goal of African RIs that is implicit in these objectives is to significantly increase trade within each integrated area and expand the area's overall trade. This goal would be achieved by carrying out agreed obligations regarding the elimination of barriers to the free movement of goods, services and factors of production, usually in accordance with a phased implementation schedule over a defined time period.

It should be noted, of course, that while most African RIs share these broad objectives, some actually espouse even more ambitious ones. The two key RIs in West Africa, i.e., the Economic Community of West African States (ECOWAS) and the Union économique et monetaire ouest africaine (UEMOA) both have objectives that go beyond the purely economic ones centred on growth through trade (Ogunkola 1994; Jebuni, Ogunkola and Soludo 1995; M'bet 1995). Similarly, the main RIs in Eastern and Southern, Africa such as the Common Market of Eastern and Southern Africa (COMESA) and the Southern African Development Community (SADC) expect to move beyond trade liberalization into deeper aspects of integration such as harmonization of broad economic policies within their areas (Lyakurwa, McKay, Ng'eno and Kennes 1993; Kasekende and Ng'eno 1995; Lyakurwa 1995). It is clear, however, that even those RIs that set their sights beyond the basic objective of promoting intra-regional trade and enhancing overall trade regard this as a key stepping stone or pre-requisite for the achievement of other and, perhaps, more ambitious goals. In any case, the performance of African RIs has generally been assessed in terms of this basic objective. One such recent evaluation is categorically negative. Specifically, Fine and Yeo (1995:3) conclude that:

"…regional integration would appear to be a dead end in terms of its ability to make a significant contribution in any substantive fashion to rapid and sustainable growth of the Sub-Saharan African economies."

This assessment reflects four salient results of various reviews of the performance of African RIs against their set objectives from the early 1970s through the early 1990s (Torre and Kelly 1992; McCarthy 1996). First, intra-regional trade as a proportion of total trade is much lower in African RI arrangements than in those of other regions, such as Asia and Latin America. Second, RI schemes in Africa have failed to boost the total exports of the areas covered by the various schemes relative to total world exports. Third, trade performance (measured by changes in trade/GDP ratios) of the "integrated" areas worsened significantly in the 1980s, in spite of the RIs and intra-regional trade actually bore the brunt of this sharp decline in trade performance. Fourth and finally, trade liberalization within the regions covered by African RI has generally been low. The widely documented failure of African RIs can be explained in terms of initial conditions, the problems of implementation, and basic design deficiency. From the beginning, African RIs lacked such success pre-requisites as pre-existing high level of intra-regional transactions, complementarities among regional partners in goods and factors of production, and potentials for product differentiation between regional partners emanating from differences in income levels and consumption patterns. In addition, most African RIs were established without strong private sector support and pressure; while many of the schemes lacked viable and well-managed mechanisms for redistributing benefits from the net gainers to the more disadvantaged regional partners.

For most of the RI schemes there were almost complete non-implementation of agreed trade liberalization schedules as well as other obligations entered into, presumably voluntarily, by members. It seems to have been fairly easy or costless to sign on the dotted line, but much more difficult to actually carry out many of the obligations. The absence of a strong and effective lobby for regional integration at the national level, and particularly in the private sector, clearly made non-implementation of integration responsibilities an easy option for most African governments.

Most African RIs were made up of countries whose basic development strategy during the 1960s and 1970s were essentially inward-oriented. The policy environment generated by this strategy contained an inherent anti-export bias. It thus had negative implications for RIs which, in principle at least, would require and thrive best in a pro-export environment. As McCarthy (1996:12) so aptly puts it: "···the acceptance at the national level of import-substituting industrial development policies created a bias against regionalism."

This is clearly a design fault. But there are others. Many of the RIs were over-ambitious, they had overlapping memberships and mandates that sometimes conflicted and were often unclear. More significantly, perhaps, the supra-national authority of the RIs was often either explicitly repudiated or implicitly ignored (Hess 1994; Oyejide 1995b; GCA 1995).

Since the early 1980s, two contrasting tendencies seem to have evolved regarding African trade policy and general economic development strategy. First, at the individual country level, many African countries have adopted SAPs, the major focus of which is outward-orientation with specific requirements for radical reform of the trade regime. In particular, outward-orientation also implies a closer integration of Africa into the world economy. Second, at the aggregate level, Africa's global and production-oriented approach to regional integration continues to reflect its attempt to create regional economies that are more independent of the world economy so as to insulate African economies from international shocks while, at the same time, promoting overall economic development through an inward-oriented import-substitution-industrialization strategy based on protected regional (and hence larger) markets.

The two tendencies are obviously inconsistent and cannot coexist permanently. It is clear that unilateral trade liberalization measures taken by individual countries under SAP effectively reduce the degree of preference enjoyed by regional partners in an integrated area. More significantly, these measures are typically articulated and implemented at the national level without reference to any existing or envisaged obligations under specific RIs.

Which tendency will prevail? Some analysts have already answered this question unequivocably. An example comes from Fine and Yeo (1995:21) in the following statement:

> Since structural adjustment policies in Africa invariably entail trade liberalization on a multilateral basis, deeper integration is more likely to occur as a consequence of these policies than through the efforts of regional and sub-regional entities especially established for this purpose. Indeed, these bodies··· could well obstruct rather than facilitate promising linkages that would otherwise emerge from trade liberalization undertaken unilaterally by many African countries.

The claim here is not that the concept of regional integration is necessarily inconsistent with outward-orientation in trade policy and development strategy. It is that African RIs need to be redesigned to reflect basic outward-orientation and acquire the characteristics of an effective framework for coordinating and harmonizing economic policies, for internalizing the regional dimensions of national-level policies, and for locking-in and thus promoting the credibility and sustainability of sound policies across the continent.

Experience with Trade Liberalization
Some Conceptual Issues
When trade liberalization occurs by deliberate design, it does so through the implementation of specific trade policy measures. But trade policy is often confused with other, and in some cases more general, macroeconomic policies. It is therefore necessary to clarify what trade policy is and what its major components and instruments are. In the context of Africa, trade policy changes constitute only one set of the many policy reforms being undertaken, usually under the general umbrella of struc-

tural adjustment programs. However, this set appears to carry substantial weight in the typical reform package. The rather sharp focus on this component may be part of the reason why trade policy is often not sufficiently differentiated, or separable, from broader macroeconomic policies. Yet, analytical convenience and the need to relate particular policies and its major components are separately identified from other policies.

In general, trade policy measures are targeted at the tradeable goods and services sector. More specifically, they influence the overall structure of incentives within this sector and thus affect the relative prices between importables and exportables. In the process, trade policy measures exert their impact through changes in the incentives for producing and consuming, as well as exporting and importing various types of tradeable goods and services (Helleiner 1992-1995). It should be obvious that while trade policy measures aim primarily at influencing the composition and levels of imports and exports, this does not preclude other factors and policies from affecting the levels of exports and imports. Clearly, exogenous changes in levels of income and/or general economic activity in a country would tend to influence both tradeable and non-tradeable sectors. At the same time, some macroeconomic policies (for example, exchange rate policy) could alter relative prices between non-tradeable and tradeable goods and hence influence the levels of exports and imports. Disentangling the effects of pure trade policy on the composition and levels of imports and exports from the effects of other policies and exogenous factors is one of the many challenges facing empirical research in this area.

Discussions of trade policy reform, especially in the context of Africa, are confined almost exclusively to a debate about the pros and cons of trade liberalization. How far and how fast should it proceed? What should be its coverage? etc. This contextual framework reflects the reality that many African countries have embraced and are implementing various forms of trade liberalization measures. Thus, in this context, trade policy reform is coterminous with trade liberalization, which in turn refers to policy changes and associated measures that are used to endow the trade regime with a more neutral incentive structure. In the process, the policies aim to reduce domestic policy distortions and open the economy to the world market. A neutral trade regime is one whose incentive structure does not discriminate between exportables and importables or between production for the domestic market and for export sales. More specifically, trade liberalization would imply transforming the trade regime from an "inward-oriented" stance that discriminates in favor of (and thus protects) import-competing activities into a "neutral" regime whose incentive structure does not distinguish between exportables and importables—or into an "outward-oriented" trade policy regime that discriminates in favor of (and thus actively promotes) exports.

It has been argued that trade liberalization that produces either a "neutral" or an "outward-oriented" trade regime confers certain productivity-enhancing and growth-promoting features on the liberalized economy (World Bank 1991). Included among these are improvement in the efficiency with which resources are allocated, increase in competition and product investment, and the creation of a favorable environment for technology transfer. In the context of a more open trade regime, technology can be transferred through at least three distinct sources: as an integral part of foreign investment, through increased trade that allows a country at the frontiers of technological development. This perspective thus suggests, first, that a trade regime that enhances import and export competition would induce increased trade and promote more rapid economic growth; and secondly, that the key to the productivity increases associated with expanded trade lies predominantly in "exporting (which) strengthens the incentive to adopt new technology by increasing the returns from innovation through expanded market opportunities" (World Bank 1991:89).

There exists, however, an alternative perspective that associates the growth-promoting effect of expanding trade with the productivity-enhancing characteristics of technology by increased imports of capital and intermediate inputs (Helleiner 1995). This perspective would, in effect, be consistent with a trade regime whose incentive structure discriminates in favor of technology-embodying imports of capital goods and intermediate inputs as a means of capturing the growth-promoting gains derivable from increased capacity utilization and the productivity-enhancing gains emanating from technology transfer. Even in the context of this perspective, however, there is a clear recognition of the important role of exports, not necessarily as the critical vehicles for technology transfer but more significantly as the primary source for financing the indispensable bottleneck-breaking and technology-bearing imports, especially in the absence of adequate capital inflows. Both perspectives suggest that the two (import and export) components of trade policy reform may have important, distinct and different implications. Hence, their differences deserve to be explicitly taken into account in the design and implementation of trade liberalization programs.

The import component of trade liberalization is of special significance in Africa for a variety of reasons. The standard instruments of trade policy, such as import tariffs and quantitative import restrictions, are often used to serve multiple objectives in this region, given the country characteristics that limit the availability and effectiveness of alternative policy handles. Hence, trade policy reform, especially as it relates to the import component, needs to be carefully designed to ensure that its full implications for the various objectives are explicitly recognized and taken into account.

The fiscal impact of changes in the form and level of import protection constitutes one of these areas of concern. Compared to other regions of the world, sources of government revenue are much narrower in Africa. This narrow base translates into an unusually heavy reliance on trade taxes. Import liberalization may lead to increased trade deficits and thus worsen the usually precarious balance-of-payment position of the typical African country. A deep, generalized and sudden import liberalization program could induce an abnormally high demand for imports, which could place unsustainable pressure on the country's balance-of-payments requiring access to substantial aid and capital inflow for a compensating exchange rate policy change. As Collier and Gunning (1992) demonstrate, although import liberalization does not inevitably worsen the balance of payments, it seems likely to do so, particularly in the short to medium term. Increasing trade deficits arising from import liberalization should worry the typical African country, which lacks flexible access to external financing.

In addition to their use for generating fiscal revenue and confronting balance-of-payments problems, various import protection measures are also used in African countries as instruments of industrial policy. These countries are justifiably concerned with the survival and future development of an efficient manufacturing sector as an important component of their development. In this context, some level of protection may need to be given to these import-competing activities, on traditional "infra-industry" (or, in this case, "infant-sector") grounds, provided it is understood that the theoretical justification for such protection also requires that it be temporary and thus strictly time-bound. This reflects the idea that while the protected activities are not internationally competitive in the short term, they have potential to become competitive in the medium term. Furthermore, import protection offered on these grounds may also be differentiated to favor particular sub-sectors of the economy that exhibit positive externalities.

These two important considerations argue the case for some non-zero import tariff levels and a non-uniform import tariff structure. At the same time, however, considerations of allocative efficiency suggest that the level of import protection should be moderate to limit the relative price distortions they can create and hence the damage they can do to the economy (Rodrik 1988). The tariff structure should be fairly simple, reflecting a limited number of tariff rates, to ensure greater transparency (Helleiner 1992).

Given its significant implications for fiscal deficits, balance-of-payments pressures and the need for some degree of industrial protection, the future direction, scope and speed of import liberalization in Africa are likely to be influenced by the relative weights attached by policy-makers to these concerns.

The export component of trade liberalization also deserves close scrutiny. This derives from the proposition that good export performance is not just a desirable goal in its own right, but is also a critical means to other important ends, i.e., deeper import liberalization and more robust overall economic growth. A successful and sustainable import liberalization program requires successful exports; or put differently, a country must export in order to import (Hachete 1991; Snape 1991).

As it relates to the export sector, trade liberalization would cover policy changes that produce a "neutral" trade regime that treats exportables and importables similarly as well as an "outward-oriented" trade regime that favors and actively promotes exports. Establishing a "neutral," much less an "outward-oriented" trade regime, as defined above, in the presence of some degree of import protection is not straightforward. Those policy measures that endow a trade regime with its import protectiveness, such as tariffs and non-tariff barriers, have an anti-export bias. More specifically, restrictions on imports translate effectively into a tax on exports; by making import substitutes relatively more profitable they increase the costs and reduce the availability of imported inputs used in the production of exportables, forcing exporters to use relatively expensive and low quality locally produced inputs. Import restrictions also subject exporters to a more appreciated exchange rate than they would otherwise have faced. These different elements of the tax on exports imposed by import restrictions combine to reduce the international competitiveness of the export sector.

To promote exports, the anti-export bias that inherently characterizes a trade regime containing a degree of import protection needs to be eliminated. This can be done through schemes that grant exporters and their suppliers unrestricted access to inputs (imported and locally produced) at internationally-competitive prices that are free of import duty and other indirect taxes. Effective systems to provide "neutral" trade regimes for exporters may be established in the framework of specially created export processing zones. Or, this can be done more simply through duty drawback or exemption schemes as well as bonded warehousing arrangements.

Given the widely recognized role of foreign exchange reserves and export performance in the initiation and sustainability of deep import liberalization, an important question is whether a stage of export promotion aimed at boosting exports should precede that of import liberalization (Shepherd and Langoni 1991; Nash 1992). This issue derives an additional relevance from the heavy emphasis of the conventional model of trade liberalization on opening the domestic economy to competition from imports (Coes 1991). Based broadly on the notion that, in foreign-exchange-constrained economies, exports can be regarded as a means of acquiring the foreign exchange with which to purchase increased imports made available through trade liberalization, it is suggested (Nash 1992:63) that "introducing other export policy reform shortly before, or at least at the same time as, import reforms permits an earlier export supply response and allows unification of the tariff structure to proceed without burdening exporters."

This principle appears to have featured prominently in the trade liberalization experiences of countries such as Brazil (Coes 1991) and Argentina (Cavallo 1991).

These and other similar experiences acknowledge the pivotal role of good export performance in sustaining import liberalization. Hence, whenever the required export performance can be more readily induced through specific export promotion measures, it would be reasonable to implement them before import liberalization is attempted. In countries with clearly defined development strategies and adequate capacity to implement them, appropriate export promotion measures can be used to achieve trade "neutrality" and thus secure outward-orientation prior to full-scale import liberalization (Levy 1993).

In any case, full-scale import liberalization is not normally achieved in one step, even in those countries (such as Chile) that choose to adopt a rapid trade liberalization process. These include a change in the form of protection, the simplification of the system of protection, lowering of tariff rates and reduction in the dispersion of protection rates. In this context, Helleiner (1995:74) refers to "the conventionally recommended sequence, beginning with the gradual reduction and removal of import quotas and licenses and their replacement with tariffs···" followed by the lowering and simplification of tariffs···.'

Replacing quantitative restriction with tariffs, or tariffication, is in itself a major element of trade liberalization. It enhances the transparency of the system of protection. Compared to non-tariff barriers, tariffs are more visible and predictable and, hence, less susceptible to corruptive manipulation. In addition, tariffs generate government revenue and should, to that extent, help to remove some of the fiscal objectives to trade liberalization. Given the clear superiority of tariffs over non-tariff barriers, countries that choose to use trade protection should rely on tariffs.

The next step in the import liberalization process can be rapid or gradual. In general, it seems reasonable to suggest that the speed with which the rationalization and lowering of tariffs should be achieved would depend on each country's initial conditions and the response of its export sector. A gradualist approach may, obviously, run the risk of "capture" and frustration by vested interest; this possibility provides the main argument for "shock therapy." Analysts who propose rapid trade liberalization tend also to indicate quite specific targets and time tables. Thus, Valdes (1993:280) suggests that a country undergoing trade liberalization should not only announce its program of tariff reductions in advance, but also accomplish its tariff reduction target "in not more than approximately three years." This rigid stance requires that the likely import surges associated with deep and rapid trade liberalization can be financed either by a comfortable level of foreign exchange earnings and reserves or through an adequate level of foreign assistance. Countries that cannot count on such assured, adequate and sustainable level of import financing should necessarily be more circumspect with regard to both the speed of tariff reduction and the ultimate target level envisaged in their trade liberalization programs.

Trade liberalization can be carried out unilaterally; it may also be implemented in the context of bilateral, regional or multilateral (global) arrangements. There is a sense in which these different trade liberalization mechanisms and modalities are related and they can be usefully integrated, wholly or partly, in a country's trade policy strategy. The strategic elements that characterize trade liberalization in the context of regional and multilateral agreements include peer pressure and reciprocal actions that help participating countries to lock-in the achievements of liberalization. Reciprocity enables each country to take liberalized policy actions that they would otherwise hesitate to take unilaterally. Such policy measures become less susceptible to pressures from local anti-reform interest groups in each country, while the fear of group penalty renders each country's policy largely irrevocable. More specifically, multilateral trade liberalization commitments entered into through GAA/WTO negotiations provide an opportunity for countries to go beyond their unilateral liberalization efforts in exchange for multilateral concessions. They can also be used to support domestic trade reform efforts as they offer an international framework for binding these reforms and thus preventing policy reversals.

An important new dimension of the strategic element to African trade policy has recently emerged (Collier and Gunning 1995; Fine and Yeo 1995). Based on the observations that trade liberalization in the context of regional arrangements has had little or no success, while unilateral trade liberalization measures implemented with donor support in the context of structural adjustment programs often lack credibility and are susceptible to reversal, it is suggested that African trade liberalization efforts may be more credible, sustainable and, hence, successful if designed and implemented in the context of a free trade area that links Africa with Europe. This reciprocal arrangement would provide as much gain as the region can expect from multilateral or global free trade, given Africa's current and prospective export markets. It would also serve the defensive purpose of protecting the region from being locked out of an emerging world of trade blocs, and provide a powerful lock-in mechanism that ensure credibility.

Focus and Extent of Trade Liberalization
Trade liberalization in many African countries has occurred in the context of a fairly broad and general phenomenon. Since the early 1980s, a marked shift from inward looking towards outward-oriented and more open trade regimes has characterized the trade policies of most developing countries worldwide. This common trend does not, of course, imply that the trade liberalization experience has been the same in different countries and across various regions. In fact, there are significant differences along such dimensions as the motivations for trade liberalization, the initial conditions from which different countries started their liberalization process as well as the scope, speed and sequencing that various countries have chosen for reforming their trade regimes. The trade liberalization experience of African countries is reviewed, in this sub-section, bearing in mind these various dimensions.

During the 1960s and 1970s, many African countries rapidly built up, in a rather haphazard manner, highly interventionist and protectionist trade regimes. These regimes were broadly characterized, on the import side, by restrictive licensing systems, fairly high tariffs, escalated or cascading tariff structures made up of several layers, varying degrees of import prohibitions, and tight foreign exchange controls. On the export side, the trade regimes featured substantial implicit and explicit taxes as well as frequent use of non-tariff barriers such as export prohibition.

These heavily protectionist trade regimes were motivated apparently, by several different concerns. The multiplicity of objectives, quite a few of which were conflicting, probably accounts for the rather haphazard, incoherent and internally inconsistent nature of the trade regimes that eventually emerged in many African countries. One of the key concerns that trade policy had to address in many African countries was the raising of government revenue. Given the heavy reliance of these countries on tariffs for revenue, it is not surprising that fiscal concern served as the primary motive for imposing high tariff rates on both imports and exports. At the same time, these countries were highly susceptible to balance-of-payments pressure, given their strong commitments to the maintenance of often unrealistic fixed exchange rate systems. In this context, trade policy often became a substitute for more appropriate policies needed to maintain macroeconomic discipline (such as exchange rate policy reform, for instance). Hence, non-tariff import controls were often applied whenever it becomes necessary to deal with recurring balance-of-payments crises. This is clearly consistent with the evidence that some 80 percent of the quantitative import restrictions notified to the GATT by African countries during the 1970s and early 1980s were justified for balance-of-payments reasons under GATT Article XVIII.B.

Some aspects of Africa's protectionist trade regimes in the 1960s and 1970s could also be traced to the desire to shelter domestic industry, as a means of enhancing overall development, in the context of the import-substitution-industrialization strategy which was quite popular at that time. Evidence in support of this include extensive exemption from duties and low tariff rates for imported inputs used by local producers, as well as the generally escalated structure of tariffs that imposed high rates on finished products and much lower rates on raw materials. It should be noted, however, that tariff rates were applied to so-called "luxury" goods thought to be consumed largely by the rich and for which no local production facility existed. Thus, high tariff rates did not always reflect the desire to protect local industry; in certain cases they reflected the desire to raise revenue based on perceived ability of the consumer to pay.

There is a strong relationship between the use of import restrictions and the appearance of balance-of-payments problems in many African countries; while the increasing significance of border taxes as a contributor to government revenue can be matched with the rising level of import tariff rates in these countries up to the early 1980s. Taken together, these relationships and trends suggest that budgetary needs and balance-of-payments concerns had much stronger impact on the evolution and structure of the trade regime in African countries than the desire to protect local industry.

Beginning around the mid-1980s, many African countries started the difficult journey of rationalizing and liberalizing their trade regime. They have been guided and prodded along on this journey by the World Bank and the IMF in the context of integrated packages of economic policy and associated reforms conventionally referred to as SAP. In spite of this common source of stimuli, progress across the countries involved has been uneven, while the sequencing of the liberalization process has been problematic. Many of the liberalization attempts have not followed the standard sequence of replacing quantitative restrictions with high (equivalent) tariff rates first before adjusting the tariffs towards lower and more uniform levels subsequently and in a phased manner.

The liberalization efforts have generally concentrated on the import side and have, to that extent, paid much less attention to the elimination of anti-export bias inherent in the trade regime. There is clear evidence, however, that "protection of import substitutes by tariffs and non-tariff barriers (NTB) in Sub-Saharan Africa as a whole has declined" (Nash 1993:38). More specifically, the level of protection has fallen by between 30 percent and 50 percent over the period from the mid-1980s to the early 1990s, and the downward trend has continued through the mid-1990s. Thus, a more recent assessment (World Bank 1995:24) finds that "since the mid-1980s, most African countries have moved from complete or nearly complete government control over imports to more open systems, and have substantially reduced the number of imports subject to quantitative barriers." It goes on to indicate that "the greatest progress has been achieved in replacing quantitative restrictions with lower and less dispersed tariff levels, more than half the countries now have average tariff rates of 15-20 percent with the highest rates set at 35-40 percent, and the number of tariff categories reduced to 4-5." But on the export side, the picture is much less favorable. According to Nash (1993:42):

"there has been little progress on establishing efficient systems to give exporters access to inputs at internationally competitive prices" since the various institutional mechanisms for achieving this objective, such as export processing zones or duty drawback and exemption schemes remain underdeveloped in virtually all African countries.

Impact of Trade Policy Reform

A descriptive policy account may provide an indication that a specific policy reform has taken place or has not. The primary reason for implementing policy reform is, of course, to influence the targeted economic variable; the corresponding change in this target variable would then serve as an indicator of policy impact. In principle, therefore, the impact of African trade policy reform can be assessed by examining the changes in appropriate target variables that can be ascribed to trade policy reform.

There are several reasons why the impact evaluation procedure described above is not as straight forward as it may, at first, appear. First, as indicated earlier, it may be virtually impossible to disentangle the effects of trade policy reform from those emanating from more general macroeconomic policies or even exogenous developments that may impact on the same set of variables. Second, trade reform episodes in particular countries may be too short to permit an evaluation based essentially on an analysis of time-series data. Third, whether a discernible impact of a trade reform episode can be found may depend quite critically on whether private agents (whose actions or inactions ultimately determine "supply response" to policy change) regard the policy as credible and, hence, sustainable. Fourth, the attribution of observed post-reform changes in the target variables to the proceeding reform effort based on the well known "before and after" comparisons is not related to appropriate counterfactuals and can be criticized on this account.

Bearing these problems in mind, the rest of this section summarizes the evidence on credibility, policy reversals and impact of trade liberalization episodes for ten African countries (Cote d' Ivoire, Ghana, Kenya, Mauritius, Nigeria, South Africa, Tanzania, Uganda, Zambia, and Zimbabwe) covered in a recent study (Oyejide, Ndulu and Gunning 1996).

This study confirms that African trade liberalization attempts are particularly prone to the twin problems of credibility and policy reversal. More specifically, credibility problems have typically resulted when the trade liberalization policy stance either generates an unsustainable balance-of-payments deficit or an unsustainable budget deficit. In either case, private agents would expect a policy reversal and act in a manner that would enable them profit from the policy reversal rather than in terms of a rational response to a credible and sustainable policy change. Of the ten countries identified, only three (i.e., Mauritius, Uganda and Zimbabwe) had credible and sustained trade liberalization episodes; although, even among these three, Zimbabwe risked a partial policy reversal when it sharply increased statutory tariffs in January 1995. Partial policy reversals accompanied by frequent policy shifts characterized trade liberalization attempts in five of these countries. This leaves two extreme cases (Nigeria and Zambia) in which total policy reversals occurred; Nigeria reversed its 1986-94 liberalization process abruptly in 1994, while Zambia executed a policy "about turn" in 1987. In virtually all cases of incredibility and/or policy reversal uncovered in this study, the corresponding trade liberalization episodes appear to have generated unsustainable balance-of-payments deficits (i.e., they were payments incompatible) and the governments concerned were apparently unwilling to use exchange rate policy aggressively enough as a means of restoring payments compatibility, in the absence of adequate and sustainable external financing.

The typical target variables in an evaluation of the impact of trade liberalization include output change, change in various components of trade, change in the performance of the manufacturing sector, and change in employment.

Based on changes in these variables, Mauritius stands out as the country with the most successful experience with trade liberalization. Its fairly long and effective trade liberalization process had positive impacts on trade, output and employments (Dabee and Milner 1995). Over the 1979-

90 period, its import/GDP ratio rose by 35 percent while the export/GDP ratio increased by 46 percent. Real GDP growth was maintained at an average annual rate of 5.6 percent; while the unemployment rate declined steadily from 28 percent in 1982 to less than 5 percent in the 1990s. It should be noted, of course, that the maintenance of macroeconomic stability, other supportive policies and a favorable external environment for exports from Mauritius played significant roles in achieving these results.

In Uganda, the import/GDP ratio increased from 9 percent to 20 percent between 1989 and 1993, but the export/GDP ratio remained roughly constant at 5 percent (Ssemogerere 1996). Only massive inflow of resources (from official and private sources) made it possible to sustain the trade liberalization process. Real GDP grew by an annual average of 6.4 percent between 1991 and 1995 while anecdotal evidence suggests positive impacts in terms of employment and performance of the manufacturing sector. In spite of the increase in the tax revenue/GDP ratio from 7 percent to 12 percent between 1991 and 1995, Uganda's liberalization could encounter fiscal incompatibility problems in the near future. In the same way, Uganda's heavy reliance on massive external financing to maintain the payments compatibility of its trade liberalization process should ring alarm bells.

Zimbabwe's trade liberalization process is more difficult to evaluate in terms of its probable impact. This is because the episode is relatively short and it largely coincides with the sharp negative drought shock that caused a drop of 7.7 percent in real GDP in 1992 alone. Overall, however, real GDP growth averaged 3.4 percent per annum in the early 1990s; while the manufacturing export sector grew by 19 percent in real terms over the 1990/91-1994/95 period. It is also estimated that employment increased by 18.4 percent between 1991 and 1995 (Ncube et al. 1995).

In spite of frequent but partial policy reversals in Kenya, its trade liberalization is assessed to have had positive effects on growth in imports, exports, output and employment (Mwega 1995). More specifically, import/GDP ratio rose from 26 percent in 1972 to 36 percent in 1993, the export/GDP ratio increased from 26 percent to 42 percent over the same period. Real GDP growth averaged 5 percent per annum in 1988 and 1989, recorded less than 5 percent annual growth rate until 1995 when it bounced back to 5 percent. Trade liberalization does not appear to have had any marked impact on the manufacturing sector, however.

A possible threat to the current liberalization process derives from its reliance on temporary balance-of-payments support in terms of external financing. But this does not appear to have affected its credibility.

Ndulu et al. (1995) demonstrates that liberalization of the exchange regime, rather than trade liberalization, has had the dominant effect on the growth of imports, exports and output. In the case of South Africa, Bleaney et al. (1995) concludes that trade liberalization prior to 1995 has focused largely on export promotion measures to eliminate the anti-export bias inherent in the protective trade regime. Import liberalization has only just started. Export promotion measures combined with exchange rate depreciation helped to produce average annual growth rate of 7.7 percent in exports over the 1984-90 period.

Ghana has had a fairly long period of uninterrupted trade liberalization experience. But, according to Tutu and Oduro (1995), the results have been less than impressive. While real GDP growth has averaged over 4 percent per annum from the mid-1980s to the early 1990s, the response of non-traditional exports has been modest. Sectoral composition of output has shifted away from import substituting and non-tradeable to tradeable sectors, as anticipated. But unemployment appears to be worsening due to the de-industrialization consequences of trade liberalization and public sector retrenchment programmes.

The results of the trade liberalization experiences of Nigeria and Zambia have some similarities. In both cases, a dominant mineral sector seems to have deflected the impact of trade liberalization; although, in both cases, the policy was reversed before it could have any real impact. In any case, Musonda et al. (1995) concludes that liberalization measures probably had only a marginal effect on the performance of the economy. During the decade from 1983, the import/GDP ratio fell from 52 percent to 33 percent, while the export/GDP ratio declined from 29 percent to 20 percent. In the case of Nigeria, Ajakaiye and Soyibo (1995) concludes that trade liberalization had no significant effect on output, employment and imports.

A few summarizing generalizations can be made on the basis of the trade liberalization experiences of this set of countries. First, there has been a shift of resources away form import-substituting and non-tradeable sectors to the tradeables. As a result, exports have responded positively, although modestly, and trade shares have generally increased. Second, some amount of de-industrialization has occurred in some countries. Trade liberalization has unleased competitive pressures that many previously sheltered and inefficient industrial firms have been unable to cope with; yet new export-oriented activities have not bloomed sufficiently to take up the slack. Third and finally, continued credibility of some of the trade liberalization processes faces serious challenge as their heavy reliance on external financing may not be sustainable.

Regional Integration and Trade Liberalization in Africa's Development Strategy

The trade regime of many African countries in the 1960s and 1970s evolved through an ad hoc and haphazard process, they reflect attempts to use one instrument (i.e., trade policy) to achieve several (and, sometimes, inconsistent) objectives. The policy reforms which began in the early 1980s and are, in many cases, still on-going have achieved some significant results but much still remains to be done especially in terms of rationalizing the trade regimes. In order that further policy reforms may be effective, they should be designed and implemented in the context of a comprehensive framework that takes account of the initial conditions in each country, its long-term development objectives and the strategy for achieving them. It is within this context that one should pose the questions: what role can trade policy play? What direction, scope, and sequencing of trade policy reform are indicated?

A broad consensus has emerged which indicates that, in the long-term, development in African countries should focus on the achievement of rapid, equitable and sustainable economic growth (World Bank 1995). Both theories and accumulated experience of successful development experience elsewhere (e.g., East Asia) suggest that a realistic overall development strategy for achieving this growth objective should stress broad outward-orientation. The peculiar characteristics of most African countries as small and historically "open" economies (in the sense of high ratios of trade to GDP) are broadly consistent with this focus. Africa's failed experiment with its largely inward-oriented development strategy of the 1960s and 1970s should also quite rightly provide an additional support for a radical change in orientation. An outward-oriented development strategy assumes that aggregate economic growth will be export-led in an environment in which macroeconomic balance and stability are maintained.

Development experience world-wide suggests that industrialization plays a significant role in the process of establishing rapid and sustained economic growth; the current marginalization of Africa in this respect would have to be addressed if the region's long-term development objective is to be achieved. The initial conditions regarding industrialization in many African countries become particularly relevant at this juncture. The region's previous import-substitution industrialization (ISI) strategy has not only failed to deliver sustainable growth of the manufacturing sector but it has also created a difficult legacy of massive inefficiency and heavy import-dependence that needs to be overcome before a new push in promoting industrialization could be worthwhile. More specifically, promoting export-oriented industrialization against the background of the ISI legacy requires (i) restructuring the existing manufacturing sector whose structure and efficiency levels are ill-suited to exporting, (ii) altering the incentive regime to provide exporters with access to inputs at international prices net of tariffs and indirect taxes, and (iii) using appropriate pro-active measures to assist manufacturers and exporters to cope with the production and learning-by-doing externalities associated with manufacturing and the difficulties of gaining access to the information and technology needed to break into export markets (Lall and Stewart 1995).

The initial conditions of Africa and its long-term development objective and strategy briefly sketched above provide the context within which future trade policy reform can be meaningfully discussed. This should begin with the recognition that credible and balanced macroeconomic policies are essential for reaping the benefits of trade liberalization. In any case, the prior achievement and continuous maintenance of macroeconomic balance and stability would release standard trade policy instruments from inappropriately chasing revenue and balance-of-payments objectives and enable them concentrate on the more legitimate role of enhancing efficiency in resource allocation and promoting long-term development via industrialization and export expansion and diversification.

Trade policy instruments would not be released from the duty of chasing revenue objectives until the importance of border taxes in the tax system is significantly reduced. This can be achieved through increased reliance on a substantially revamped domestic indirect tax system which, for example, replaces the existing patchwork of sales taxes with a more comprehensive value-added-tax. Similarly, a more flexible and responsive exchange rate policy backed with adequate financing is necessary if trade liberalization is to be consistent with the maintenance of balance-of-payments stability. Deep import liberalization cannot be sustained and would not be viewed by private agents as credible in the absence of these conditions.

Following the establishment of macroeconomic balance and stability, full import liberalization would be the first best way to eliminate anti-export bias and this promotes exports. But, given the existence of some manufacturing capacity in many African countries part of which could be restructured, revived and transformed into export-oriented local production facilities, a certain degree of protection should be retained. The inherent anti-export bias in the resulting trade regime should be compensated for by providing direct and indirect exporters access to imported inputs at international prices through the effective use of such institutional mechanisms as export processing zones, in-bond warehousing, and duty drawback and exemption schemes.

Future trade liberalization programs in Africa would be more credible and effective if they emanate from clearly articulated and coherent strategies that explicitly recognize specific objectives and address both the import and export sides of the trade "coin." Similarly, the context in which trade liberalization programs are implemented could have significant implications for their success and sustainability. Most of Africa's recent trade liberalization efforts have been based on SAPs supported by the World Bank. The basic model that underpins these programs assumes that adjusting countries would "receive balance of payments assistance equivalent to 20-50 percent of exports earnings" during their liberalization process (Harrold . 1995:89). As Collier and Gunning (1992) convincingly argue, such aid-induced trade liberalization attempts may not be credible and therefore suffer reversals because the "inducement" is not permanent and its temporary nature is quickly recognized by private agents. In competition, credibility of trade liberalization is enhanced when it is implemented in the context of a "reciprocal-threat" arrangement in which the certain reactions of partners make unilateral policy reversal costly. Hence, the credibility of future trade liberalization attempts in Africa is likely to be enhanced to the extent that they are "GATT-bound" or are implemented within the terms of effective regional integration (including north-south reciprocal) schemes.

Works Cited

Ajakaiye, D.O. and Soyibo, "Trade Liberalization in Nigeria, 1970-93: Episodes, Credibility and Impact." Mimeo, AERC, Nairobi (1995).

Balassa, B., "Exports, Policy Choices and Economic Growth in Developing Countries after the 1973 Oil Shock," *Journal of Development Economies*, 18 (1985):23-26.

Bleaney, M., M. Holden and C. Jenkins, "Trade Liberalization in South Africa." Mimeo. Nairobi: AERC (1995).

Cavallo, D.F., "Argentina: Trade Reform, 1976-1982." In: G. Shepherd and C.G. Langoni, eds., Trade Reform: Lessons From Eight Countries, San Francisco: ICEG (1991).

Coes, D., "Brazil: Precedents and Prospects in Foreign Trade." In: G. Shepherd and C.G. Langoni, eds., *Trade Reform: Lessons From Eight Countries*, San Francisco: ICEG (1991).

Collier, P. and J.W. Gunning, "Aid and Exchange Rate Adjustment in African Trade Liberalizations," *Economic Journal*, 102, (1992):925-939.

_____. "Trade Policy and Regional Integration: Implications for the Relations between Europe and Africa." *The World Economy*, 18 (1995):387-410.

Debeen B. and Milner, "Evaluating Trade Liberalization in Mauritius." Mimeo. Nairobi: AERC (1995).

De la Torre, A. and M.R. Kelly, 1992, *Regional Trade Arrangements*, Occasional Paper 93, IMF, Washington, D.C.

Diaz-Alejandro, C.F., "Trade Policies and Economic Development." In: P. Kenen, ed., *International Trade and Finance*, Cambridge University Press (1980).

Evans, H.D., "Alernative Perspectives on Trade and Development." In: H. Chenery and T.N. Srinivasan, ed., *Handbook of development Economics*, North-Holland, Amsterdam (1989).

Fine, J. and S. Yeo, "Regional Integration in Sub-Saharan Africa: Dead End or a Fresh Start?" Mimeo. Nairobi: AERC (1995).

Fosu, A.K., "Exports and Economic Growth: The African Case." *World Development*, June (1990).

Global Coalition for Africa (GCA), "Rationalization and Strengthening of Regional Integration Institutions in Africa: Background and Recommendations." Mimeo. Washington: GCA (1995).

Hachette, D., "Chile: Trade Liberalization since 1974." In: Shepherd, G. and C.G. Langoni, eds., Trade Reform: Lessons from Eight Countries, San Francisco: ICEG (1991).

Harrold, P., M. Jayawickrama, and D. Bhattasali, *Practical Lessons for Africa From East Asia in Industrial and Trade Policies*, World Bank Discussion Papers, Africa Technical Department Series, Washington, D.C.: World Bank (1995).

Helleiner, G.K., "Industrial Organization, Trade and Investment, A Selective Literature Review for Developing Countries." Mimeo, Toronto (1985).

_____. "Outward Orientation, Import Instability and African Economic Growth: An Empirical Investigation." In: S. Lall and F. Steward, ed., *Theory and Reality in Development*, London: Macmillan (1986).

_____. "Trade Strategy in Medium-Term Adjustment." Mimeo. Helsinki: Wider (1988).

_____. "Introduction." In: Helleiner, G.K., ed., *Trade Policy, Industrialization and Development: New Perspectives*, Oxford: Claredon Press (1992).

_____. *Trade, Trade Policy and Industrialization Reconsidered*, World Development Studies No. B, Helsinki: Wider (UNU) (1995).

Helpman, P. and P.R. Krugman, *Market Structure and Foreign Trade: Increasing Returns, Imperfect Competition and the International Economy*, Cambridge: MIT Press (1985).

Hess, R., "Rationalization and Strengthening of Integration Institutions in Africa." Mimeo. Washington: GCA (1994).

Jebuni, C.D., E.O; Ogunkola, and C.C. Soludo, "A Review of Regional Integration Experience in Sub-Saharan Africa: Case Study of ECOWAS." Mimeo. Nairobi: AERC (1995).

Kasekende, L.A. and N. Ng' eno, "Regional Integration and Economic Liberalization in Eastern and Southern Africa." Mimeo. Nairobi: AERC (1995).

Krueger, A.O., "Trade Policy as an Input to Development." *American Economic Review*, 70, 2, (1980):288-292.

Lall, S. and F. Stewart, "Trade and Industrial Policy in Africa." North South Roundtable, Johannesburg (1995).

Levy, B., "An Institutional Analysis of the Design and Sequence of Trade and Investment Policy Reform." *The World Bank Economic Review,* 7, 2, (1993):247-262.

Lewis, W.A., "The Slowing Down of the Engine of Growth." *American Economic Review,* 70, (1978):565-564.

Lyakurwa, W.M., *Trade Policy and Promotion in Sub-Saharan Africa.* Special Paper, No. 12, Nairobi: AERC (1993).

_____. "Regional Integration and Trade Liberalization in Sub-Saharan Africa: Regional Case Study of SADC." Mimeo. Nairobi: AERC (1995).

_____.A. McKay, N. Ng' eno and W. Kennes, "Regional Integration in Sub-Saharan Africa: A Review of Experience and Issues." Mimeo. Nairobi: AERC (1993).

M' bet, A., "CEAO and UEMOA within ECOWAS: The Road Ahead Towards West African Economic Integration." Mimeo. Nairobi: AERC (1995).

McCarthy, C., "Regional Integration in Sub-Saharan Africa: Past, Present and Future." Mimeo. Nairobi: AERC (1996).

Michaely, M., "The Lessons of Experience: An Overview" in Shepherd, G. and C.G. Langoni, eds., *Trade Reform: Lessons From Eight Countries,* San Francisco: ICEG (1991).

Mwega, F.M., "Trade Liberalization, Credibility and Impacts: A Case Study of Kenya; 1972-94." Mimeo. Nairobi: AERC (1995).

Nash, J., "An Overview of Trade Policy Reform, with Implications for Sub-Saharan Africa," in Fontaine, J.M., ed., *Foreign Trade Reforms and Development Strategy,* London and New York: Routledge (1992).

_____. "Trade Policy Reform Implementation in Sub-Saharan Africa: How Much Heat and How Much Light? Mimeo. Washington, D.C.: World Bank (1993).

Musonda, F.C., Adam, and P.A. Andersson, "Trade Liberalization in Zambia, 1970-1995." Mimeo. Nairobi: AERC (1995).

Ncube, M., P. Collier, J.W. Gunning and K. Mlambo, "Trade Liberalization and Regional Integration in Zimbabwe." Mimeo. Nairobi: AERC (1995).

Ndulu, B.J., "Economic Reforms and Development in Sub-Saharan Africa: What have we Learned and What are the Challenges Ahead?" Mimeo. Nairobi: AERC (1995).

_____. J. Semboja, and A.V.Y. Mbelle, "Trade Liberalization in Tanzania: Episodes and Impacts." Mimeo. Nairobi: AERC (1995).

Ogunkola, E.O., "An Empirical Evaluation of Trade Potential in the Economic Community of West African States." Final Research Report, Nairobi: AERC (1994).

Oyejide, T.A., "Exports and Economic Growth in African Countries." *Economic Internationale,* 28 (1975).

_____., "Regional Economic Integration in Sub-Saharan Africa." In T.A. Oyejide and M.I. Ibadan, eds., *Applied Economics and Economic Policy,* Ibadan: Ibadan University Press (1993).

_____, "Trade and Trade Policies in African Development: Some Critical Research Issues," mimeo, CODESRIA, Dakar, (1995).

_____, "Rationalizing and Strengthening African Integration Institutions:

The Case of West Africa," mimeo, GCA, Washington, D.C., (1995b).

_____, B.J. Ndulu, and J.W. Gunning, 1996b, "Trade Liberalization in Selected African Countries: Introduction and Overview," mimeo, AERC, Nairobi.

_____, I.A. Elbadawi, and S. Yeo, "Case Studies of Regional Integration in Sub-Saharan Africa: Introduction and Overview," mimeo, AERC, Nairobi.

Rodrik, D., "Imperfect Competition, Scale Economies, and Trade Policy in Developing Countries," in R.E. Baldwin, (ed.), *Trade Policy Issues and Empirical Analysis*, University of Chicago Press, Chicago (1988).

Shepherd, G. and C.G. Langoni, (ed.), *Trade Reform: Lessons From Eight Countries*, ICEG, San Francisco (1991).

Snape, R.H., "East Asia: Trade Reform in Korea and Singapore," in Shepherd, G. and C.G. Langoni, (ed.), *Trade Reform: Lessons From Eight Countries*, ICEG, San Francisco (1991).

Srinivasan, T.N., "International Aspects: Introduction to Part 5," in H. Chenery and T.N. Srinivasan, (eds.), *Handbook of Development Economics*, Vol. II, North-Holland, Amsterdam (1989).

Ssemogerere, G. and D. Fielding, "Trade Liberalization and Regional Integration: Case Study of Uganda, 1990-94," mimeo, AERC, Nairobi.

Taylor, L., "Trade and Growth," *The Review of Black Political Economy*. (1986).

Tussie, D. and D. Glover, (ed.), *The Developing Countries in World Trade: Policies and Bargaining Strategies*, IDRC, Ottawa (1993).

Tutu, K.A. and A.D. Oduro, "Trade Liberalization and Regional Integration: Case of Ghana," mimeo, AERC, Nairobi (1995).

Valdes, A., "The Macroeconomic Environment Necessary For Agricultural Trade and Price Policy Reforms," *Food Policy*, August, pp. 272-283 (1993).

World Bank, *Accelerated Development in Sub-Saharan Africa: An Agenda For Action*, World Bank, Washington, D.C. (1981).

_____, *World Development Report*, Oxford University Press, New York (1987).

_____, *World Development Report: The Challenge of Development*, Oxford University Press, New York (1991).

_____, *Trade Policy Reforms Under Structural Adjustment*, World Bank, Washington, D.C. (1992).

_____, *Adjustment in Africa: Reforms, Results and the Road Ahead*, Oxford University Press, New York (1994).

_____, *A Continent in Transition: Sub-Saharan African in the mid-1990s*, World Bank, Washington, D.C. (1995).

Chapter 4

Industrialization and Growth in Sub-Saharan Africa:
Is the Asian Experience Useful?

Charles Chukwuma Soludo

Introduction

> Let me now turn to the question of future research. In very broad terms, I see two possible directions. One avenue would take as its starting point the view that there is still a lot to learn about the East Asian experience, and that the focus should be on the still unanswered questions. The other approach, while acknowledging that there are a lot of unanswered questions, would take as its initial position the view that the real issue is the relevance of the East Asian experience for other developing countries. While there is undoubtedly scope for research in both areas, I would like to suggest that emphasis ought to be on the second... now the East Asian study is completed, the research agenda lies more in Africa and other developing countries than it does in East Asia (Squire 1993).

In three major Reports, the World Bank has sought to achieve three separate but interrelated tasks: develop the principles of the "market-friendly" approach to economic development (1991); "prove" that the high-performing Asian economies (HPAEs)—Japan, Hong Kong, South Korea, Singapore, Taiwan, Malaysia, Indonesia, and Thailand—industrialized and grew rapidly *because* they adopted the approach (1993); and "demonstrate" that Africa's economic stagnation was due to failure to apply the principles and that those economies which showed significant signs of "learning from Asia" were performing better (1994).

Expectedly, the equivocations, omissions, and misinterpretations in these reports have been the sources of interminable controversies which have emitted more smoke than light. This chapter aims to summarize major strands of the debate, clarify some of the issues pertaining to Africa, and in particular, examine the opportunities for, and constraints to, drawing any useful lessons from the Asian experience.

No issue has attracted more attention in the development literature recently than the spectacular growth of the HPAEs which was sustained through a virile, diversified, and competitive industrial complex. Notwithstanding evidence of widespread and heavy government interventions and perhaps also a more benevolent external environment, the *Asian Miracle* (World Bank 1993) insists that the HPAEs' rapid industrialization and growth was essentially because of their "market-friendly" approaches to development. First, the Report argues that the main characteristic of government policy in these economies is that they "got the fundamentals right" —ensuring macroeconomic stability, building up human capital; creating effective and secure bank-based financial systems; limiting price distortions and maintaining the relative prices of traded goods close to international prices; being open to foreign trade and technology; and developing agriculture. Second, it asserts that interventions have been designed to rectify market failures, modest in scope (making them consistent with macroeconomic discipline), and implemented with performance standards which ensured "contest-based competition."

Implicit in these interpretations, and consistent with the overall thrust of Structural Adjustment Programmes (SAP) is a denial of any significant role for the external environment, a diminution of the role of the State in the process, and a denial that any other approach could have led to the same or better outcomes. External factors are considered largely unimportant in explaining performance because "all developing economies faced the same external environment." Therefore, domestic policies are all that countries require to make or mar their industrial growth. On the nature of domestic policies, the major intellectual framework of the World Bank (1991 Report 1) argues that, "competitive markets are the best way yet found for efficiently organizing the production and distribution of goods and services. Domestic and external competition provides the incentives that unleash entrepreneurship and technological progress."

What has generated heated debate is the seemingly unsuccessful attempt to re-construct the East HPAEs' success to fit into the above framework, and the obdurate insistence that other developing countries have no option but to duplicate this contestable version of the Asian experience. The veracity of the claims and empirical findings by the BWI regarding these HPAEs has been vigorously challenged by a number of economists (often referred to as "revisionists"). For virtually every aspect of the World Bank's assertion about the causes of the Asian success, the revisionists have produced a matching evidence in the opposite direction.

So, which of the interpretations of the Asian industrialization story should Africans believe? More fundamentally, can the Asian experience (whatever the dominant interpretation of that experience) be duplicated in Africa under the Uruguay GATTified world, the SAP framework, high debt-burden and pervasive external shocks, the peasant domestic economic structures, and the changed and continually changing international environment? The central goal of this chapter is to illuminate some of the issues/ questions raised above.

In this chpater, we present some stylized facts about Africa in a global economic context; we evaluate the two contrasting interpretations of the East Asian miracle, and draw out useful lessons from the experiences. The major constraints to a duplication of the Asian experience in the African context are examined as well.

Africa in the Global Economy: Some Stylized Facts

The position of aggregate sub-Saharan African (SSA) economies in the world economy is very marginal. In economic terms, the SSA of about 500 million people is about the size of Belgium (a small European country of about 10 million people). The SSA accounts for about one percent of world output, and less than two percent of total trade.

Its marginal position would not be worrisome if the region were experiencing any improvements in its socio-economic performance relative to other regions. Rather than any noticeable progress to catch up with the rest of the world, the region has the sad distinction of being the only part of the world that experienced zero average per capita growth over the last thirty years, including average annual negative growth rates, -0.35% over the last two decades. Other developing regions have, in contrast, posted more satisfactory records: the East Asia and the Pacific achieved a spectacular average annual per capita real GDP growth rate of over 5% during the 1965-90 period; while Latin America recorded average rate of almost 2% per annum (Elbadawi 1995:1). From available records, out of the 20 poorest countries of the world, 16 (80%) are from the SSA. The SSA is the most illiterate region, has the least investment (as a proportion of GDP) in education as well as in other social sectors, technology, research and development, etc. Africa also has the highest population growth rate relative to other developing regions.

A graphic picture of Africa's level of underdevelopment is captured by table 4.5 (see appendix for all tables), which compares the means of some social and economic indicators for Africa in 1990 and three southeast Asian countries in 1965. From the table, Africa in 1990 was very much comparable to these Asian economies in 1965. However, Africa was worse-off than these Asian economies were 25 years earlier in terms of GDP per capita, school enrolment, mortality rate, and life expectancy.

An examination of the long-term trends in real per capita growth of individual SSA countries shows a persistently low and declining per capita income growth. The decline has persisted since the 1970s with median average growth rates falling from 1.24% in the 1970s to -1.31% in the first half of the 1990s. Moreso, the proportion of SSA countries with positive per capita income has fallen from 66% in the 1960s to 62% in the 1970s, and further down to 48% and 31% in the 1980s and 1990s respectively (Elbadawi and Ndulu 1995:3). The performance of the agricultural and industrial sectors over the 1980—92 period is also dismal. The inclusion of Nigeria in the SSA sample biases the results to give the impression of relative improvements between the 1980-87 and 1987-92 periods. For example, average growth rate of agriculture improved from 1.3% in 1980-87 to 1.8% in 1987-92 period for aggregate SSA. When Nigeria is excluded from the sample, the growth rate actually declined from 1.8% to 1.1% in the two periods. The startling nature of this result can be appreciated when it is recalled that about 65% of Africa's labor force is engaged in agriculture, and that this sector accounts for more than 30% of GDP. Similarly, the industrial sector growth rate declined from 3.4% in 1980-87 to 1.6% when Nigeria is excluded from the SSA. As a proportion of GDP, the industrial sector has shrunk from 33% in 1980 to 25% in 1992 (see Table 4.1). Productivity of resources is lowest in Africa, and over half of its population is said to live in absolute poverty.

There are two major strands of thought when it comes to explaining the dismal performance of the SSA relative to the rest of the world. One stresses the "external factors," while the other blames the "internal policy distortions." The debate is largely unsettled, and we would return to some aspects of it later in this chapter. Whatever were the causes of the structural weaknesses of the SSA, it is intuitive that such weaknesses adversely hurt its competitiveness and performance in the global trade. Svedberg (1995:21-22) summarizes the deterioration in Africa's trade performance. The SSA's share of world exports declined from a little over 3 percent in 1950 to barely one percent in 1990. Between 1950 and 1970, Africa increased its share of all developing countries' exports, even though its global share declined. The decline in SSA's share of global exports ceased during the 1970s mainly due to the two consecutive real oil price increases. After the collapse of the second oil boom (1979-81), the SSA share of both world and LDC exports declined dramatically in the 1980s. In the non-oil products, the SSA held its position in world markets up until the mid-1970s. Between 1970 and 1988, the SSA share declined by almost two-thirds. In some ways, this decline might reflect the sluggish growth of world non-oil primary commodity markets whose share in global exports fell from 26% in 1970 to 16% in 1988. On the other hand, the declining share of SSA exports is also a consequence of failure to boost alternative exports—especially the manufactures exports. According to Svedberg, the most noticeable phenomenon is that the African countries as a group failed to maintain their shares in the stagnating primary commodity world markets. Between 1970 and 1988 this share fell from 7.0 percent to 3.7 percent (see Tables 4.2 and 4.3). It is therefore not surprising that by the beginning of the 1990s, the SSA share of world exports was down to one percent and that of LDC exports to five percent. Specifically, "export earnings in current dollars fell from US$50 billion in 1980 to about $36 billion at the beginning of the 1990s. The whole of sub-Saharan Africa, forty-five countries with almost 500 million inhabitants, now has export revenues less than half of those of Hong Kong, a nation of 6 million people" (Svedberg 1995:21).

The deterioration in SSA trade and economic performance occurred in spite of the spate of SAPs since the early 1980s. The necessity for diversifying the export base away from over-arching dependence on primary commodities to manufactures is echoed in most policy documents. The East Asian experience is cited profusely (and sometimes carelessly) as "the model" to follow, in order to transform the fragile and miniscular industrial base into a competitive and prospering one. Under SAP, the SSA is called upon to "compete and grow" —a la the Asian way.

What makes the HPAE stand out among the developing countries is the rapid, sustained growth of their economies since 1960 combined with the most equitable income distribution among all developing countries. Indeed, "since 1960, the HPAEs have grown more than twice as fast as the rest of East Asia, roughly three times as fast as Latin America and South Asia, and twenty-five times faster than sub-Saharan Africa. They also significantly outperformed the industrial economies and the oil-rich Middle East—North Africa region" (World Bank 1993:2). Life expectancy in these countries improved from 56 years in 1960 to 71 in 1990, while the proportion of people living in absolute poverty dropped from 58% in 1960 to 17% in 1990 in Indonesia, and from 37% to less than 5% in Malaysia during the same period. Several other social and economic indicators (education, appliance ownership, etc) improved rapidly in the HPAEs and are at levels that sometimes surpass those in industrial economies.

When compared with the dismal picture about SSA performance during the same period, the miraculous triumph of the HPAEs becomes even more startling. For example, in 1965 Indonesia's GDP per capita was lower than Nigeria's and Thailand's lower than Ghana's. Indonesia was very comparable to Nigeria in many respects—heterogeneous but large population, and reliance on oil as the mainstay of the economy. Like Ghana, Thailand was a poor agricultural country. By 1990, Indonesia's GDP was three times that of Nigeria, and Thailand had become one of the world's best performing economies while Ghana was struggling to regain its former income level. At another level, Malaysia borrowed the palm technology and palm seedlings from Nigeria in the 1960s. As at 1990, Malaysia had become about the world's largest exporter of palm produce while Nigeria became a net importer. What explains the seemingly miraculous transformation of these HPAEs from underdevelopment into the exclusive club of the rich and the powerful, while Africa stagnated?

The Asian Industrial and Economic Growth: Two Alternative Interpretations

Several alternative interpretations abound in the literature pertaining to the dominant causal factors in explaining the east Asian industrial and economic success. Confronted with the same sets of facts about the Asian economies, analysts have differed widely about the causal relationships. Two dominant strands of interpretations stand out in the literature, namely, the neoclassical view and the so-called revisionists' counter explanation. Each of the two views represents a school of thought within which there are diverse shades of opinions about the finer details and degree of "state versus market" interventions. The salient arguments of each school are examined below.

The Neoclassical Interpretation

> In the organization of a liberal order, pride of place is given to market rationality. This is not to say that authority is absent from such an order. It is to say that authority relations are constructed in such a way as to give maximum scope to market forces rather than to constrain them. Specific regimes that serve such an order, in the areas of money and trade, for example, limit the discretion of states to intervene in the functioning of self-regulating currency and commodity markets. These may be termed "strong" regimes, because they restrain self-seeking states in a competi-

tive international political system from meddling directly in domestic and international economic affairs in the name of their national markets... (Ruggie 1982:381).

The neoclassical interpretation of the Asian triumph is rooted in the effort to validate the neo-liberal order as the dominant development paradigm. Liberalism itself has undergone important transformations—essentially from the laissez-faire liberalism of the 19th century to the post world war order which constrained liberalization of trade, payments, capital controls, etc by the requirements of domestic stability. An unresolved question pertains to the nature of state interventions required to make the liberal order work effectively. Does an ultra domestic liberal order have to accompany a liberalized external economic relations?

At the start of SAPs in the early 1980s, laissez-faire policies at both the external and domestic fronts were recommended. "Getting prices right" was the dominant slogan. This has been modified in the face of stiff criticisms, and the emerging "Washington consensus" is articulated in the World Bank (1991) as the "market-friendly" approach. Under this approach (which is a variant of the neoclassical thesis), the presence of market failures are acknowledged but rejects most corrective interventions as likely to boomerang. In essence, "government failure" dominates market failure. "The appropriate role of government is to ensure adequate investments in people, provide a competitive climate for private enterprise, keep the economy open to international trade, and maintain a stable macroeconomy. Beyond these roles, governments are likely to do more harm than good, unless interventions are market friendly" (World Bank 1993: 10). Specifically, governments should:

> Intervene reluctantly—Let markets work unless it is demonstrably better to step in... **Apply checks and balances**—Put interventions continually to the discipline of the international and domestic markets.. Intervene openly—Make interventions simple, transparent, and subject to rules rather than official discretion. (World Bank 1991:5)

The 1991 World Development Report (which articulates the above framework) has emerged as the intellectual backbone of the "new Washington consensus." In the subsequent two major volumes—the 1993 Asian Miracle, and the 1994 Adjustment in Africa—enormous efforts are devoted to *proving* that the "new consensus" framework explains the performance of "successful" economies, and stagnation in others is due mainly to non-compliance with the tenets of the framework. More specifically, it is no surprise that the semi-official World Bank policy study, the Asian Miracle, did little more than attempt to demonstrate that the Asian experience conforms to the "market-friendly" thesis. We summarize key conclusions of the 1993 Report—which represents an extension of the neoclassical interpretation.

According to the Report, the HPAEs achieved high growth by "getting the basics right." The principal engines of growth were private domestic investment and rapidly growing human capital. Also, high levels of domestic financial savings sustained the high investment levels. Agriculture experienced rapid growth and productivity improvement, while population growth rates declined more rapidly than in other developing countries. They also had a head start because of a better educated labor force and a more effective system of public administration. Thus, according to the Report (p.5), "there is little that is 'miraculous' about the HPAEs' superior record of growth; it is largely due to superior accumulation of physical and human capital." Each of the HPAEs "maintained macroeconomic stability and accomplished three functions of growth: accumulation, efficient allocation, and rapid technological catch-up," and these were accomplished with "combinations of policies, ranging from market oriented to state led, that varied both across economies and over time" (p.10).

Two kinds of such policies can be distinguished: fundamentals and selective interventions. Some of the most important fundamental policies are those that encourage macroeconomic stability, high investments in human capital, stable and secure financial systems, limited price distortions, and openness to foreign technology. On the other hand,

> these fundamental policies do not tell the entire story. In most of the economies, in one form or another, the government intervened—systematically and through multiple channels—to foster development, and in some cases the development of specific industries. Policy interventions took many forms: targeting and subsidizing credit to selected industries, keeping deposit rates low and maintaining ceilings on borrowing rates to increase profits and retained earnings, protecting domestic import substitutes, subsidizing declining industries, establishing and financially supporting government banks, making public investments in applied research, establishing firm—and industry-specific export targets, developing export marketing institutions, and sharing information widely between public and private sectors. Some industries were promoted while others were not" (World Bank 1993:5-6).

After noting the methodological difficulty in establishing statistical links or causality relationships between specific state interventions and economic performance, the Report relies on "essay in persuasion," based on analytical and empirical judgements to make certain inferences. It asserts that "our judgement is that in a few economies, mainly in Northeast Asia, in some instances, government interventions resulted in higher and more equal growth than otherwise would have occurred" (p.6). The Report however quickly qualifies its inference by defining the pre-requisites of effective government intervention—which limits the lessons of such experience for other developing countries. First, it notes that governments in Northeast Asia developed institutional mechanisms which allowed them to establish clear performance criteria for selective interventions and to monitor performance. Such interventions have taken place in an unusually disciplined and performance-based manner. Furthermore, the Report argues that for the interventions that attempt to guide resource allocation to succeed, they must be geared towards correcting "market failures," and that was what the HPAEs did. Second, it is argued that the costs of interventions, both explicit and implicit, did not become excessive and damaging in these economies. For example, when selective interventions have threatened macroeconomic stability, the governments of HPAEs are said to have reacted on the side of the latter, and price distortions arising from selective interventions were also less extreme than in many developing countries.

On the nature of instruments used by the HPAEs to achieve success, the Report argues that there was no dogmatic insistence on defined and immutable instruments. Rather, "they used multiple, shifting policy instruments in pursuit of more straightforward economic objectives... Pragmatic flexibility in the pursuit of such objectives—the capacity and willingness to change policies—is as much a hallmark of the HPAEs as any single policy instrument" (pp. 11-12). Whatever the nature of specific instruments employed in each of the HPAEs, the Report asserts that an overriding consideration and feature of such instruments was "competition via either markets or contests." It is argued that the HPAEs have gone beyond market-based competition by creating contests that combine competition with the benefits of cooperation among firms and between government and the private sector. Such contests range from very simple non-market allocation rules to very complex coordination of private investment in the government-business deliberation councils of Japan and Korea. "The key feature of each contest, however, is that the government distributes rewards—often access to credit or foreign exchange—on the basis of performance, which the government and competing firms monitor" (p.11).

Such competitive atmosphere was efficiently nurtured in the HPAEs by a competent and impartial referee—strong institutions. The HPAEs had high-quality civil service that was insulated from political interference, and coupled with transparent and efficient legal system. In such circumstance, monitoring performance was enhanced and enforcement of contracts was effective. The Report diminishes the importance of "developmental states" in the HPAEs but emphasizes the central role of "government-private sector cooperation" and efficient institutional framework. The countries have built a business-friendly environment that was generally hospitable to private investment and enterprise. Through the "deliberation councils" and other mechanisms, the HPAEs enhanced effective communication between business and government. In addition, several programmes were effectively implemented in these economies to ensure equitable access to the gains of growth—massive public housing programmes in Hong Kong and Singapore; comprehensive land reforms in Korea and Taiwan; rice and fertilizer price policies to raise rural incomes in Indonesia; and assistance to workers' cooperatives and establishment of programmes to encourage small and medium-size enterprises in many countries.

Perhaps, the most controversial aspects of the Report relate to its interpretation of the industrial and trade policy regimes of the HPAEs. Industrial policy is narrowly defined "as government efforts to alter industrial structure to promote productivity-based growth" (p.304). Such productivity-based growth derives mainly from learning, technological innovation, or catching up to international best practices. The Report empirically examines whether industrial policy helped to accelerate productivity in the HPAEs by addressing two sets of related issues:

> • First, did industrial policy alter the sectoral configuration of industries in ways that we would not predict based on factor intensities and changing relative factor prices? If changes in the sectoral composition of output are largely market-conforming, industrial policies must have failed in at least one of their objectives to guide industrial development along paths that it would not take if it were guided by market forces.
>
> • Second, what were the rates of productivity change in industry in the HPAEs, and what were their sectoral patterns? If rates of productivity change in industry are low overall or in promoted sectors, prima facie, industrial policy did not meet its productivity-enhancing objective.

In both cases, the Report concludes that industrial policies were largely ineffective in the HPAEs. In most of the HPAEs, sector-specific policies to promote industrial growth and productivity were employed to varying degrees. Japan led the way with its heavy industry promotion of the 1950s and South Korea and others followed the lead. Important aspects of these policies included import protection as well as subsidies for capital and other imported inputs. The empirical estimates confirm that, on average, rates of productivity change in industry in Japan (before 1973), Korea, and Taiwan were high by international standards, and that productivity-based catching up was taking place. The bad news however, is that despite the gamut of sectoral (preferential) interventions, there is little evidence that industrial policies have affected either the sectoral structure of industry or rates of productivity change. Except for Japan (1960-79) in the chemicals and metalworking machinery complex, the Report's evidence shows that productivity change has not been higher in promoted sectors. It argues that the industrial structures in Japan, Korea, and Taiwan "have evolved during the past thirty years as we would expect given factor-based comparative advantage and changing factor endowments" (p.21). The industrial policies are believed to have produced mainly "market-conforming results." This is principally because while these governments selectively promoted capital and knowledge-intensive industries, they also took steps to ensure that they were fostering profitable, internationally competitive firms. Furthermore, their industrial policies incorporated a large amount of market information and used performance, usually export performance, as a yardstick. With exports as the performance criteria (which was also rewarded by access to credit and foreign exchange), even firms which benefited from higher-than average rates of protection in the domestic market understood that in the near future they would be forced to compete in world markets.

Indeed, the Report's major thesis is that the HPAEs industrial growth and competitiveness were strongly anchored on an export-push strategy. Such export-push strategy—the winning mix of fundamentals and interventions to encourage rapid export growth— "was the HPAEs' most broad based and successful application of selective interventions," and this is the strategy that "holds the most promise for other developing economies" (p.358). The hallmark of the strategy was diversification into, and promotion of, manufactured exports. All the HPAEs (except Hong Kong) passed through an import-substitution phase, with high and variable protection of domestic import substitutes but these periods ended earlier than in other economies. Rather than conserve foreign exchange through import restrictions, the HPAEs sought additional foreign exchange by increasing exports. The HPAEs achieved this by adopting strategic pro-export policies that established a free trade regime for exporters and offered a range of other incentives for exports.

The pro-export incentives and strategies took varied forms in different countries and at different times. In Hong Kong and Singapore trade regimes approximated free trade, with domestic prices linked to international prices; export credit was made available even though they did not subsidize it, and Singapore made extra efforts to attract foreign investment in exporting firms. Japan, Korea, and Taiwan adopted mixed regimes that were largely free for export industries. These three northern economies halted the process of import liberalization for extended periods while heavily promoting exports— "the promotion of exports coexisted with protection of the domestic market" (p.22), or conversely, "protection was combined with either compulsion or strong incentives to export" (p.359). Export incentives were not neutral among firms or industries: efforts were made to promote specific industries. Korea employed firm-specific export targets; whereas in Japan and Taiwan, access to subsidized export credit and undervaluation of the currency acted as an offset to the protection of the local market.

The Report (p.359) further argues that "In the HPAEs that intervened selectively to promote exports, contests based on performance in global markets played the allocative role that is normally ascribed to neutral exposure of both import-substituting and exporting industries to international competition." These contest-based incentive structure required high government institutional capability—civil service and public institutions staffed by competent and honest civil servants. A key to the success of the HPAEs was the government' s ability to "combine cooperation with competition." Export performance was the success index, and the Report argues that every other consideration was subordinated to this over-arching criterion. Thus, the emphasis on export competitiveness gave businesses and bureaucrats a transparent and objective system to gauge the desirability of specific actions or policies.

In the area of external debt and management of external shocks, the Report argues that the difference between the HPAEs and other developing countries is that the former was more agile rather than luckier. It notes however that external debt was not a major constraint for the HPAEs as only Indonesia, Korea, Malaysia, and Thailand had public or publicly guaranteed external debt. None of these indebted economies has faced a debt crisis in the sense of having to reschedule debt—ostensibly because increases in debt have been matched with rapid adjustment, and the high export-GNP ratio and rapid economic growth have enabled them to maintain creditworthiness (p.113).

With regard to the magnitude of external shocks, the Report argues that while the shocks to the East Asian economies were sometimes smaller than shocks to other developing countries, more often they were about the same or larger. The difference in terms of the impacts of those shocks lies in the "agility" and "capacity" of the HPAEs to manage them. For example, by limiting distortions and tightly supervising banks, the HPAEs' governments reduced the spillover from the real sector into the financial sector. Furthermore, flexible labor and capital markets enabled the real sector to react quickly to government initiatives, setting off new growth cycles that eased the recessionary impact of stabilization measures (p.117). These features enabled the HPAEs to recover more quickly from macroeconomic shocks than other economies.

While some analysts maintain that the HPAEs were rather "lucky" with respect to a more benevolent external environment and external shocks, the Report insists that they faced the same or sometimes worse external environment and that the key to their success remains their "superior, market-friendly, domestic policy environment." On the summary explanation for the main factors that contributed to the HPAEs' superior allocation of physical and human capital to high yielding investments and their ability to catch-up technologically, the Report insists that:

> Mainly, the answers lie in fundamentally sound, market-oriented policies. Labor markets were allowed to work. Financial markets, although subject to more selective interventions to allocate credit, generally had low distortions and limited subsidies compared with other developing economies. Import substitution... was quickly accompanied by the promotion of exports and duty-free admission of imports for exporters. The result was limited differences between international relative prices and domestic relative prices in the HPAEs. Market forces and competitive pressures guided resources into activities that were consistent with comparative advantage and, in the case of labor-intensive exports, laid the foundation for learning international best practice and subsequent industrial upgrading.

These conclusions which draw on the "market-friendly" framework of the 1991 World Development Report seem to form the foundation for the "three guiding principles" for "African development made easy" as embodied in the World Bank (1994) study. These principles (or commandments?) can be summarized as: Get macroeconomic policies right; Encourage competition; and Use scarce institutional capacity wisely (World Bank 1994:9). These "commandments" are said to be based on "successful experience elsewhere" —which is a tacit reference to the Asian experience. In the light of these, the crucial lessons that Africa can learn from the Asian experience is to "get the fundamentals right and adopt the export-push strategy under a market-friendly framework." A large body of literature is however emerging which ostensibly challenges both the interpretation of the Asian experience and the advertised lessons for other developing countries. Such challenge and alternative interpretation of the experiences constitute our task in the next section.

Revisionists' Counter View[1]

> Revisionists argue that East Asian governments "led the market" in critical ways. In contrast to the neoclassical view, which acknowledges relatively few cases of market failure, revisionists contend that markets consistently fail to guide investment to industries that would generate the highest growth for the overall economy. In East Asia, the revisionists argue, governments remedied this by deliberately "getting the prices wrong" —altering the incentive structure—to boost industries that would not otherwise have thrived... The revisionist school has provided valuable insights into the history, role, and extent of East Asian interventions, demonstrating convincingly the scope of government actions to promote industrial development in Japan, Korea, Singapore, and Taiwan, China (World Bank 1993:9)

In contrast to the neoclassical interpretations, the revisionists have mustered enormous evidence to demonstrate that the Asian miracle was principally due to activist, "developmental states" which had clearly defined visions of their economies and pragmatically devised and implemented well-honed interventions to wheel the markets to desired directions. Such views are provided as the anti-thesis of the dominant paradigm (the neoclassical view) and also tendered as alternative policy lessons for developing countries. According to Gore (1996:41) "attempts to interpret East Asian development in the terms of the dominant paradigm entail omissions and over-simplifications, semantic slippages and narrative incoherence... The cost is further delay in proper understanding of the many important policy innovations which have been created in East Asia, innovations which, with adaptation, could do much to improve development policy design in poorer countries." To clarify issues and inform the debate, much of the revisionists' response pertain to a re-interpretation of certain conceptual issues, and also marshalling evidence to show that the state interventions "to shape the efficient operations of the market" largely explain the Asian ascendancy.

One major area of controversy relates to the nature and role of the state versus the markets. A fact acknowledged by both schools is that the private sector was the nucleus of economic activity in the HPAEs. Equally true was the pervasive and activist state intervention in the working of the markets. This blend between markets and state has earned the Japanese system a description of "a plan-oriented market economic system." In this sense, defining the debate in terms of market versus state constitutes a misrepresentation. Both played critical roles, but their relative importance is the bone of contention. There is little doubt that leaders of the HPAEs have tended to be either authoritarian or paternalistic. What the neoclassicals question, is the revisionists' view that governments in these economies actively and effectively promoted and directed the markets in order to achieve well-defined national development objectives.

The main logic of government intervention in the HPAEs (perhaps with the exception of Hong Kong) lies in the strong priority given to promotion of "national interests" with strong emphasis on rapid and sustained industrialization. In the case of Japan prior to the 1970s, the strategic orientation was to catch up with the more industrialized economies, and policy orientation was towards production rather than consumption. Policy was consciously guided by international comparisons between production patterns, techniques and organizational structures in Japan and in more industrialized economies. In other HPAEs of Taiwan, Korea, and Singapore, national survival and security also influenced the industrialization drive.

The size of the state in the economies of the HPAEs was small relative to that of the private sector (about 10 percent of domestic investment). However, the size had little to do with the activist public policy. While relying on the creativity and dynamism of the private sector to drive growth, the States "orchestrated the actions of private-sector firms, stimulating investment, innovation, and productivity improvement, and guiding and coordinating entrepreneurial decisions. This was far from a *laissez-faire* economy" (Gore 1994:9). This kind of goal-oriented approach to economic policy has led many analysts to describe the HPAEs as having "developmental states." Correctly interpreted, this does not imply that States supplanted the markets, but rather entailed the search for, and implementation of so-called "market-conforming" policies: "not telling business managers what to do, but seeking to increase the capabilities of firms by providing them with better means of production, to expand the set of market opportunities open to firms, to diminish risks associated with seizing opportunities, and to alter the relative risks and rewards of particular courses of action. Such policies are not market-replacing, but rather market-augmenting and market-accelerating" (Gore 1996:16).

This state – market relationship is rooted in a theory concerned with the requirements of catching up by late industrializers. Unlike the Latin American dependency theory which directed attention to how international centre-periphery relations blocked or distorted progress in the peripheral countries, the Asian theory (developed in, and popularized by, Japan) analyses how late industrializers can catch up with earlier industrializers through well-honed national policies which use the existing international relationships. The Asian theory does not assume that pursuit of national interest should necessarily involve protectionism, and excessive controls on capital movements and trade. Equally, it does not prescribe a blind adherence to the tenets of free trade and capital movements: pragmatism is the key word, and its content at a particular time and circumstance is tailored to serve "national interests." For example, the norms of a global literal order are not totally ignored by the Asian practice, since multilateral obligations have required phased implementation of certain principles of international economic relations. These are however delayed as far as possible if they conflict with the achievement of national development goals. To the extent that these policies responded to correct market failures, they could be described to be "market friendly" of some sort. But in these economies, especially Japan, market failure was not defined in relation to efficiency of resource allocation, but rather in relation to the ability of the market mechanism to achieve the goals set by the government. Contrasting the concept of, and responses to, market failures in the United States and Japan, Okimoto (1989:11) observes that:

> ...the United States' concern can be described as reactive, ad hoc, and focused on market failures without reference to industry-specific goals. By contrast, Japan's approach is anticipatory, preventive, and aimed at positively restructuring the market in ways that improve the likelihood that industry-specific goals will be achieved. There is a fundamental divergence in expectations and objectives and hence in policy actions. Whereas Americans are content to let the chips fall where they may, the Japanese prefer to remove as much uncertainty as possible from the market processes. Their disposition to bend, twist, and shape the market is analogous to their practice of using ropes, wires, and strings to bend and twist the trunks and branches of trees into shapes that fit the Japanese aesthetic composition of a landscape, garden, or bonsai plant.

This belief and practice that with the right policies, one could literally "twist" the market to behave as desired have informed much of government policies in several aspects of the HPAEs. No where are these pervasive interventions more noticeable than the industrial sector. Akamatsu (1961, and 1962) formalized the industrialization strategy of Japan, and provide the first theorization of how growth occurred in newly industrializing countries. One of the papers written in English was entitled "A theory of unbalanced growth in the world economy." He related the growth process in emerging countries to analysis of the trends towards differentiation and uniformization in the international economic structure as such countries developed. According to Akamatsu (1962:209), domestic production of imported consumer goods is identified as "the takeoff stage in the wild-geese-flying pattern." This would occur through "a struggle of economic nationalism" in which "there should be fostered a domestic consumer goods industry powerful enough to win in the competition with imported consumer goods and to recover the home market from the hands of foreign industries" (Akamatsu 1961:13).

The theory shows that national economic policy is important to promote this through protectionist measures, and to promote the accumulation of capital and the technological adaptability of the people in the country seeking to industrialize. Then as these consumer industries grow, they develop into export industries, and at that point a further process of import substitution begins with regard to capital goods industries—which in turn become export industries. As for the markets, Akamatsu theorizes that for both consumer and capital goods, the less industrialized countries initially provide important markets, but as production progresses from crude and simple goods to complex and refined goods, more advanced countries become significant market outlets. Over time however, exports of simple consumer goods begin to decline as other developing countries themselves begin to produce these goods, and competing with the early "newly rising countries."

This sequential process of industrialization and growth requires activist State policies to guide the economy along the way: guiding the process by which firms improved productivity and gained international competitiveness through firm-level policies which sought to improve production capabilities within particular industries. But governments all over the world in one way or another intervene in the operations of the industrial sector. For example, in the United States, the interventions take a variety of forms: anti-trust laws, industrial standards, pollution regulations, labor laws, infrastructural and defence expenditures, etc. The difference between the US and those of the HPAEs is that in the latter, policies affecting industry are coordinated and viewed as a coherent whole, and the governments have a strategic view of their countries' industrial development relative to the rest of the world. Such conscious efforts to wheel the economy to desired direction (especially in Japan) are exemplified by presentation of the Vice-Minister of Japanese Ministry of International Trade and Industry (MITI)—Yoshishisa Ojimi—to the OECD industrial Committee in 1970:

> There was a great outgrowth of industries that depended on low wage labor during the pre-war and post-war transition when Japan was plagued by shortages of capital. At the same time, these industries enjoyed an advantage from the viewpoint of the theory of comparative advantage..... Should Japan have entrusted its future to the development of those industries characterised by the intensive use of labor?... If Japan had adopted the simple doctrine of free trade and chosen to specialize in this kind of industry, it would have sentenced its population to

the Asian pattern of stagnation and poverty. Japan would have remained a weak link in the free world,...

The MITI decided instead to promote heavy industries that require intensive employment of capital and technology, industries such as steel, oil refining, petrochemicals, automobiles, aircraft, all sorts of industrial machinery, and electronics, including electronic computers. In terms of the comparative cost of production, these industries should be the most inappropriate for Japan. From a short run, static viewpoint, promoting their development would seem to conflict with economic rationalism, but from a long-range viewpoint, these are precisely the industries where the income elasticity of demand is high, technological progress is rapid, and labor productivity rises fast. Without such industries it would have been extremely difficult to employ a population of 100 million and raise their standard to that of Europe and America. Logical or not, Japan had to have these heavy and chemical industries. (OECD 1972)

Ojimi's presentation points to the conclusion that the purpose of industrial policy was Japan's domestic development, which grew out of a conviction that comparative advantage can be "created and directed" by conscious government policy. Even though the industrialization process was essentially driven by dynamic and innovative firms seeking long-term profits, growth in their market share, and competitive advantage, the government did everything it could to orchestrate both the pace and direction of change. As Shinohara (1982:23) summarizes it, "The success of guidance from above was only made possible by dynamism in industrial circles." A gamut of incentives/ policies were employed to affect the investment conditions. Some of them include: fund procurement, land availability and construction costs, low and sectorally administered interest rates, development finance institutions which provided preferential interest rates to specific industries at early phases of development; specific fiscal measures—reduced effective tax on corporate earnings, tax exemptions and special depreciation allowances; etc. In general,

... the governments in Japan, the Republic of Korea and Taiwan Province of China did not intervene either reluctantly or transparently in any of these economies. Specifically, in their periods of fast economic growth, the governments of Japan (1950-73) and the Republic of Korea used a wide array of interventionist instruments including: import controls; control over foreign exchange allocations; control over multinational investment and foreign equity ownership; heavy subsidization and "coercion" of exports, particularly in the Republic of Korea; a highly active State technology policy; restrictions on domestic competition and government encouragement of a variety of cartel arrangements in the product markets; promotion of conglomerate enterprises through mergers and other government measures (the Republic of Korea). The governments in these countries, not only intervened at the sectoral level, but also far more intrusively at the level of the individual firm through so-called "administered guidance" (Singh, 1995:6).

Particularly in the case of Japan, several of these incentives were provided in the context of "industrial rationalization" programmes designed by the MITI in cooperation with the firms/industries concerned. The idea was that "all aspects of a firm in any given industry have to be rationalized if the firm is to compete with other firms and survive in domestic and overseas markets." This involved an intense effort to reduce production costs and increase productivity at the firm level, a process which has been characterized as "the attempt by the State to discover what it is individual enterprises are already doing to produce the greatest benefits for the least cost, and then, in the interest of the nation as a whole, to cause all the enterprises of an industry to adopt these preferred procedures and techniques," coupled with measures, at the industry level, to improve the environment of firms (e.g., industrial location and physical infrastructure planning) and to promote the appropriate degree of competition and cartel-like cooperation (see Gore 1994:25). Despite substantial (though reluctant) liberalization of the economy under the weight of international pressures, the Japanese traditional development policy—based on protection of the home market and promotion of domestic industries through a variety of means—continues till date. "The MITI has to use more indirect instruments as well as moral persuasion to a far larger degree than it did before. "Administered guidance," MITI's close links with industry and trade associations and, importantly, business practice, are the essential policy tools in the new context" (Singh 1995:14).

The governments of the HPAEs also actively managed industrial competition at home and strategically managed the economies' interaction with the global economy. The industrial framework in the HPAEs can be described as that of competition that is tempered by cooperation among the firms, and the government actively encouraged these arrangements. In the specific case of Japan and Korea, oligopolistic competition among domestic firms was broadly encouraged, but managed in a pragmatic manner which tempered competition if it was adjudged "excessive" or if it conflicted with the achievement of specific objectives. In the sphere of the financial sector, governments in these economies strongly regulated competition.

At the domestic level, foreign competition was restricted through the instruments of the Foreign Trade and Exchange Control Law (1949) and the Foreign Investment Law (1950) in Japan. All foreign exchange earned in the economy of Japan was surrendered to a government agency which decided on how such would be used for imports. Also, foreign investors wanting to license technology, acquire stock, share patents, or enter into contracts that provided them with assets in Japan had to be licensed. Much of the competitive environment in Japan was laid during the US occupation after the world war: the few large business groups that controlled the industrial sector prior to the war (the zaibatsu) were dissolved during the occupation; anti-monopoly law was introduced in 1947; while business reforms were undertaken which introduced the business practice of divorcing ownership from management.

When the US occupation ended, the Japanese government revised the Anti-Monopoly Law—which among other things permitted the formation of cartels. A major result was the emergence of loose but largely effective business groups (kereitsu). Competition in the heavy and chemical industries was largely oligopolistic, while free competition reigned in other sectors, especially textiles. Active government policies were evident in attempts to grant certain firms exemptions from the Anti-Monopoly Law to form cartels. For example, in the early 1950s, special exemptions were made for small businesses, for exporting, and to promote rationalization ("rationalization cartels") and to offset the effects of recession ("depression cartels"). The latter could make agreements on production quantities, sales quantities and price, and were authorized under certain conditions "if supply exceeds demand, prices fall below average costs, and where a number of producers are likely to go out of business." Rationalization cartels, on the other hand, are meant to make concerted efforts to exchange or restrict technology, standardize goods produced, work out specializations of product line, or make common use of transport or storage facilities (see Caves and Uekesa, 1976:486).

Another example that some HPAEs did not follow a policy of maximum domestic competition or unfettered market-determined entry or exit of firms is the experience of South Korea. The Korean government surpassed the Japanese in actively helping to create large business conglomerates, promoting mergers, and directing entry and exit of firms according to the requirements of technological-scale economies and world demand conditions. The result is that Korea has one of the largest market concentrations in the manufacturing sector in the world. Commenting on the industrial structure in Korea, the UN (1993) observes that:

> Such a structure is the deliberate creation of the Government, which utilized a highly interventionist strategy to push industry into large-scale, complex technologically demanding activities while simultaneously restricting FDI inflows tightly to promote national ownership. It was deemed necessary to create enterprises of large size and diversity, to undertake the risk inherent in launching high-technology, high skill activities that would remain competitive in world markets. The chaebol acted as the representative and spearheads of the Government's strategy: they were

supported by protection against imports and TNC entry, subsidized credit, procurement preference and massive investments in education, infrastructure and science-technology network.

The rules of behavior established by the State in Korea regulated competition among the domestic firms. Big businesses were prevented from colluding. As Cho (1987) notes, after 1975, inter-group competition in Korea became more fierce as each *chaebol*, or diversified business group, tried to qualify for generous subsidies to establish a general trading company by meeting government performance standards in terms of minimum export volumes and number of products.

One other major issue of controversy concerns the extent of "openness" of the HPAEs. Evidence suggests that the HPAEs (except Hong Kong) did not adhere to the tenets of openness, international competition and close integration with the world economy as propagated by the various World Bank reports and policy documents. Japan in particular, was in every sense a very closed economy in terms of its imports of manufactured goods (estimated at only 2.4% of GDP in 1979) during its catching up period. Specifically, during the 1950-70s, the economy operated under various forms of draconian import controls—both formal and informal. Both Korea and Taiwan had similar experience. These protectionist measures were used by the HPAEs' governments to alleviate the balance of payments constraints, and protection also played very important positive role in promoting technical change, productivity growth and exports in these economies. Protection provided the firms with a captive home market leading to high profits, which enabled the firms to undertake higher rates of investment, to learn by doing and to improve the quality of their products. Given the demand for exports as a measure of success and eligibility criterion to continue the enjoyment of protection and incentives, the profits in the protected domestic market not only increased investment but also greatly aided the export performance.

In terms of the extent to which domestic prices in the HPAEs deviated from international prices, Japan, Korea, and Taiwan can be shown to be very closed economies. The Japanese government, for example, maintained strict exchange controls and kept a steady nominal exchange rate with respect to the United States' dollar over almost the entire period of 1950-73. Purchasing power parity calculations, using Japanese-US price indices, show a 60 percent real appreciation of the exchange rate between 1950 and 1970 (see Singh 1995:20-22). Such level of real appreciation is certainly anything but how to "get prices right." With relative prices in Korea, Taiwan, and Japan more distorted than in several other developing countries such as Brazil, India, Venezuela, etc, further explanations pertaining to their international competitiveness need to be proffered by the World Bank economists.

A major lesson from the experience of the HPAEs is that "openness" / outward-orientation, and international competition are terms that should be interpreted with caution. Various World Bank reports and studies by neoclassical economists have generally equated outward or export oriented economies as "based on private enterprise, managed by market forces and operating under a virtual free-trade regime—at least as far as their production was concerned" (Alam 1994:3). It is generally argued that an outward-oriented strategy allows countries to reap the benefits of specialization according to comparative advantage, permits the realization of economies of scale, and provides the spur of competition which induces technological change. A common conclusion is that the more open the economy and the closer its integration with the global economy, the faster its rate of growth. Paradoxically, the HPAEs are presented by the neoclassical economists as examples par excellence of what outward-orientation as described above can achieve.

Heterodox economists (see for example, Singh 1995, Amsden 1993, Alam 1994, Gore 1996, etc) dispute such characterization and argue instead that the evidence from the HPAEs is at variance with the neoclassical interpretation. In the first instance, the revisionists argue that during their periods of rapid growth, instead of a deep and unconditional integration with the global economy, the HPAEs evidently sought "strategic" integration: they integrated up to the point where it was as much in their interest to do so as to promote national growth. The empirical definition of an optimal degree of openness is controversial, but even some neoclassical economists accept that it does not necessarily imply "free trade." An alternative theoretical perspective is provided by Chakravarty and Singh (1988). According to their analysis, "Openness" is a multi-dimensional concept; apart from trade, a country can be "open" or not so open with respect to financial and capital markets, in relation to technology, science, culture, education, inward and outward migration. Moreover, a country can choose to be open in some directions (say trade) but not so open in others, such as foreign direct investment or financial markets (see Singh 1995:23). Such analysis implies that there is not a unique optimum form or degree of openness which holds true for all economies at all times. Some factors which could affect the desirable nature of openness include the world configuration, past history of the economy and its stage of development. The authors also observe that the timing and sequence of opening are very critical as errors in these could inflict irreversible losses to the economy. The HPAEs' selective but "strategic" integration with the global economy accords with this analytical perspective.

Another source of disagreement pertains to the conception of outward or export-orientation by the neoclassicals as implying the absence of inward-looking or import-substitution (IS) strategy. In other words, the dominant paradigm sees both strategies as perfectly symmetrical: exporting does not require a preceding or simultaneous period of import substitution. The IS strategy involves the bias of incentives in favor of production for the domestic market over production for exports. On the other hand, outward-orientation implies absence of bias against exports: technically, a trade regime is outward-oriented if, on the average, incentives are neutral, biased neither for nor against exports. Such neutral trade regime could be in place in a situation of free trade, or alternatively, it could coexist with situations where import tariffs are combined with, and offset by, export subsidies. As Krueger (1978:89) notes, "a regime could be fully liberalized and yet employ exceedingly high tariffs in order to encourage import substitution."

Evidence from the HPAEs supports the view that export-led growth and import substitution are not two separate strategies but rather an organic, inseparable whole. Many of the HPAEs' exports that emerged in the 1980s had a long gestation period under State promotion: exporting did not begin immediately after the industry was established (see Amsden 1993: 4). Kim (1993) summarizes the experience of South Korea by noting that:

> The average effective rate of protection was atypically high for the "strategic" industries. In some industries, protection quickly became redundant with firms experiencing a rapid *rite de passage* from an infant to an exporter, but in others, where technology was complex and marketing was more elaborate, protection had been relatively long lasting in order to provide a longer period of incubation.

A probable reason why the HPAEs did better than other developing countries in the 1980s was because they had emphasized for a longer time period both the import-substitution and export promotion, rather than just the IS strategy. In the HPAEs, industries that benefited from the IS strategy were also required to meet specified export performance. This pressure to also export made the critical difference between HPAEs and other developing countries. As Amsden (1993:4) concludes, "the lesson for other countries from East Asia's trade and industrial policy is not necessarily to abandon subsidized import substitution—otherwise exports may fail to become more diversified and knowledge- and capital-intensive. Instead, the lesson is to subject every import-substitution industry to various forms of discipline, including possibly some export target, however modest."

The economies of Japan and Korea are also noted to have greatly discouraged foreign direct investment (FDI). One reason for this is that it would have been difficult for the authorities in Japan or Korea to use "administrative guidance" to control foreign firms with as much success as it did with domestic firms. Again, it is argued that there is a strong link between the local ownership of the large firms (chaebols) in Korea and their extraordinarily high investment in research and development (R&D). Korea was reputed as having the highest expenditure on R&D among developing countries.

A pertinent question at this stage concerns the extent to which we could attribute the rapid industrialization and economic growth of Japan and other HPAEs to the "active industrial and trade policies" discussed above. The World Bank economists (in the Asian Miracle study) reject the effectiveness of these policies on two major hypotheses:

(a) That the industrial structure which emerged in these economies was not all that different from what it would have been had these countries not had an industrial policy (i.e., the observed industrial structure was *ex-post* market conforming and accorded with the changing relative factor intensities and prices); and
(b) That the total factor productivity growth of the industrial-policy-favored sectors was no different from that of the unfavored sectors.

Heterodox economists have argued that the conclusion of policy ineffectiveness based on these two hypotheses is very inadequate. Hypothesis (a) above is fraught with serious methodological flaws. Such a conclusion is very difficult to establish empirically as it requires the construction of a credible counterfactual with all its notoriously difficult choice of assumptions. Alternatively, it would have required a comparison of HPAEs and other countries with similar characteristics as the HPAEs but which did not have active industrial policies. Specifically for Japan, it could be necessary to compare the periods before-and-after the industrial policies (i.e., before and after the World War II); or assess the extent to which the industrial policy objectives were accomplished. The Asian Miracle does not apply any of these methodologies. Its hypothesis (a) could be said to be at best "speculative."

Other empirical evidence on the effectiveness of the HPAEs' industrial policies have produced mixed, but largely positive results. Friedman (1988) and Okimoto (1989) have shown that in some instances, the policies did not achieve their set goals, and even had counter-productive effects. On whether the policies actually affected the growth and competitiveness of the industrial sector, Boltho (1985a and 1985b) and Magaziner and Hout (1980) have provided detailed and careful evidence (based on case studies of specific industries) which strongly suggest that the industrial policies were successful in propelling the targeted industries into pre-eminence in international competition. Equally problematic is the methodological plank on which hypothesis (b) rests. Its major difficulty is that of attempting to isolate the total factor productivity growth (TFPG) of individual industrial sectors in a setting where "spillover effects" are very high. As Singh (1995:17) argues, such a test overlooks the effects of industrial policy on a country's balance of payments and its long-term rate of growth of domestic demand. For example, by confining its attention only to the supply side effects of productivity growth and technical change (as predicted by the TFPG approach), the Asian Miracle study erroneously hypothesizes that "spillovers" of these activities will be confined only to the favored sectors or their close sub-sectors within the two digit industrial classification the Study has analyzed. But, "to the extent that industrial policy helps to relieve the balance of payments constraint, most sectors will benefit from higher rates of growth of production and hence productivity (by Verdoorn's Law), and not just the favored sectors. In other words, the spillovers will be universal." A serious flaw of the Asian Miracle study is that it focused on the period 1960-1990—thereby avoiding the deep analysis of the development experience in the 1950s, which is crucial for understanding how rapid growth and industrialization got started in the first instance.

Conceptually, it is difficult to assess the type of industrial policies employed by the HPAEs (especially Japan) since they worked through the actions of the market, and entailed the shaping of market institutions. Disentangling the effects of the markets and policies is an extremely difficult task. Indeed, one can argue that "for this type of policy, which seeks to expand private sector capabilities and market opportunities, the more successful the policy, the less its success is apparent" (Gore 1994:39). This raises the apparent futility of attempting to polarize the analysis of the HPAEs into those who glorify the "industrial policy" (state action) and those who elevate the markets. Both, synergistically played significant roles that are difficult to isolate. A more fruitful line of enquiry would rather involve an understanding of how government policy and corporate strategy worked together to enhance the production capabilities and productivity of firms in the HPAEs, and their competitiveness within the global context.

Other major issues which have attracted strong counter interpretations from the revisionists pertain to the nature and role of human capital formation, why the HPAEs escaped external debt overhang, and the impacts of external shocks on the economies. Human capital formation was an integral component of the industrial policy designed to enhance the national strategy for technological development. The HPAEs, as late industrializers, realized the prominent role which education at all levels (especially tertiary education) and technological infrastructure—human and physical—could play in their catching-up strategy. Several World Bank reports stress the importance of primary and secondary education for achieving growth, but ignores the heavy emphasis of the HPAEs on higher education. After the Second World War, the Japanese adopted a national technological system which involved active collaboration among the government, firms, universities, and indeed the entire society. Higher education especially in science and engineering was very central to the process. This experience was adopted by other HPAEs, and currently several Asian countries have higher annual output of graduate engineers per hundred thousand of population than Japan. Singh (1995:44) neatly sums up the lessons from HPAEs in terms of higher education by arguing that:

> ...the World Bank's emphasis on early education would not appear to be an adequate way of enhancing the international industrial competitiveness of semi-industrial countries. To compete in the world industrial economy, it is also essential to have higher educational institutions, scientists, technologists and engineers. Universal primary and secondary education is a worthy goal in its own right, but it does not provide the wherewithal to compete in the international market place. It is undoubtedly far more expensive on a per capita basis to provide higher education than to provide primary or secondary schooling. The former is also necessarily elitist, but this is the price that has to be paid for seeking international competitiveness, a price that the East Asian countries have been willing to pay.

The other major issue of debate is the types and magnitudes of external shocks faced, and how the HPAEs managed them. In the early 1980s, developing countries were subject to four different kinds of external shocks: a demand shock (resulting from recession in the OECD economies); a terms-of-trade shock; an interest rate shock (due to monetarist policies in the OECD); and a capital supply shock (which resulted from the cessation of lending to several developing countries). Given a world of imperfect wage-price flexibility, balance of payments disequilibria, capital rationing and foreign exchange constraints, all the four shocks were relevant. The impact of these shocks on the HPAEs was relatively inconsequential, and the interpretation of this phenomenon is the subject of a raging debate. According to the World Bank documents, the HPAEs faced similar shocks and of the same magnitude as other developing countries. The reason why the HPAEs escaped the deleterious effects of the shocks is attributed to their "agility" : openness to international trade and "sound fundamentals" which increased the HPAEs' resilience.

On the contrary, the revisionists advance arguments which largely attribute the HPAEs' "escape" from the crisis to "good luck." First, the World Bank's thesis that "openness" of the economies was responsible for their "agility" is punctured by evidence. As preceding discussions in this chapter show, several East Asian economies implemented fairly strict exchange controls, and thus their financial markets were less open. In comparison with many other developing countries which were badly affected by the debt crisis (particularly in Latin America), some HPAEs were less open. Furthermore, it has been shown that even the least open Asian economies—China, India, and Pakistan were also able to cope effectively with the shocks. More convincing explanations would therefore need to be proffered. One plausible explanation posits that the magnitude of the shocks to the HPAEs was much smaller than those which befell other developing regions. For example, available evidence shows that the terms of trade of the South Asian economies and East Asian NICs either remained the same or improved in the 1980s, while those of Latin American countries deteriorated by more than 15% and those of sub-Saharan African countries fell by nearly 30% (see Singh 1995).

Furthermore, as Singh (1995:38) argues, "perhaps the most important single reason why the Asian countries escaped the debt crisis of the last decade was that they were not subject to the capital supply shock." Despite the fact that the East Asian economies had worse current account deficit-GDP ratio than the heavily indebted Latin American countries, and Korea's external debt, fiscal and current account deficits were very comparable to those of Brazil and Mexico, the banks continued to lend to Korea and other HPAEs but ceased lending to several other developing countries. What explains this preferential treatment of the HPAEs? Some analysts attribute this to "good luck." For Williamson (1985), the differences in the banks' treatment of Mexico and Brazil on the one hand, and of the Republic of Korea on the other, can be explained by a "herd" instinct on the part of the banks and by a "contagion" effect. He argues that had South Korea been a Latin American country, it would have also been subject to the contagion effect, and might not have been so "lucky." Thus, while other developing regions (Africa and Latin America) continued to face or confront credit-rationing and experience reverse capital flows to the industrial countries throughout the 1980s, the East Asians continued to experience positive net capital inflows. These are certainly non-trivial issues and they could have made all the difference between the Asians and other regions in the 1980s.

Summary of Prospective Lessons from the Revisionists' Interpretation

The foregoing review and recent studies on the experiences of the HPAEs point to several interesting lessons about the kinds of policies and institutions required to overcome underdevelopment in the shortest possible time. Of course, a country by country review of the experiences easily reveals that Asian countries differed markedly in terms of the character and intensity of the policies and institutions employed. In broad terms, some analysts point to the differences between the north eastern countries of Japan, Hong-Kong, South Korea, and Taiwan province of China on the one hand, and the south eastern countries of Malaysia, Thailand, and Indonesia on the other. The latter group is claimed to have followed a less interventionist policy regime than the former. The differences might be ones of degree rather than of kind, and it is still possible to summarize the broad principles that characterized the rapid economic transformation of these economies.

One common thread in all of the countries is the pre-eminent emphasis on the ability of the State to consciously articulate the society's long-term development vision and using a battery of instruments to mobilize the country's productive energies to attain such vision. Rapid investment promotion was seen, not only as a prerequisite but also a guarantee of rapid productivity growth. Because of perceived "market failures" of various kinds, the HPAEs sought to provide active "guiding hand" to the operations of the market. For them, market failure was not defined in terms of efficient allocation of resources, but rather in relation to the ability of the market mechanism to achieve specific development goals set by the government. Intervention was not, unlike the experiences of many other countries, geared "to constrain the business sector as a whole in the interests of other classes, and still less to replace private enterprise with centralized state control; nor has it been a system for extending favors to certain individual interests" (UNCTAD Secretariat 1996:5). In these economies, the primary purpose of policy was to promote the interests of the business sector as a whole in a manner consistent with a broader set of national interests. A second major lesson pertains to the role of strong and efficient institutions in the process of development. As the experiences of the HPAEs show, these institutions themselves can be consciously "created" to serve the development needs, but they need to evolve within the context of each country's socio-historical and cultural milieu. A third major lesson is that activist policies can still succeed, despite the constraints of the external environment and this depends on the dexterity and skill of government-business initiatives in interpreting and maneuvering the GATT and other rules of international economic cooperation.

More specifically, the broad range of principles that underlie the success of the Asians can be summarized in four categories (see UNCTAD Secretariat 1996:11-12):

A. Policies and institutions to promote profit-investment nexus, which entailed:

(i) providing a stable political environment and pro-investment macroeconomic policies to sustain the investors' confidence;
(ii) providing both general and specific investment incentives by using measures to create artificially high profits. This was achieved in two ways. First, a range of fiscal instruments were used to supplement corporate profits and to encourage their retention in order to accelerate capital accumulation; tax exemptions and special depreciation allowances were applied both on a general and targeted industries. Second, a set of trade, financial and competition policies were used to create "rents" which boosted corporate profits and thus provided investible resources available to corporations. Such policies included a mix of selective trade protection, controls over interest rates and credit allocation, and various strategies to manage competition.
(iii) disciplining entrepreneurs by closing off channels for unproductive investments and capital flight, and restricting luxury consumption. Restrictions on consumption were done directly through restricting the import and domestic production of luxury consumption goods, and indirectly through high taxation and restrictions on consumer credits.

B. The promotion of an export-investment nexus by:

(i) initially promoting traditional exports (primary commodities and labor-intensive industries) to maximize foreign exchange receipts to buy capital goods that embody more advanced technologies;
(ii) promoting more demanding industries identified on a number of criteria (e.g. productivity growth potential, conformity with domestic technological capabilities, demand prospects) through the creation, manipulation and timely destruction of rents; and
(iii) upgrading through strategic integration with the international economy, using the disciplinary power of international markets alongside measures of export promotion, and using FDI selectively and strategically to access more advanced technologies abroad.

C. The creation of a strong government-business network:

(i) promoting an independent economic bureaucracy, including a network of agencies at the sectoral level;
(ii) establishing strong links between industrial firms and the financial sector in ways to promote productive investment.

D. Addressing the danger of marginalization:

(i) supporting small-scale producers both in the rural and industrial sectors through targeted public investment, subsidized credit and appropriate advisory services;
(ii) supporting upgrading by linking small producers to large firms and public research institutes.

As indicated earlier, differences exist among the HPAEs in terms of the breadth and depth of application of these principles. The hallmark of their development experience is that these economies had their destiny securely in their hands, and freely experimented with alternative instruments as changing circumstances dictated. Development was thus a dynamic process, requiring strategic interventions to pre-empt the market, or to correct observed distortions that endangered the attainment of state goals. A major question that remains is whether these major lessons are still valid for other developing countries to replicate in the light of changed and continually changing international environment.

Constraints to Learning from Asia: Why Africa Is Different

From the foregoing reviews of the neoclassical and revisionist interpretations of the Asian success, there seem to be as many points of consensus as there are disagreements about the important lessons for other countries. Both schools of thought agree that the private sector was the linchpin of economic activity: they disagree on whether such virile private sector was deliberately and actively cultivated and promoted by the State; they agree that the HPAEs maintained macroeconomic stability, built strong institutions, invested heavily in the acquisition of human and physical capital, and actively promoted exports, but disagree intensely about the mechanisms for accomplishing them. Furthermore, they disagree on the nature and activism of the State, the nature and effectiveness of industrial policies, correct interpretation of the trade regime and the HPAEs'
"strategic" or "laissez-faire" integration with the global economy, and whether or not the HPAEs were "lucky" with respect to the impacts of the

external environment. The debate is raging and the emergence of "consensus lessons" from East Asia as blue-print for other developing countries is a largely unfinished business. Even the World Bank (1993: 102) admits that:

> The search for policy explanations for East Asia's success has not been completely successful. Each of the broad views of the relationships between public policy and rapid growth—neoclassical, revisionists, and market-friendly—adds important elements to our understanding, but none fully captures the complexity of public policy and rapid growth in the HPAEs.

In essence, there is still a lot of puzzle—large unexplained aspects of the HPAEs' performance. This confounds the debate, and cautions against any hasty attempt to embrace any one of the schools as having "the" correct interpretation. However, increasingly the "revisionist" interpretation is being accepted as the "more correct" evaluation. For SSA, the key question is whether there are constraints and peculiar circumstances that would diminish the efficacy of the Asian lessons. We address issues pertinent to the question below.

External Shocks Are Decisive for Africa

A limitation of the debate between the neoclassicals and revisionists is its narrow pre-occupation with the "nature and role" of domestic policies. Implicitly, the logic is that once the "correct" domestic policies are implemented, any developing country has great chances of mimicking the Asian success. This line of reasoning is guilty of what Gore (1996) calls "methodological nationalism" —by which economic performance is wholly attributed to national characteristics (particularly, domestic policies). It ignores the nature of interrelationships between the content and effectiveness of domestic policies and the external environment.

For example, in its 1994 Report (as in most other earlier reports), the World Bank still maintained that for Africa, "the main factors behind the stagnation and decline were poor policies—both macroeconomic and sectoral—emanating from a development paradigm that gave the state a prominent role in production and in regulating economic activity" (p.20). On page 135, it notes, albeit with equivocation, that "much of the evidence... shows that the external environment has not been a major direct factor in Africa's long-term growth. But short-run growth can be affected by large and sudden changes in the external environment—that is, changes in net transfers or the terms of trade." The report contains other equivocating references to the impacts of the external environment. On the debt issue, it notes that "Africa needs debt relief because it cannot service its current debt obligations, but external debt has not yet been a major obstacle to growth since net external transfers...were positive for most countries in the region." Then on the terms of trade, it observes that "The adding-up problem—the decrease in commodity prices from having major producers of primary commodities expand their supply simultaneously—is an obstacle for exporters of several African major primary commodities: cocoa, and to a lesser extent, coffee, sisal, tea and tobacco." Yet, neither the terms of trade nor the external debt burden significantly affected growth in the regression analysis conducted by the study. A major conclusion of the report seems to be that such factors could only have significant impact in the "short-run" but become insignificant in the long-run. Thus, to promote growth and development which are long-run phenomena, external factors do not matter.

We resist couching the discussion in terms of the old, and seemingly sterile, debate about the relative importance of domestic policies and external factors in explaining Africa's economic crisis. Several authors (see in particular, Soludo 1993) have illuminated this controversy and we do not intend to rehearse it here. Elbadawi and Ndulu (1995) have summarized more recent evidence on this debate. According to some studies (see Easterly et al., 1993; Fischer 1993; Barro, Sala-i-Martin 1995; and Mendoza 1994) there is a strong link between external shocks and growth performance especially in the 1980s: once external shocks are properly accounted for, policy variables become less significant in explaining economic performance, suggesting a strong link between the two groups of variables. Of the external shocks, the terms of trade and its variability have the most significant negative impact on growth.

The usual response of those who deny the impact of external factors is to argue that since all developing countries faced and continue to face the same shocks, domestic policy should be the focus of analysis. But it is not true that the HPAEs faced similar shocks as Africa. During the decisive period of the HPAEs' catch-up strategies (1950-1970s) the terms of trade for developing countries (especially for primary commodities) was more favorable. African, Asian and Latin American developing countries benefited in varying degrees. The shocks were most volatile and disruptive in the 1980s, and as we noted earlier, the terms of trade for the HPAEs "either remained the same or improved in the 1980s while those of Latin American countries deteriorated by more than 15 percent and those of sub-Saharan countries fell by nearly 30 percent" (see Singh 1995: 39). The volatile external shocks were most pervasive when the Asians had developed virile industrial, financial, and institutional structures to adjust and manage the shocks better.

On the contrary, what has amplified the magnitude and consequences of the shocks in the case of SSA is not only the absence or weaknesses of institutions to deal with them, but more so, their types and transmission mechanism to the economy. It was, indeed, the apparent pervasiveness of the external shocks in crippling development efforts in Africa that led to the summoning of an extra-ordinary session of the United Nations General Assembly in 1986 to consider Africa's development crisis. This resulted in the adoption and partial implementation of the "United Nations Programme of Action for African Economic Recovery and Development, 1986-90" (UNPAAERD). In the "Final Review and Appraisal of the Implementation of UNPAAERD," the United Nations Secretary-General, among other things, noted that "the combination of depressed commodity prices, the crushing debt servicing burden and capital flight, estimated at some $30 billion in the last five years, added up to $80 billion in potential domestic resources that the continent could not utilize...In 1986 alone, Africa lost $19 billion as a direct result of falling commodity prices, with more than $50 billion of export earnings lost during the entire UNPAAERD period." While the HPAEs have reduced their reliance on the volatile primary exports from 67% of total exports in 1970 to 26% in 1993, the SSA countries merely reduced from 83% in 1970 to 76% in 1993 (Killick et al., 1995). This high concentration of exports on primary commodities in the face of unprecedented volatility in their prices partly explains Africa's vulnerability. As Elbadawi and Ndulu note, the terms of trade shocks faced by SSA in the 1970-90 period account for an income loss of 3.8 percent of GDP which is twice the 1.9% loss for other developing countries. For the SSA that is about the size of the Belgian economy, the losses are non-trivial.

On the structural weaknesses of SSA that make it more vulnerable to the shocks, Elbadawi and Ndulu argue that a critical one is the structural dependence of African economies on imports for production and investment which, in turn, depend on uncertain revenue from exports on the one hand, and on autonomous external resources to finance them on the other. Domestic savings in Africa (however high) cannot be converted into investment or productive capacity utilization without foreign exchange to import capital goods and intermediate inputs. Foreign exchange availability thus constrains both capacity utilization and output expansion. Availability of foreign exchange is ostensibly exogenous: dependent on the volatile export performance, as well as the uncertain bilateral aid and conditional multilateral and bilateral loans.

"It is the significant exogeneity of the determinants of the foreign exchange constraints that determines the relative exogeneity of growth performance and its sensitivity to exogenous shocks in Africa" (Elbadawi 1995: 19). Furthermore, most SSA countries depend on trade-related taxes for government revenue (over 40% of revenue in most cases). It is therefore little surprise that shocks affecting the trade performance not only worsen the current account deficits, but also have severe consequences on the government budget in Africa.

Compounding the apparent exogeneity of foreign exchange sources and their constraint on domestic investment and production, is the excruciating external debt overhang. The meagre and uncertain foreign exchange required for effective industrial take-off is devoted to servicing debt, and its magnitude alone saps confidence and scares investors. In effect, "...the debt crisis has halted development..., no policy for growth in the short-term could be contemplated without resolving the debt problem at the same time" (see Elbadawi 1995:16). It has become a decisive crisis and the World Bank has acknowledged that export growth cannot solve the debt crisis of the low (and even medium) income SSA countries. These countries would require sustained export growth rates of no less than 15% annually (for at least five years) to bring the debt - export ratio below the critical threshold of 200%. Recent experience of actual export performance points to exports growing at an average annual rate of around 3%. Thus "actions to reduce significantly the African debt overhang are urgently required and must aim at substantial reductions in the total stock of debt. Without such debt reduction, it is not realistic to expect public and private investment, important for development, to expand" (Centre for Development Research 1995:70). Green (1993:74) dramatizes the SSA's dilemma by noting that, "Most SSA governments (and economies) have few degrees of freedom left. They are hemmed in by present resource levels, by poor external economic environment prospects and limited probability of any large increases in the inflow of net resources—as well as by the need to deliver perceived benefits to voters."

Furthermore, a critical issue that is little understood is the extent to which domestic policy instruments are endogenous to the external shocks. This is amplified in the special case of Africa because of the weaknesses or absence of pertinent institutions, and therefore "limited degrees of freedom" in terms of policy instruments at the disposal of policymakers. Even as early as 1987, the Managing Director of the IMF drew attention to the limitations of domestic policies in adjusting countries given the increased globalization. According to him;

> Effective policies in the developing countries—central as they are to a successful growth-oriented adjustment programme—are not sufficient... Industrial countries can and should provide crucial support by following sound monetary and fiscal policies that are compatible with healthy, non-inflationary growth of world demand, lower international interest rates, and an appropriate pattern of exchange rates; by rolling back protectionism; and

by providing increased official development assistance and adequate official export credit. Clearly, it will be more difficult to make progress in reducing debt burdens if the real interest rate on debt exceeds the growth rate of developing-country exports, and if the incentives to adopt more "outward-looking" policy reforms are sapped by protectionist barriers. To work well, the adjustment process must be symmetric. We cannot have two standards of adjustment—one for industrial countries and the other for developing countries. (Camdessus 1987:10)

More recently, Singh (1995:39) provides further illustration of this endogeneity of domestic policies to the external environment by citing the case of Latin America. He observes that;

...it was not the case of the Latin American countries being incompetent or unaware of the desirability of balanced budgets, etc., but that the economic shocks many of them suffered were so gigantic that their social and political institutions simply could not deal with the ensuing redistributive struggles over a diminished national cake. Hence, many of them experienced episodes of hyper inflation and extreme macroeconomic instability. This has led to capital flight....

Africa has a worse scenario given its peculiarities of dependence on the external sector for foreign exchange and government revenue. There is also the dependence on the uncertain foreign aid even to finance the bulk of government budget in many countries. Any shock such as the perennial terms of trade or the unpredictable aid resource flows therefore reverberates throughout the economy. It is easy to show that much of the fiscal crisis in many SSA results from the struggle by the authorities to preserve the critical minimum of State activities in the face of a volatile revenue base. There is therefore a sense in which much of the macroeconomic instability is endogenous to exogenous shocks. With policymakers preoccupied with the chase of this moving target—macro stability—there is little effort for creating the growth fundamentals as the Asians did.

So what does SSA learn from the HPAEs in this regard? Unfortunately, not much can be learnt since the East Asians never experienced the predicaments of pervasive and unpredictable external shocks and a debt burden that has halted sustainable development. Even South Korea that had comparable external debt burden as Mexico and Brazil was spared the turbulent experience of the 1980s through uninterrupted and increased external resource flows while other developing countries in Africa and Latin America witnessed reverse capital flows. If the HPAEs were confronted with similar shocks and debt burden in their early years of industrial catch-up, it is arguable whether they could have been able to escape the strangulating effects of these shocks to achieve the spectacular growth associated with them. Thus, the inference in several World Bank documents that the SSA could duplicate the East Asian experience of the 1960-1970s simply because Africa in 1990s is comparable to Asia of the 1960 is largely a *non-sequitur*. The circumstances the SSA faces *currently* are evidently different from the ones Asia faced in its historic road to success.

External Environment and "Rules of the Game" Have Changed

A very important issue that is often glossed over in discussions of the HPAEs' rapid industrialization is that they faced a more benevolent external environment that was more favourable to their quest than what Africa faces currently. The flying-geese industrialization strategy of Japan which soon spread to other HPAEs encouraged the rapid flows of FDI from Japan (as the hegemon in the region) to the neighbouring countries with conducive investment climate. Rising wages in the fast growing Japan (the lead goose) squeeze profits in its low-skill labor-intensive industries. Such industries move to the next row of geese—South Korea and Taiwan—initially setting off FDI flows and subsequently intra-industry trade flows. "The process repeats, with Japan expanding its newer, higher-tech activities and hiving off older, lower-tech ones to the Republic of Korea and Taiwan, where industrial development and wage pressures in time induce FDI and industry relocations to ASEAN countries further back in the flock" (Felix 1994: 19). The other countries provided a gamut of incentives to attract and retain the FDI flows from Japan and the other early geese. The result of this model was the formation of an integrated regional economic block. Intra-regional trade in recent years about equals in value US—Japanese trade, and such trade is heavily intra-industry. The higher level of intra-regional economic integration the HPAEs built up with the aid of dirigiste policies has enabled liberalization to proceed much more gradually and less disruptively of inter-industry linkages in East Asia than elsewhere—Latin America and Africa (see Felix 1994). Africa on the other hand, has lacked and may continue to lack, such a hegemon or growth pole which Japan provided in the case of the HPAEs.

Furthermore, the other external factor that provided high growth impetus was the massive US technical assistance, aid, and investment—especially in the military infrastructure—during the cold war era. The American market was also open to exports from the HPAEs. These efforts were to build up some Asian economies as a bulwark against communism. In the effort to contain communism, the US and other western countries rather supported the evidently authoritarian or paternalistic State of the HPAEs. Such state with vision and muscle to ram through its dirigiste policies through the society was supported. As Mongula (1994:84) has observed,

> ...the problem with the blind admiration of the performance in the countries of South East Asia is that it did not consider the specific conditions of those countries, including their colonial and neo-colonial relations with the US and Britain, their small geographical sizes, their dependent and foreign controlled economies, and their social instability and repression. In fact, the political context under which those countries achieved their economic growth is quite different from the one which the donors are recommending for sub-Saharan Africa.

Times have changed and the new slogan under the "new world order" is democratization. Asia is currently democratizing when it has built strong institutions that could withstand the travails of experimentation with a new system that is based on popular participation. Interminable controversy still surrounds the question of the right kind of state for African development in the bid to "catch-up." Should it "learn" from the HPAEs and encourage autocratic regimes provided they are "developmental," or should the current wave of democratization—with the seemingly divisive (ethnic-based) politics of bread and butter (with weak institutions) that has overwhelmed much of Africa—be promoted? It is not clear that "learning" from Asia in this regard would be the best option in today's world.

There are several other ways that the global trading and economic playing field has changed tremendously from the one Asia faced. Even the World Bank (1993: 360-368) admits that "times have changed." The early export drives of most HPAEs took place amid the expanding world economy of the 1950s to the mid-1970s. The industrial economies expanded fast, barriers to trade were declining, and global efforts to spur free trade under the General Agreement on Tariffs and Trade (GATT) were prevalent.

All that has changed. Economic growth in the industrialized economies has weakened, and these are the traditional markets for developing countries' exports. Regional trade groupings are becoming more and more prominent—with all the protectionist tendencies. The global markets are becoming more competitive. "Each economy seeking to emulate the East Asian export-push model will encounter difficulties because many other economies are trying to do the same" (World Bank, 1993:362). Studies have estimated that Eastern Europe and republics of former Soviet Union will increase their share of world trade from 10 to 23 percent, relying initially on agricultural commodities and labor intensive goods that would compete directly with the products of other developing countries.

Also, the GATT rules have changed. Over the years, developing countries have invoked Article XII to claim and receive exemptions from GATT rules. Such exemptions were used to safeguard their balance of payments, and Article XVIII allowed promotion of infant industries. Developing countries were then free to implement trade policies of their choice at home while benefiting from the openness of industrial markets. This was the regime during much of the period of Asian ascendancy. With the shift of GATT's earlier narrow focus on tariffs to a broad agreement on rules, developing countries will find it increasingly difficult to enjoy the exemptions. Equal and compensatory treatments among trading partners characterize the new era. Developing countries that wish to increase trade and attract investment are also under pressure to open their economies and bring their practices more closely in line with those in industrial countries. According to the World Bank (1993:365), "those that attempt to use more interventionist versions of the export-push model will risk retaliation from industrial-economy markets or punishment under the GATT." Indeed, industrial economies have already begun to retaliate against allegedly unfair export practices by imposing quotas or countervailing duties. Furthermore, for other policy options, such as financial repression, increasing integration with global markets will limit the scope for intervention.

It is evident from the foregoing that African economies face a different set of "rules of the game" in the 1990s than the HPAEs faced in the 1950s through the mid-1970s. If we accept that the interventionist and activist policies of the HPAEs which subordinated every other consideration to "national interests" were key to the Asian triumph, then African economies are up against the tide in terms of useful lessons. Most analysts would readily concede that for late industrializers such as SSA, a more "development-oriented" State, with strategic and aggressive promotion of "national interests" would be required to escape the throes of vicious stagnation. The State would literally have to create and promote "comparative advantage." But such a policy bent which propelled Asians to greatness is no longer tolerated in the "new world order." Even if Africa can duplicate the aggressive industrial policies of the Asians, the current "adding-up" problem whereby most developing countries pursue similar policies is a serious handicap. With the increasing waves of regionalism all over the world, where lies the markets for Africa's industrial products assuming they become significant? It is very difficult to see how the Asian experience up to the 1970s can help Africa out of the present dilemma.

Added to the "adding-up" problem with respect to production and exports, Africa faces a much more intense competition for scarce investible resources today than any region has faced in the past. The end of the cold war and desperate bid by the East European countries to compete in the world capitalist order, together with increasing globalization, has raised the bidding price for resources. Globalization has meant that financial resources are also highly mobile. Donor fatigue plagues Africa, and FDI to the region is the least in the world. Given the binding foreign exchange constraint in the investment and production process in SSA, attaining rapid and sustainable growth through industrialization is a tall task. This is compounded by the debt problem which not only drains off scarce investible resources (in the midst of declining export earnings) but also portends future tax on investment. Investors therefore demand front loading of incentives as a response or prefer the "waiting option." Moreover, pervasive ignorance about Africa and its potentials is commonplace in the West, as well as reluctance (for whatever reasons) to invest there. For example, while western investors are willing to make minimal investment in "information gathering" about investment potentials in eastern Europe, Asia and some parts of Latin America, Africa literally has to "lobby" them to "visit Africa and see things for themselves." Many African countries have provided comprehensive and generous industrial policies under SAP, to attract FDI, but the inflow is still insignificant. The image of Africa as the "dark" continent where nothing works and backwardness reigns is itself a problem. Western investors are content to investing in communist China, Vietnam, and other Asian countries but not in the best performing African economies. It is not clear how the Asian experience can help in these matters.

SAP as the Key Stumbling Block to Learning from Asia

Asia did not industrialize by having the BWIs literally take their economic policy-making institutions hostage. The Asian Miracle study admits that "there is no unique Asian experience" as each country did what it considered to be of prime national interest in order to industrialize and grow. Under SAP, with its neoclassical monoeconomic underpinning, most of SSA is being handed a common recipe that neither takes cognisance of individual country peculiarities nor appreciates the extent of constraints imposed by the external environment. Given the enormous leverage exercised by these institutions, the SSA countries still grudgingly swallow the bitter pills of adjustment even without any hope of a lasting cure. The several refinements of, and add-ons to, the various SAP frameworks demonstrate that the learning curve is rather cumbersome, and that the internal consistency of the instruments is questionable.

A major handicap of the SAP package is that it is an incomplete theory of development, and does not embody a comprehensive and consistent strategy for industrial development. For example, it took nearly one and half decades of SAP for the World Bank to admit that "adjustment alone cannot put countries on the path of sustainable long-term development." According to the World Bank (1994:35-39):

> With today's poor policies, it will be forty years before the region returns to its per capita income of the mid-1970s.... Even with good policies, catching-up is not easy. So far, low-income countries have had less success than middle-income countries in growing quickly. Poor policies explain some of the lag, but poor endowments might also limit the potential for rapid growth in the short term, underlining the importance of investing in human capital and strengthening institutional capacity.... Adjustment programs... are no substitute, however, for the long-term development efforts needed to build the capabilities of Africa's people.

Implicit in the Bank's position is that Africa should have two sets of economic programmes to address the crisis: the SAP on the one hand, and "long-term development efforts" on the other. But the World Bank (1994) gives the mistaken impression that there is a sequencing process between adjustment and development, with the former preceding the latter—which is contrary to the Asian experience. Aside from casual references to the imperatives of long-term development—investment in human and physical capital—there is no conscious attempt to reconcile such with the SAP instruments, the debt burden, dwindling resource base (falling export revenues and requirements of "fiscal discipline"), etc[2]. Unconstrained by external debt or foreign exchange, and with rapidly growing economies, the Asians effectively invested in education, infrastructure and technological acquisition. Where are the resources for the SSA countries to make comparable investments in these areas (as percentage of GDP) as the Asians did? SAP merely sermonizes these issues without any concrete effort to address them. Even more worrisome is that the World Bank is known to preach that primary and secondary education should be emphasised at the expense of tertiary education, contrary to the experience of East Asia. Asia had the resources and it invested in all sectors of education because technological and industrial catching-up could not have been successful without a crop of high level manpower. The real challenge to SAP and its protagonists is how the SSA can effectively operationalize the Asian lesson of investing in infrastructure and human capital within the framework of SAP.

Furthermore, the inadequacy of current policies to usher in industrial growth is amplified by the fact that in Africa, the results from the World Bank (1994) study show that countries with the best adjustment records are not necessarily the ones with best industrial performance. The results (World Bank 1994:95-98, 247-8) assess growth performances in industry and manufacturing sectors over two periods—1981-86 and 1987-91. Uganda had the best performance, recording 17.3% and 23.6% improvements in industry, and manufacturing respectively. Uganda is reported to have achieved this success by having limited interventions in the markets but with only small improvement in its macroeconomic stance. Gabon follows in terms of manufacturing sector performance (15.3% turnaround) and yet it had "deterioration" in its macroeconomic policies, and maintained "medium" level interventions in the markets. On the other hand, The Gambia had both a "large improvement in macroeconomic policies" and "limited" intervention in the markets—the best preferred option according to the study—and yet the industrial sector witnessed a decline (minus 2.3%). Even the Central African Republic which was judged to have had a "deterioration" in policies, and maintained "heavy" market interventions performed better than Zimbabwe and Burkina Faso which had "large improvements" in policies.

In otherwords, if improvements in macroeconomic policies and limited intervention in the markets are the requirements for industrial competitiveness and growth, then Africa has a lot more lessons to learn than these World Bank prescriptions under SAP. In any case, the obdurate insistence on "limited market interventions" as a mechanism for industrial "catch-up" is largely at odds with the Asian experience. A major lesson from Asia is that Africa cannot industrialize by taking dictations from outsiders. It must be free to take the initiative—to do whatever it deems fit (within the bounds of international order) to compete and win. No country has industrialized by being arm-twisted in the ways and manner to do so. Indeed, the "Pragmatic flexibility in the pursuit of such objectives—the capacity and willingness to change policies" which the World Bank (1993:11-12) identified as the hallmark of the HPAEs' success is not available to Africa under the SAP framework.

Africa Is Too Fragile to Compete

Industrial competitiveness within an increasingly globalized world order would continue to be the central plank of economic growth in the foreseeable future. By this, various countries would continue to employ a battery of competition policies and other micro-industrial-efficiency measures to increase their market shares or revealed comparative advantage in critical industrial goods. Market penetration and power depend largely on the relative size and structure of the market (e.g., number and productive edge of competitors, geographical and commodity trade patterns, ease of entry, contestability extent, trade barriers, and availability of present or potential substitutes). The extent of success of a firm/industry or even country in this global contest would be aided by the absolute size of the firm/industry, its links and access to inputs and other output producing industries, cost structure, and its influence in, and by, the international market[3].

Aside from the key factors enumerated above as determining market penetration and competitiveness, other important determinants of competitiveness that are not seriously incorporated in standard price and trade theory pertain to the space-bridging (transport), and transaction costs. This has become inevitable in the light of increasing globalization which has made productive resources highly mobile—seeking locations with safety and highest returns. According to Kasper (1994) any realistic analysis of the East Asian experience must explicitly include the roles of: (i) space, transport and communications costs, and international factor mobility, (ii) the information, transaction, and organization costs of doing business and the role of institutions in economizing on these costs, and (iii) oligopolistic competition and competitive evolution. This is because, in modern economies with advanced division of labor and trade over vast distances, transport and transaction costs tend to make up at least 40 percent of producing the national product (North 1992). Thus, the competitive edge of many firms/businesses often depends on how well they cope with these costs and how well the society in which they operate manages to reduce these costs.

Technological prowess, conducive policy framework, and a level international playing field seem to be at the heart of the competitiveness game. Evidently, the playing field is high-tech: very capital intensive and breath-taking technological changes. In terms of prospects for developing countries, standard growth theory predicts that low-income countries will grow faster than high-income countries, because they can borrow technologies from the rest of the world and increase the marginal productivity of capital more rapidly than advanced countries. But this requires actively taking advantage of the technology, knowledge, and experiences of other nations (World Bank 1994). As Frieschtak (1989:1) suggests, "the least developed countries need to build up their industrial endowments, markets, and institutions before they can use competition to the fullest advantage as a tool of industrial policy." Unfortunately, Africa is decades behind the rest of the world in these matters, and does not have the institutions, capital, and entrepreneurial class. The very limited hard infrastructure—roads, ports, communications facilities, etc—has deteriorated, and the "soft" infrastructure—appropriate institutions—is largely absent.

An important lesson from the rapid industrialization of the HPAEs is that the process was propelled by a vibrant and innovative private sector. Such a private sector did not emerge over-night: it took a combination of environmental and historical factors as well as deliberate policies to nurture the sector to maturity. On the contrary, the private sector in much of Africa is still very fragile.

Gibbon (1994) provides a detailed characterization of the African private sector. The "purely" or "largely" private sector in Africa comprises mainly the tiny "person" or "person and a family member" operations undertaken mostly for survival purposes. These enterprises face serious structural constraints in attempting to expand their operations and these include: their primitive technical level, shortages of skilled labor, difficulty in obtaining credit, and continuous competition for household and enterprise resources with other household members and other income-generating activities. There is also the problem of markets for the private sector. Markets in several areas of operation of these kinds of enterprises comprise small concentrations of poor consumers. Thus, demand is constrained and, given that the populations are highly dispersed in an environment of poor infrastructure, the cost of moving beyond these market constraints is enormous. Beyond this peasant/ small scale enterprises, the private sector in Africa mainly consists of enterprises dependent upon different degrees of state connection. As against the pervasive illicit kinds of connections, the open connection entails state assistance with regard to subsidies, access to credit, inputs and state contracts, and implicit or explicit protection via tariff barriers, confinement policies, guarantees of market monopoly, etc. There is however a thin line dividing the various kinds of enterprises, as one type easily transforms to the other, or even straddles between "purely" private nature and enjoyment of illicit state connections.

The dominant group of the private sector activities is the category with state connections as well as the outposts of major multinational corporations. Evidence from a sample of SSA countries is that many of such enterprises have either collapsed, or are choking under the yoke of stabilization and import liberalization. As Gibbon (p. 23) observes, "the consequence has been that, following trade liberalization, consumption of locally manufactured clothing has in most places been mainly displaced not by consumption of cheaper (and possibly more efficiently-produced) new imports, but by imported second-hand clothing." Furthermore, most of the enterprises have proved to be incapable of competing in export markets. "Even the minority of enterprises (mainly the largest companies) who produced goods of competitive quality at competitive prices had very little basic knowledge of export markets or marketing. There were no state interventions which would have either overcome these obstacles, or helped smaller enterprises to reach the stage where they would have confronted them." In the 1990s, Africa's share of the global trade in manufactures has not exceeded a pitiable 0.43 percent. Such is the state and dilemma of private sector and industrial development in Africa.

In Nigeria for instance, many of the industries established under the import-substituting regime have collapsed, while in some sectors, some of the surviving ones have experienced improved capacity utilization. The booming sectors are mostly the food, drinks, beverages, and pharmaceuticals. The critical but nascent capital goods sector is still sprawling and its performance is far below the level prior to the commencement of SAP. Much of the private sector activities has been driven underground into the commercial sub-sector: informal petty and wholesale trading, smuggling, and speculative activities. Furthermore, some analysts have argued that one of the major reasons for the lack-lustre privatization programmes in some African countries is the weakness of the entrepreneurial class. If these weaknesses of, or even retrogression in, the private sector virility can be generalized for many countries in Africa, then such an environment is ripe for anything but industrial revolution. Such was not the experience of the HPAEs.

Conclusions

In the 1960s through early 1980s, Africa followed the IS/ structuralist strategy which had its own theoretical and empirical basis—and which were, just as the current neoclassical policies, the dominant orthodoxy for a majority of developing countries. The HPAEs ostensibly followed a mixture of the IS/ structuralist model and aggressive outward-orientation. On the other hand, the western capitalist and socialist countries followed different paths to industrialization and development. *Ex-ante*, no one was sure which of the contending schools would dominate, but the *ex-post* triumph of the neoclassical experiment (at least currently) is a historical fact. Today, in retrospect, other schools of thought are said to have been "mistakes" : had the countries followed the neoclassical path, they would have been ushered into the hall of fame as industrial giants. Moreso, there is a certain revisionist bent of the neoclassical ascendancy, that is, the attempt to re-construct the historical experience of every seemingly successful economy as having followed policies consistent with the neoclassical paradigm. Such has been the experience with the controversial attempts by the neoclassical economists (particularly the World Bank staff) to interpret the development experience of the HPAEs, and moreso, the questionable inference that other developing regions such as the SSA could duplicate such experience. The review of strands of interpretations regarding the causal factors in the HPAEs' successful industrialization points to several dark and grey areas of the experience for which neither school has unambiguously provided the answers. If there is any useful conclusion from the exercise, it is that gaining a comprehensive and correct understanding of the HPAEs' success story is largely an unfinished business.

From the limited light thrown so far on the HPAEs especially by the revisionists, we contend that several of the useful lessons from either of the interpretations would have limited applicability in SSA under its present circumstances. This is in the light of the changed and continually changing global environment, current terrain of international industrial competition, external shocks afflicting the economies and endogeneity of domestic policy handles, the constraints of the SAP package, as well as the structural weaknesses of the SSA economies. Even with the best of policies, neither the World Bank nor other independent analysts are very optimistic. As Green (1993:82) concludes,

> The chances of SSA breaking out into Asian NIC-type economic dynamism by the year 2000 are nil. Even one such case would be surprising. Neither the unlimited cheap capital, easy access to export markets for newcomers, nor the human capital bases for such a transformation exist now or can be foreseen for the coming decade.

So where does that leave Africa in the quest for rapid industrialization as the means to sustainable growth in the new world order? In retrospect, many analysts and even African policymakers acknowledge their share of the "past mistakes" in failing to track the Asian policy stance and performance during the past decades. The pertinent question now is where do we go from here? A certain faulty analogy seems to pervade most of the World Bank analysis: that the past is *always* the best guide to the future, and therefore if it can be proved that other successful economies followed a certain path to development, then intending developers would necessarily have to track that path. But every empirical economist knows that inference from historical data often becomes suspect in the presence of strong and frequent "regime shifts" or "structural breaks." The terrain of the changed and continually changing international environment, the nature and dynamics of the shocks afflicting the SSA economies, as well as the SAP policies foisted on these economies constitute a major source of the "regime shift." Under such a fundamental alteration of both the "playing field" and "rules of the game," it becomes doubtful that Africa's industrialization catch-up strategy should be hinged on attempting to go back 30-40 years to re-enact the HPAEs' experiment. Perhaps, the importance of the historical review of Africa-Asia past development experiences is to enable us not to repeat "past errors," but the evident "structural breaks" caution against any blind imitation of the "past correct" policies. Care should be taken therefore, so that some of the commentaries being sermonized by the World Bank as "the gospel truths" of development, do not divert attention from the fundamental constraints to Africa's recovery and integration into the world economy.

It is evident that the development challenge facing Africa requires major strategic, creative thinking. The SSA needs to do certain things very differently, but the question is how? A dominant view is that the best chance of sustained development in Africa should be pursued within the framework of a growth-promoting and market-accelerating model, but which is human-centred to address the pervasive poverty of the region. The specifics of such a model are still controversial and so are the strategies for breaking out of the stranglehold of the external environment. Useful ideas are emerging about the imperatives of the international community reaching a contract with Africa. This contract would be a re-invigorated variant of the UN-PAAERD. Several suggestions abound in the literature pertaining to the role of the international community in terms of: cancellation/ reduction of external debt by creditors as first step in "levelling the playing field"; a Marshall Plan for Africa; unlimited access of Africa's manufactured exports to the industrial country markets; increased resource flows targeted at institution building, human and physical capital development, and diversification of the economic base and export promotion; etc. On the other hand, some ideas are emerging about the nature of domestic policy directions in SSA to enable it compete: integration of the endogenous growth literature with the broader analytical framework of development literature (poverty reduction, etc); nurturing more activist but democratic "developmental states"; strengthening of regional cooperation/ integration with strategic linkages to other regional groupings rather than preoccupation with national SAPs; investments, policies, and building of institutions which promote competition and are market-accelerating; etc. A more rigorous analysis of the details and integration of these broad issues into a consistent but viable development strategy is beyond the scope of this paper, but the issues point to potential areas of fruitful enquiry.

With regard to specific area of strategies for rapid industrialization, several questions emerge. For example, where would Africa's comparative advantage lie in the future, and how can such comparative advantage be "created" —*a la* the Asian way—without being constrained by the GATTified world? Should Africa start small (with little things that have lost competitiveness, and as Ojimi argued in the case of Japan—areas that have lowest productivity and lowest returns), or, should it go high-tech. and seek to compete all the way? In the latter case, how would SSA do it, and with what technology and market penetration strategies? It has become fashionable to suggest diversification of the export base into manufactures, but the challenge is to demonstrate how such an option is viable and that chosen sectors/products are potentially competitive. In essence, while it is important to learn how the Asians dealt with the *past*, it is even more important to understand how some of the countries (especially those in "transition") are coping with current constraints and opportunities. For example, a lot of lessons could be learnt from other emerging industrializers (e.g., China, Vietnam, etc.) in terms of their innovations in interpreting and exploiting the provisions of the GATT/WTO and their strategies for "economic diplomacy." Evidently, the challenges to Africa's industrial competitiveness and economic growth are gigantic even as the options seem gloomy.

Notes

1. The review in this section benefits from the works of Gore (1994 and 1996); Singh (1995); Felix (1994); Kasper (1995); and Amsden (1993).
2. See Soludo (1996b) for more detailed analytical exposition on this issue, and the reconciliation of the SAP and development requirements within a more holistic theoretical framework.
3. See Frischtak (1989), Lieberman (1990), McGeehan (1968), for elaborate discussions of the determinants, and major components of the policy and institutional context of a competitive industrial environment.

Works Cited

Akamatsu, K. "A Theory of Unbalanced Growth in the World Economy." *Weltirtschaftliches Archiv*, 86, (1961):196-215.

Akamatsu, K. "A Historical Pattern of Economic Growth in Developing Countries." *The Developing Economies* 1, 1 (1962):3-25.

Alam, M. S. *Governments and Markets in Economic Development Strategies: Lessons from Korea, Taiwan, and Japan*; New York: Praeger (1994).

Amsden, A. H. "Structural Macroeconomic Underpinnings of Effective Industrial Policy: Fast Growth in the 1980s in Five Asian Countries," UNCTAD Discussion Papers, 57 (1993).

Barro, R. and Z. Sala-i-Martin *Economic Growth*, McGraw Hill (1995).

Boltho, A. "Was Japan's Industrial Policy Successful?," *Cambridge Journal of Economics*, 9, 2 (1985a).

Boltho, A. "Japan's Industrial Policy," in Z. A. Silberston and A. M. Schaefer, eds, *Industrial Policy and International Trade*; Philadelphia: University of Pennsylvania Press (1985b).

Borensztein, E., M.S. Khan, C.M. Reinhart, and P. Wickham "The Behavior of Non-Oil Commodity Prices," Occasional Paper No. 112; Washington, DC, International Monetary Fund (1994).

Calvo, G.A., L. Leiberman, and C.A. Reinhart "Capital Inflows and Real Exchange Rate Appreciation in Latin America: The Role of External Factors," *IMF Staff Papers*, 40, 1 (1993).

Camdessus, M. "Opening Remarks" in V. Corbo, M. Goldstein, M. Khan, eds, *Growth-Oriented Adjustment Programs*, Washington, DC, International Monetary Fund (1987).

Caves, R.E. and M. Uekesa "Industrial Organization," in H. Patrick and H. Rosovsky, eds, *Asia's New Giant: How the Japanese Economy Works;* Washington, DC: The Brookings Institution (1976).

Centre for Development Research, "Structural Adjustment in Africa: A Survey of the Experience," Copenhagen, March (1995).

Chakravarty, S. and A. Singh (1988), *The Desirable Forms of Economic Openness in the South*, Helsinki; WIDER.

Cho, S. *The Central Trading Company: Concept and Strategy*, Lexington, MA: Lexington Books (1987).

Deaton, A. "International Commodity Prices, Growth and Politics in sub-Saharan Africa," Paper presented at the Plenary Session of the Research Workshop of the African Economic Research Consortium, Nairobi: May/June (1995).

Easterly, W., and R. Levine "Is Africa Different?: Evidence from Growth Regressions," Mimeo; Washington, DC, The World Bank (1993).

Elbadawi, I. *World Bank Adjustment Lending and Economic Performance in Sub-Saharan Africa*, Policy Research Working Paper, No.1001, Washington, DC: The World Bank (1992).

Elbadawi, I.A. "Consolidating Macroeconomic Stabilization and Restoring Growth in Sub-Saharan Africa," Paper presented at the Overseas Development Council's Conference on "Africa's Economic Future," Washington, DC: April 24-25 (1995).

Elbadawi, I.A. "Market and Government in the Process of Structural Adjustment and Economic Development in Sub-Saharan Africa," Paper presented at the Institute of Economic Research of Hitotsubashi University Symposium on "The World Economy in Transition—Market and Government: Foes or Friends?," Tokyo: February 8-10 (1996).

Elbadawi, I.A, and B.J. Ndulu "Growth and Development in Sub-Saharan Africa: Evidence on Key Factors," Invited paper presented at the 1995 World Congress of the International Economic Association, Tunis, Dec. 17-22 (1995).

Felix, D. "Industrial Development in East Asia: What Are the Lessons for Latin America?," UNCTAD Discussion Papers, No.84 (1994).

Fischer, S. "The Role of Macroeconomic Factors in Growth," *Journal of Monetary Economics*, 32, 3 (1993):485-512.

Friedman, D. *The Misunderstood Miracle: Industrial Development and Political Change in Japan*, Ithaca, NY: Cornell University Press (1988).

Frischtak, C.R. "Competition Policies for Industrializing Countries," Policy and Research Series, The World Bank, Washington, DC (1989).

Gibbon, P. "Structural Adjustment and Structural Change in Sub-Saharan Africa: Some Provisional Conclusions," Mimeo, Centre for Development Research, Copenhagen (1996).

Gore, C. "Development Strategy in East Asian Newly Industrializing Economies: The Experience of Post-War Japan, 1953-1973 (1995).

Gore, C. "Methodological Nationalism and the Misunderstanding of East Asian Industrialization," UNCTAD Discussion Paper, No. 111 (1996).

Green, R. "The IMF and the World Bank in Africa: How Much Learning? in T.M. Callaghy and J. Ravenhill, eds, *Hemmed In: Responses to Africa's Economic Decline*, Columbia University Press, New York (1993).

Guasch, J.L. and S. Rajapatirana "The Interface of Trade, Investment, and Competition Policies: Issues and Challenges for Latin America," Policy Research Working Paper, No. 1393; Washington, DC: The World Bank (1994).

Hadjimichael, M.T., D. Gura, M. Muhleisen, R. Nord, and E. Murat Ucer "Effects of Macroeconomic Stability on Growth, Savings, and Investment in Sub-Saharan Africa: An Empirical Investigation," IMF Working Paper, WP/94/98; Washington, DC (1994).

Jamal, V. "The African Crisis, Food Security and Structural Adjustment," in *International Labor Review*, 127, 6 (1988).

Kasper, W. "Global Competition, Institutions, and the East-Asian Ascendancy," An International Center for Economic Growth Publication; ICS Press, San Francisco (1994).

Killick, T. *A Reaction Too Far: Economic Theory and the Role of the State in Developing Countries*, ODI Development Policy Studies (1989).

Kim, M. "National system of Industrial Innovation: Dynamics of Capability Building in Korea," in R. R. Nelson, ed., *National Innovation System: A Competitive Analysis*, Oxford University Press (1993).

Krueger, A. O. *Foreign Trade Regimes and Economic Development: Liberalization Attempts and Consequences*, Cambridge, MA: Ballinger Pub. Co. for NBER (1978).

Lieberman, I. "Industrial Restructuring: Policy and Practice," Policy, Research and External Affairs; The World Bank, Washington, DC (1990).

Lipumba, N.H.I. *Africa Beyond Adjustment*, Overseas Development Council, Washington, DC (1994).

Magaziner, I.C. and T.M. Hout, *Japanese Industrial Policy*, Berkeley, CA: Institute of International Studies, University of California (1981).

Marsden, K. "African Entrepreneurs: Pioneers of Development," Discussion Paper No.9; Washington, DC: International Finance Corporation (1990).

McGeehan, J.M "Competitiveness: A Survey of Recent Literature," *The Economic Journal*, LXXVIII, (1968):243-262.

Mendoza, E.G. "Terms of Trade Uncertainty and Economic Growth: Are Risk Indicators Significant in Growth Regressions," Mimeo; Washington, DC: International Monetary Fund (1994).

Mongula, B.S. "Development Theory and Changing Trends in Sub-Saharan African Economies 1960—89," in U. Himmelstrand, et al, eds, *African Perspectives on Development: Controversies, Dilemmas and Openings*, E.A.E.P, Nairobi (1994).

North, D.C. *Transaction Costs, Institutions, and Economic Performance*, Occasional Paper No. 30; San Francisco: International Center for Economic Growth (1992).

OECD, *The Industrial Policy of Japan*, OECD, Paris (1972).

Okimoto, D. *Between MITI and the Market: Japanese Industrial Policy for High Technology*, Stanford, CA: Stanford University Press (1989).

Putterman, L. and D. Rueschemeyer, eds, *State and Market in Development: Synergy or Rivalry?*, London, Lynne Rienner Publishers (1992).

Reinhart, C.A., and P. Wickham "Commodity Prices: Cyclical Weakness or Secular Decline?," *IMF Staff Papers*, 41, 2 (1994)

Ruggie, J.G. "International Regimes, Transactions and Change: Embedded Liberalism in the Post-war Economic Order," *International Organization*, 36, 2, (1982):379-415.

Shinohara, M. "Patterns and Some Structural Changes in Japan's Postwar Industrial growth" in L. Klein and K. Ohkawa, eds, *Economic Growth: The Japanese Experience Since the Meji Era*, Homewood, III: Richard D. Irwin Inc (1968).

M. *Industrial Growth, Trade and Dynamic Patterns in the Japanese Economy*. Tokyo: University of Tokyo Press (1982).

Singh, A. "How Did East Asia Grow So Fast? Slow Progress Towards an Analytical Consensus," UNCTAD Discussion Papers, No. 97 (1995).

Soludo, C.C. *Growth Performance in Africa: Further Evidence on the External Shocks Versus Domestic Policy Debate*, Development Research Paper Series No. 6, United Nations Economic Commission for Africa, Addis Ababa (1993a).

Soludo, C.C. *Implications of Alternative Macroeconomic Policy Responses to External Shocks in Africa*, Development Research Paper Series DRPS No.8, United Nations Economic Commission for Africa, Addis Ababa (1993b).

Soludo, C.C. "Multilateralism, Regionalism and the Effectiveness of Nigeria's Trade Policy," in A.H. Ekpo, ed., *External Trade and Economic Development in Nigeria*, Nigerian Economic Society (published volume of the 1995 conference papers), Lagos (1995a):313-334.

Soludo, C.C. "A Framework for Macroeconomic Policy Coordination Among African Countries: The Case of the ECOWAS Sub-Region," *African Economic and Social Review*, 1, 1 and 2; (1995b):1-31.

Soludo, C. C. "Intertemporal Fiscal Policy Rules, Deficit Reduction Plans and Macroeconomic Performance in Nigeria," Paper presented at the senior national policy workshop organised by the Nigerian Economic Society (March 6-8, 1996), Lagos: forthcoming in A.H. Ekpo, ed., *Monetary and Fiscal Policies Under Structural Adjustment in Nigeria* (1996a).

Soludo, C.C. "Theoretical Framework of SAPs in Africa: Some Better Ideas from Dead Economists," in *Beyond Adjustment: The Management of the Nigerian Economy*, Nigerian Economic Society, Published conference proceedings (1996b).

Soludo, C.C. "In search of Alternative Analytical and Methodological Frameworks for an African Economic Development Model," final research report submitted to CODESRIA under the Project on "African Perspectives on Structural Adjustment" (1996c).

Sorsa, P. "Competitiveness and Environmental Standards: Some Exploratory Results," Policy Research Working Paper, No.1249; Washington, DC: The World Bank (1994).

Squire, L. Remarks in "Proceedings of the Symposium on the East Asian Miracle," Symposium hosted jointly by the World Bank and the OECF, Tokyo: December (1993).

Svedberg, P. "Trade Compression and Economic Decline in Sub-Saharan Africa," in Blomstrom, M. and M. Lundahl, eds, *Economic Crisis in Africa: Perspectives on Policy*, Boutledge, London (1995).

UN *Transnational Corporations from Developing Countries*, New York: United Nations (1993).

UNCTAD Secretariat, "East Asian Development: Lessons for a New Global Environment," Report presented at a conference in Kuala Limpur, Malaysia, 29 Feb.-1 March (1996).

UNIDO "Private Sector Development and Privatization in Developing Countries: Trends, Policies and Prospects," Mimeo (1993).

Wade, R. *Governing the Market: Economic Theory and the Role of Government in East Asian Industrialization*, Princeton, NJ: Princeton University Press (1990).

Williamson, J. "Comment on Sachs," *Brookings Papers on Economic Activity* 2 (1985).

World Bank "The Challenge of Development," *World Development Report*, Washington, DC: The World Bank (1991).

World Bank *The East Asian Miracle: Economic Growth and Public Policy*, New York: Oxford University Press (1993).

World Bank, *Adjustment in Africa: Reforms, Results, and the Road Ahead*, New York: Oxford University Press (1994).

Table 4.1: Trends in Sub-Saharan Africa's economic

Indicator	Economic Performance Indicators		
	1965-73		1973-80
	SSA Total	Excl. Nigeria	SSA Total
GNP per Capita	2.9	1.2	0.1
Agriculture	2.2	-1.9	-0.3
Industry	13.8	-	4.3
Export	15.1	5.7	0.2
Import	3.7	3.4	7.6
Indicator	Structural Economic Change and		
	1965		1980
	SSA Total	Excl. Nigeria	SSA Total
Agriculture (% of GDP)	43	40	30
Industry (% of GDP)	18	20	33
Gross Domestic Investment (% of GDP)	14	15	20
Gross Domestic Savings (% of GDP)	14	15	22
Government Consumption (% of GDP)	10	12	13
Govern. deficit incl. grant (% of GDP)			6.8
External Balance (% of GDP)			(1.5)
Net ODA (% of GDP)			3.6
Total Debt Services as % of Exports			11.6
Terms of Trade (1987 = 100)			154

Source: UNCTAD(a) 1979, 1989 and 1991, Table 1.1
(see Svedberg, 1995: 22)

performance, structure and macroeconomic balances, 1965-92

(average annual change in %)				
1980-87		1987-92		
Excl. Nigeria	SSA Total	Excl. Nigeria	SSA Total	Excl. Nigeria
-0.7	-2.8	-1.2	-2.0	-0.4
0.7	1.3	1.8	1.8	1.1
1.8	-1.2	3.4	2.4	1.6
2.0	-1.3	1.7	3.5	3.4*
2.9	-5.8	-1.5	0.8*	2.0*
Macro Economic Balances				
	1987		1992	
Excl. Nigeria	SSA Total	Excl. Nigeria	SSA Total	Excl. Nigeria
34	34	35	30	29
25	28	24	25	23
20	16	16	17	17
15	13	11	12	10
16	16	17	14	16
4.7	5.5	4.7	8.4	8.4
(7.0)	(4.2)	(5.1)	(3.7)	(6.1)
6.6	8.0	9.6	11.3	14.0
	27.5		25.7	
128	100	100	80	79

Table 4.2: Value of exports from world, less developed countries and Sub-Saharan Africa, 1950-90

Region	1950	1960	1970	1980	1990
Billions of current dollars					
(1) World	60.7	129.1	315.1	2,002.0	3,415.3
(2) LDCs	18.9	28.3	57.9	573.3	738.0
(3) SSA	2.0	3.8	8.0	49.4	36.8
Share of LDCs–(%)					
(4) World exports	31.1	21.9	18.4	28.6	21.6
Share of SSA-(%)					
(5) World exports	3.3	2.9	2.5	2.5	1.1
(6) LDC exports	10.6	13.4	13.8	8.6	5.0

Source: UNCTAD(a) 1979, 1989 and 1991, Table 1.1 (see Svedberg, 1995: 22)

Table 4.3: The share of Sub-Saharan Africa in world exports of major product category (percent)

Product category	Share of product category in world exports		Sub-Saharan Africa's share of world exports		Share of product category in SSA exports	
	1970	1988	1970	1988	1970	1988
Crude oil (SITC 331)	5.3	6.0	6.5	6.9	14.0	34.5
Non-oil products (SITC 0-9 LESS 331)	94.7	94.0	2.2	0.	86.0	65.5
Primary commodities[a] (non-oil)	25.9	16.3	7.0	3.7	73.8	50.4
Agricultural commodities	7.0	2.8	6.3	3.6	12.3	8.2
Minerals & ores	7.5	3.8	9.7	4.2	30.0	14.6
18 IPC commodities[b]	9.1	4.3	16.1	10.0	59.1	35.6

Source: Derived on the basis of data from UNCTAD (b) 1984, 1986, 1988 and 1989: various tables; UNCTAD (a) 1980: Tables 1.1 and 1.2. (see Svedberg, 1995:23)
Notes: [a] SITC – 0, 1, 2-(233, 244, 266, 267), 4, 68, and item 522.56
[b] What UNCTAD labels the integrated Programme Commodities (IPC) (which supposedly are of greater importance to developing countries): bananas, cocoa, coffee, cotton and cotton yarn, hard fibres and products, jute and jute manufactures, bovine meat, rubber, sugar, tea, tropical timber, vegetable oils and oilseeds, bauxite, copper, iron ore, manganese, phosphates and tin.

Table 4.4: Characteristics of agriculture in SSA and other developing regions, 1982-84 (unless otherwise stated)

Region	Arable land per agricultural worker (ha) (1)	% of arable land irrigated (2)	Fertilizer consumption per cultivated ha (kg) (3)	Cereal yield (kg/ha) 1975-77 to 1982 (4)	Tractors per 1,000 cultivated ha, 1978-80 (5)
SSA[1]	1.6	2	9	800	1.4
Asia	0.8	27	46	1900	2.0
Near East/North Africa	2.8	20	72	1400	6.6
Latin America	4.9	7	53	2000	4.5

[1] Excluding South Africa
Source: Columns 1-4 from FAO, 1987, table 7: column 5 from FAO, 1984, table 7. (see Jamal, 1988:657)

Table 4.5: Sub-Saharan Africa in the 1990 vs. Southeast Asia in 1965. Differences in means of economic & social indicators

Characteristics	Mean Value for	
	For Asia 1965 Indonesia, Malaysia, Thailand	For Africa 1990 Selected countries
GDP per capita (at 1985 US$, ICP)	1320.0	1030.0
Agriculture share GDP (%)	37.0	35.3
Manufact. share GDP (%)	10.3	11.0[a]
Saving share GDP (%)	17.0	16.0
Investment share GDP (%)	16.0	16.0
Exports (gnfs) share GDP (%)	21.0	29.0
M2 share GDP (%)	19.6	19.45
Urban Pop. share (%)	18.3	29.0
Primary school enrolment (% eligible pop.)	80.0	68.0
Secondary school enrolment (% eligible pop.)	18.0	17.0
Adult illiteracy (% adult pop.)	35.3	53.0
Population (1000) per physician	15.0	23.5
Infant mortality (per 1000)	90.3	107.0
Life expectancy (years)	52.5	51.0
Population Growth Rate	2.7	3.1

Note: a: Data are for 1989.
Source: The World Bank (1994: 38)

Table 4.6: Asia's share of merchandise exports (percent of GDP)

	1970-1975	1976-1980	1981-1985	1986-1990	1991-1993
Newly Industrializing Economies					
Hong Kong	...	60.8	80.0	107.6	122.2
Korea, Rep of	17.8	26.0	30.9	31.1	24.6
Singapore	...	136.2	134.0	144.3	135.1
Taipei, China	35.0	46.0	48.2	46.8	40.6
China, People's Rep of	...	...	7.9	11.5	16.5
Southeast Asia					
Indonesia	17.8	24.4	22.7	22.0	25.8
Malaysia	38.3	48.1	47.5	60.5	71.5
Philippines	15.9	16.0	15.2	17.5	19.6
Thailand	13.2	18.0	17.9	24.7	29.4
Vietnam	...	...	...	...	...
South Asia					
Bangladesh	4.6	5.9	5.8	6.3	7.6
India	3.9	5.0	4.6	5.2	7.8
Pakistan	9.5	9.1	9.0	12.0	14.2
Sri Lanka	14.0	24.0	21.5	20.5	27.1
G7 Countries					
Canada	20.6	23.5	25.4	24.2	24.8
France	13.8	15.7	17.6	16.8	16.8
Germany	19.3	21.9	25.6	25.9	15.5
Italy	14.0	18.2	18.3	15.8	9.9
Japan	10.0	11.0	12.8	9.5	8.8
United Kingdom	16.5	20.7	20.6	18.3	18.5
United States	5.2	6.9	6.3	6.3	7.3

Source: World Bank, Stars Database (September 1994)

Chapter 5

Impact of Structural Adjustment on Industrialization and Technology in Africa

Samuel M. Wangwe and Haji H. Semboja

Introduction

The post-independence period was perceived by African countries as an opportunity to develop their economies. Industrialization was perceived as an integral part of the development agenda. Industrialization was expected to facilitate transformation of economic structure from predominantly agricultural economies to modern industrial economies. The share of industry in the economy was expected to rise, generate opportunities for employment, raise levels of productivity and raise incomes and standards of living of majority of the population.

In the external sector, it was envisaged that industrialization would bring about restructuring of the predominantly primary sector to a more diversified export sector in which exports of industrial products would increasingly play an important role. Industrial exports were associated with dynamic products in which specialization and technological learning were attained. On the import side two concerns were raised. First, the balance of payments problems were associated with deteriorating terms of trade. Second, economic independence was sought through substitution of imported goods by domestic production of those goods. These two considerations, drawing from the Latin American experience constituted the logic of the import substitution industrialization (ISI) model that the post independence African countries adopted.

The failures of the ISI model and the introduction of the structural adjustment programmes (SAP) underpinned by the neoclassical framework, led to the strong focus on import liberalization and almost a "no industrial policy" posture. Import liberalization was intended to provide corrective signals and incentives to the manufacturing and other sectors, increase the level of competition and technical efficiency, and stimulate factor productivity. The results so far show that the expected benefits of the new policy regime under SAP have not materialized.

In this chapter, we examine the industrialization experience of African countries, assess the performance over time and proffer explanations for the lack-lustre performance. Finally, we examine the major challenges facing the industrialization process in Africa, and chart the road ahead. The rest of the chapter is organized as follows: In the remaining parts of this section1, we evaluate the industrialization strategy of the 1960s and reasons for its failure. This is followed by an evaluation of the industrial performance under the SAP regime, with a view to understanding why the promises of SAP are not realised. Lastly, we provide a detailed evaluation of the challenges ahead and the road to more effective industrialization in Africa.

Industrial Development in the 1960s and 1970s

The industrialization agenda in this period followed the logic of the import substitution model. The Import Substitution Industrialization model was implemented following the logic of the Harrod-Domar model in which capital investment is the driving force of economic growth. The prevalence of the infant industry arguments investments in industry were accompanied by protective measures. The implementation of the ISI model was driven by investments. Although technology in the form of capital goods was embedded in imports there was little effort made towards technological learning.

The production of goods which were formerly imported was accompanied by imports of capital goods and intermediate inputs which were required to maintain desired levels of capacity utilization. Two implications of this phenomenon may be identified. First the desired production levels could only be maintained if adequate levels of imported inputs were tenable. This implied that foreign exchange earnings had to cope with the requirements of imported inputs to keep industry operating at the desired levels. Second, the industrial investments and operations were not necessarily associated with strong linkages with the rest of the economy.

Manifestation of the crisis

Towards the end of the 1970s, with variations across sectors, there were indications that the high expectations from industrial development were under threat. The main indications of this threat were low returns, inefficiency, low capacity utilization and low labour productivity. In fact, industries were not growing up as envisaged. There are variations across countries. However, specific country cases largely exhibited poor performance.

For over twenty years, Africa's industrial performance has not been satisfactory. The growth of manufacturing value-added (MVA) over the period spanning 1980 to 1993 was only 3 percent per annum in real terms, and the rate declined steadily over time, from 3.7 percent in the first half of the 1980s to 2 percent between 1989 and 1994. Several studies have suggested that this trend persisted even in the later period. The MVA growth rate fell from 3.3 percent during 1989-90, to 0.4 percent in the 1991-92 period, and registered only a modest recovery, reaching 1.7 percent in the 1992-94 period. Moreover, the growth performance conceals the continued stagnation or actual falls in MVA in many countries, particularly those of Sub-Saharan Africa (SSA) as some countries suffered from a sustained "de-industrialization" during the past decade and a half. Compared with anywhere else in the developing world, Africa's was the most serious manufacturing capacity loss.

Besides industrial activities conducted by those doing minimal processing of local natural resources, only a few of the industrial activities promoted have "matured" to become fully competitive. "Value-added" export-oriented activities that have driven many dynamic developing economies (such as those of the South East Asian countries) are conspicuously absent in Sub-Saharan Africa. Despite the low wages, only a few labor-intensive activities aimed at world markets have taken root in Africa. Linkages between local industries remain minimal and most superficial. The technological level of the existing industrial activities remains generally low. In broad terms, therefore, industry in Sub-Saharan Africa has failed to achieve structural transformation and export diversification. Yet, this was the expectation of the governments of these countries, and indeed, is the purpose that industry serves in many other parts of the developing world.

There are variations across countries. However, specific country cases largely exhibited poor performance. For instance, in the case of Tanzania a period of de-industrialization followed during 1980-1984, when the economy recorded a marginal Gross Domestic Product, (GDP) growth of 0.8 percent per annum while the manufacturing sector declined at annual average rate of 5 percent. A significant but temporary growth was recorded during 1985-1989 when the real GDP grew at 3.9 percent per annum and manufacturing grew at 2.3 percent. Official statistics indicate that the contribution of the industrial sector to the GDP declined from 10.2 percent between 1980 and 1984, to about 8 percent between 1990 and 1994.

The case of Kenya is better than that of Tanzania but there too the picture is far from rosy. Although the industrial sector is competitive relative to the neighboring countries (CBS 1961-1995a; Coughlin and Ikiara 1988), Kenya's industrialization process is still in an early phase. The sector lacks a strong indigenous technological base. The country has already taken advantage of the most obvious economic opportunities for import substitution and now wishes to find ways to sustain rapid industrial development. However, this may be hampered by:

- its dependency on imported capital and intermediate inputs;
- an uncertain future potential export market in the neighboring countries; and
- the relatively small domestic market.[1]

Industrialization in Kenya has taken place behind the wall of relatively heavy protection[2] against foreign competition (CBS 1961-1992a; Coughlin and Ikiara 1988 and 1991; Semboja 1993). This was based on a relatively over-valued exchange rate that kept imported capital goods and intermediate inputs relatively cheap, subsidized interest rates that made domestic investments attractive, direct government participation in industry, interventions in the form of the provision of direct loans and equity capital and access to foreign exchange for imported inputs and remittances at subsidized official rates. These pervasive, protection[3] and government-sponsored industrial incentives have had a strong impact on the total economy and particularly on the price relationships.

With the relatively high level of effective protection present in the domestic markets, the industrial sector became inward oriented. Many inefficient firms were able to reap high profits in the domestic market, while operating low capacity utilization levels. Under the circumstances, such firms have seen no reason to take the risks inherent in the competitive export markets.

In the case of Zimbabwe between 1982 and 1987, all subsectors (except foodstuffs and chemicals) registered negative growth rates. The highest decline occurred in metals, followed by wood and paper, and textiles and clothing. In general, the GDP growth rates were much lower after independence than during the first ten years of UDI. For the whole sample period (1967-87), total manufacturing posted an average annual rate of 5 percent.

In 1991, manufacturing in Zimbabwe accounted for 25 percent of the real GDP, while agriculture and mining contributed 12 percent and 7 percent respectively. Prior to the adoption of ESAP, the Zimbabwean economy was characterized by an import-substituting and highly protected economy that included a large, diverse, manufacturing sector that was inter-and intra-linked. Although the manufacturing sector's share of the GDP grew and dominated over the years (from 12.5 percent in 1945 to 16.6 percent in 1960 to 24 percent in 1970, 25 percent in 1990 and 22.9 percent in 1994), it was heavily reliant on agriculture.

Employment within the manufacturing sector increased from 159,400 employees in 1980 to a peak of 205,400 employees in 1991. In 1993, the number fell to 187,700. For the half-year period ending June 1994, employment recovered to 206,900 which was a 10 percent increase over 1993 levels. The fall in employment levels since 1992 is the result of company closures due to the adverse effects of ESAP and the relaxation in labor regulations—particularly those relating to the hiring and firing of workers.

While it is often suggested that the industrial performance of African countries in the 1960s and 1970s was dampened by the import-substitution strategies adopted, a reversal (i.e., import liberalization), has not convincingly answered the problem. Initial industry-related research in SSA attributes the poor performance of this sector to the import liberalization policies that constituted an important element in the World Bank/IMF structural adjustment packages which many African countries adopted in the 1980s. It is widely accepted that the response to structural adjustment programs in SSA in the mid-1980s has generally been poor, and that the manufacturing industry in these countries has particularly failed to expand in terms of outputs, productivity and exports.

The initial concerns about undynamic export structure and the role of industrialization in dynamizing the export structure is as valid as ever.

Apparently, the unchanging structure of exports suggests that the export of traditional primary commodities paints a pessimistic picture and has not seriously been considered in the design and implementation of SAPs. The first period of export pessimism was based either on declining terms of trade for primary products (Prebisch 1954) or on the notion that the absorptive capacity of foreign markets was low (i.e., elasticity pessimism) as argued by Nurkse (1959). The future of most of these traditional exports is rather bleak, considering the low-income elasticities of demand associated with these products as markets for many of these commodities are showing signs of saturation. A recent study has examined Africa's main primary exports and concluded that the path followed in Africa, of exporting mineral and agricultural resources, is a dead end because of the poor prospects of these resources (in their non-VA forms) in the world market (Brown and Tiffen 1992). For instance, studies of the world cocoa market, in which exports from Africa comprise 55 percent of the traded volume have shown that over a 20-year period spanning 1960/65 to 1980/85, cocoa consumption increased by a mere 40 percent with negligible increase in major consumer markets such as the EU (12 percent) and the USA (19 percent) over the same period (ITC 1987). Consumption of Robusta coffee which is the dominant export from Africa has stagnated at 12.5 million bags since the mid-1970s, while world exports of Arabica coffee doubled between the 1960s and 1980s (Brown and Tiffen 1992). For coffee from Africa, these trends are aggravated by problems of product quality, delays and unreliability.

The empirical evidence that is available on the terms of trade in Sub-Saharan Africa seems to suggest that the export pessimism thesis continues to hold. A World Bank study, taking the year 1980 as the base, has indicated that the terms of trade index declined to 91 in 1985 and 84 in 1987 (World Bank 1989). In a recent, more general survey of empirical studies on this subject, Killick (1992) has concluded that "there is now wider acceptance than was formerly the case of the declining real commodity price thesis." If the importance of technological change is taken into consideration, the problem of export structures is even more serious, since the production of most traditional exports is associated with relatively limited technological dynamism.

Recent efforts to revive exports within the traditional setting have provided further evidence that exports cannot provide an engine for growth unless the export structure changes. Evaluating the impact of structural adjustment programs in Sub-Saharan Africa, Hussain (1994) has indicated that between 1985 and 1990, the export volumes of nine major export commodities in countries which had undertaken adjustment programs increased by 75 percent as compared with the 1977-79 averages. Yet, (and most interestingly), export earnings from these exports had fallen by 40 percent over the same period, due to deteriorating barter terms of trade (Hussain 1994).

The current position of Africa in world trade is characterized by two main features; first, it has a small and declining share in world trade and second, its presence in world trade is largely confined to primary exports and the importation of non-primary products (Wangwe 1995).

Africa's share in world trade is not only small, it has been declining; ranging between 4.1 percent and 4.9 percent of the world trade during the 1960-1965 period, it fluctuated around 4.4 percent during the 1970s (it was 4.7 percent in 1975), and declined to 2.3 percent in 1987 (UNCTAD 1993).

Between 1980 and 1987, while world exports were growing at 2.5 percent per year, Africa's exports were declining at an annual rate of 7.5 percent.

The share of non-oil primary exports declined even more dramatically, from seven percent to four percent over the same period (Sharma 1993). Manufactured exports, though small, have exhibited a similar trend. The share of manufactured exports from Sub-Saharan Africa (SSA) in world trade declined from 0.38 percent in 1965 to 0.23 percent in 1986 (Riddell 1990). In relation to other developing countries, the share of Africa's manufactured exports declined from 5.2 percent in 1975 to 2.6 percent in 1985 and further to 2.5 percent in 1990 (UNIDO 1993). A preliminary study of the impact of the Single European Market has indicated that SSA countries lost their share mainly to other developing countries between 1987 and 1991, in spite of the preferential market access accorded to Africa through the Lomé Convention (UNCTAD 1993b). These trends suggest that Africa has lagged behind in competitiveness relative to the rest of the world. This is an indication that relative to other regions, growth in productivity and technological acquisition and the development of innovation in the export sector in Africa have been low. Africa's lack of competitiveness in traditional and non-traditional exports, is a challenge that needs to be met if Africa is to improve its position in world trade.

Africa's position in the new global trade relations will largely be determined by what action is taken in two important directions: first, increasing the regional and international competitiveness of its production activities by changing the structure of exports towards more dynamic, non-traditional products (in terms of their demand prospects and their potential to effect technological change); and second by tackling the structural bottlenecks inherent in the entire system of governance and those specifically addressing export promotion and industrial development. The need to effect change in the export structure will inevitably bring discussions of policy issues relating to export diversification, the transformation of production structures, and industrialization, back on the development agenda.

Apparently, import substitution industrialization was intended to provide correct signals and incentives to the manufacturing and other sectors, increase the level of competition and technical efficiency, and stimulate total productivity. However, for some countries this possibility is remote. Internal structures and infrastructural rigidities resulted in high production costs, eventually making it difficult for local industries to compete favorably with cheap imports in the domestic market. This impaired possibilities of securing an export market. However, import liberalization is just one side of the story, the other is institutional capability.[4]

Explanation of Poor Performance

The Policy Framework

The Berg report of 1981 took the lead in putting forward the explanation of the poor industrial performance taking the position that policy inadequacies were responsible for this situation. It was argued that these took the form of wrong incentives, distorted prices, overvaluation of the exchange rate, depressed interest rates, urban bias and overemphasis on industry at the expense of agriculture. The Berg report identified what it called three critical domestic policy inadequacies:

- trade and exchange rate policies had overprotected industry;
- little attention had been given to administrative constraints in mobilizing and managing resources for development and the public sectors became extended;
- consistent bias against agriculture in price, tax and exchange rate policies.

The focus of the analysis was on the efficiency with which resources were used. The report refers to distorted industrial development brought about by trade and exchange rate policies which provided incentives to develop import-intensive industry and offered protection which gave no incentive to productivity growth.

The key policy prescriptions of the Berg Report (1981) were: correction of overvalued exchange rates; improved price incentives for exports and for agriculture; lower and more uniform protection for industry; reduced use of direct controls; and privatization. The report was largely silent on non-price factors and institutional support infrastructures.

The Berg report suggests that exchange rate and interest rate policies made capital goods imports cheap and provided incentives to invest in capital intensive industries. In the same report, however, it is also pointed out that due to extra expense on transport and construction costs industrial projects in Africa typically require investment costs that are 25 percent higher than in developed countries and for some industries the margin may be as high as 60 percent (World Bank 1981). This suggests that structural and infrastructural constraints were recognized but were not given sufficient attention in the analysis. Long term issues relevant to industrialization were given very little attention.

Long term issues that were discussed by the report were rather narrowly confined to rapid population growth and urbanization; resource planning regarding soil, forestry and firewood supplies; and encouraging more regional integration to get round the problem of small markets.

The Berg report buttressed in the neoclassical paradigm postulated that Africa had comparative advantage not in industry but in agriculture. In fact the report argues that countries which had put resources in agriculture performed better. However, these countries (e.g., Malawi) have not been able to sustain what appeared to Berg to be dynamic growth.

Industrial Performance

Evaluation

Over time, the World Bank has increasingly recognized the importance of structural and institutional factors in explaining industrial performance in Africa. The World Bank's long term perspective study (World Bank 1989) correctly pointed out that "it is not sufficient for African governments merely to consolidate the progress made in their adjustment programs. They need to go beyond the issues of public finance, monetary policy, prices and markets to address fundamental questions relating to human capacities, institutions, governance, the environment, population growth and distribution, and technology" (World Bank 1989:1). It is suggested that the attention to human resources, technology, regional cooperation, self-reliance and respect for African values provide the main focus of the proposed strategy. The proposed strategy stresses the:

- the need to put in place an enabling environment for infrastructure services and incentives to foster efficient production and private initiative;
- enhanced capacities of people and institutions; and
- that the growth strategy must be both sustainable and equitable.

It is recognized that industrialization in Africa has achieved little indigenous technological development and proposes greater emphasis on acquiring the necessary entrepreneurial, managerial and technical skills. Incentives should focus on training and technological adaptations (World Bank 1989).

Proposing a new start to industrialization, the report correctly identifies the challenge as one of building on the base of technical skills and industrial experience that is already in place to achieve the industrial transformation envisioned in the Lagos Plan Action.

Fostering African entrepreneurship is envisaged through improving the business regulatory and policy environment, expanding access to credit, encouraging self-sustaining services (e.g. technical services, subcontracting) and stimulating local markets.

Charting out a strategic agenda for the 1990s the report (World Bank 1989) suggested the following in relation to industrialization:

- adjustment programs must take fuller account of investment needs to accelerate growth and the goal of macroeconomic balance must be to fundamentally transform Africa's production structures;
- human resource development;
- institutional reforms for capacity building;
- fostering private investment;
- providing efficient infrastructure services;
- fostering regional integration and coordination;
- continuation of special programs of assistance to Africa.

This promising progress towards understanding the poor industrial performance in Africa was countered by a subsequent World Bank report (1994) better known as the "Road Ahead." The "Road Ahead" was in many ways a retrogression compared with progress that had been made in the 1989 World Bank report.

The "Road Ahead" has suggested that there has been improvement in the performance of industrial output especially among the countries that made large policy improvement.

The "Road Ahead" starts with the words "in the African countries that have undertaken and sustained major policy reform, adjustment is working." It goes on to argue that the impact of adjustment has been particularly marked in industrial and export growth. Its sample consists of 29 countries in Africa that have undergone SAPs. Since many of these adjusting countries have not yet implemented fully the policy reforms recommended, the report divides them into three groups to assess the effects of SAPs—6 with "large improvements" in macroeconomic policies; 9 with "small improvements"; 11 with a "deterioration"; and 3 are unclassified. Changes in growth performance, in GDP, exports, industrial production and agriculture are all assessed with reference to these categories, and are separated between a pre- and post-adjustment period (1981-86 and 1987-91 respectively). The premise is that the first group shows the effect of structural adjustment most and the third, the least; and, similarly, that the differences between the second and the first periods show the effect of adjustment.

The effect on manufacturing of SAPs measured in this way shows that, on average, countries with the most adjustment have enjoyed the largest improvement in manufacturing performance; and those with the least have had the least improvement, this purports to show that SAPs are good for industrial growth.

While these findings appear to support a position pointing to the benefits of SAPs for industry, the following points need to be made:

First, the groupings according to improvements (large or small) and deteriorations in policy have little or nothing to do with adjustment. They are based entirely on macroeconomic policy and not on adjustment in the sense of "getting prices right" through imports and other forms of liberalization. If they show anything, it is that improving internal and external balances is conducive to growth. The impact on resource allocation in response to market orientation cannot be assessed.

- Medians for groups have no statistical significance if individual variations within the groups are larger than the variations between groups.
- The deteriorating policy groups have higher manufacturing growth rates in both periods than the other groups; and the large improvement group has the lowest growth rate in the later period of adjustment. One conclusion could be that adjustment, as such, had no special effects on manufacturing growth; and that the differences in growth rates were caused by other factors.

The statistical association between economic reforms (as defined by the three aspects of policy improvement) and performance must be interpreted with caution for the following reasons:

- Statistical associations do not necessarily imply casual relationships;
- Other factors could have influenced performance (e.g. terms of trade, weather, political stability);
- Variations in implementation of reforms occured in terms of timing, intensity, speed and consistency;
- Fragility and reliability characterized the data base.

Industrial Performance by Grouping: Casting Doubts

Adopting the grouping of countries according to the "Road Ahead" it is shown that the relationship between industrial performance and policy improvement is different to the picture painted by the report as demonstrated by Lall and Stewart (1996).

Various aspects of industrial performance in Africa were recalculated using more recent data from the African Development Bank, as well as from the World Bank and other sources. The coverage was extended to 50 countries in Africa, including 21 that had not undergone adjustment (ADB 1995).

The periods used were also slightly different. A longer overall period, 1980-93, was taken, with the later sub-period 1990-93 to capture the more recent effects of policy reforms. The highest rate of growth in both periods is for the adjusting countries with improved macroeconomic policies. The lowest is for the countries with policy deterioration for 1980-93 and for the non-adjusting low-income countries during 1990-93. This suggests, again, that macroeconomic policy improvement helps economic growth in general. However, there is no significant statistical difference between any of the group growth rates. This indicates that, the differences in group performance could be caused by factors other than macroeconomic policy differences. The conclusions reached in the "Road Ahead" have not survived the test of time.

Reforms in the exchange rate were geared towards shifting resources to the more productive sectors and stimulating exports. Trade liberalization has meant that clients had options to import many items. Devaluation dominated the export incentive shift and the rise in export profitability compared to domestic sales. Export recovery, which was recorded in some of the sample countries could not be sustained. The initial recovery was a temporary phenomenon, which was followed by decline. This outcome is demonstrated by the case of Tanzania and Ghana, countries which have been categorized as relatively successful adjusters (World Bank 1994). For instance in the case of Tanzania, devaluation accompanied by resumptions of foreign aid inflow induced some output recovery and export expansion. Much of this output recovery originated from better utilization of existing capacities rather than from new investments.

The performance of the industrial sector in respect of foreign exchange earning is still small. For the period 1987-1992, the country's cumulative value of exports remained small. The share of the sector to total exports was 18.1 percent in 1987 having increased to 26 percent in 1990, before decreasing to 18.2 percent in 1991 and 16 percent in 1992. Textile and textile products were among the products that mainly contributed to the export earnings. According to the USAID-CTI (1994) report, the share of industrial products to total exports has been on the low side due to unsatisfactory prices, a tax structure not conducive to the promotion of exports and an inadequate export drive. Furthermore, poor technology in the packing of industrial export products exacerbated this poor performance. Lack of a packaging industry in Tanzania has led to the importation of packing materials.

Although the expansion of exports was certainly encouraging, it has to be viewed from the context of starting with a very low performance base that typified the economic crisis period of the early to mid-1980s. The expansion rate has not been sustainable and signs of stagnation have since appeared (from 1990/91). There was a significant decline in manufacturing exports and in non-traditional exports (NTE), during the 1990-1994 period. The decline in NTE suggests that the process of elimination of market distortions is incomplete.

In the case of Ghana the story is similar. Manufacturing value-added did rise rapidly after 1983, when imported inputs were made available to existing industries that were suffering from substantial excess capacity. However, as liberalization spread to other imports and excess capacity was used up, the exposure to world competition led to a steady deceleration of industrial growth. Thus, the rate of growth of MVA fell to 5.1 per cent in 1988; 5.6 per cent in 1989; 1.1 per cent in 1990; 2.6 per cent in 1991; and 1.1 per cent in 1992.

This performance does not suggest that Ghanaian manufacturing is responding well to liberalization. Employment in manufacturing has fallen from a peak of 78,700 in 1987 to 28,000 in 1993. There has been a rise in the number of small enterprises, but this is in low-productivity activities aimed at local markets. Foreign investment has not responded well to the adjustment; and most of it is concentrated in primary activities, rather than in manufacturing; and domestic private investment has not picked up sufficiently to dynamize the manufacturing sector.

At the same time, large sections of the manufacturing sector have been devastated by import competition. It is obvious that the long period of import-substituting industrialization, with the lead taken by state-owned enterprises, left a legacy of inefficiency and technological backwardness. It may also have left some technological capabilities, but not at the level that rapid liberalization could stimulate to reach world levels. The adverse impact of liberalization has, therefore, been strongest in more modern, large-scale part of the industrial sector, which had the most complex technologies and so suffered most from the lack of technological capabilities. Industrial survivors and new entrants are basically in activities that have "natural" protection from imports—very small-scale enterprises, making low-income or localized products.

What this ignores is that even the ability to compete internationally in low-technology labor-intensive industries requires a level of productivity and managerial and technical skills that is, at present, lacking in SSA. The few relatively well-managed firms that exist are largely foreign owned; and among local enterprises, the better ones have entrepreneurs that are well educated. The typical local firms, on the other hand, has entrepreneurs with low education, a poorly skilled workforce and no methods for raising their technological capabilities; and lacks of ability even to perceive and define their technological problems.

This suggests that the recovery was not based on improvement in the core capabilities, which are necessary for firms to make gains in international competitiveness.

More advanced technology has been imported through capital goods imports, joint ventures (largely with TNCs), turnkey projects, technical management agreements and to a lesser extent, direct foreign investment. However, the degree of technology innovation, adaptation and assimilation has been low. The capability of providing own technological requirements, i.e., technological self-reliance has remained low. Benefits from the use of imported technology have been reduced by wrong choices and/or the lack of clear awareness of a domestic technological capacity that foreign technology could complement. Moreover, foreign technology has not been found to (selectively) complement the domestic technology system.

Although the importation of foreign technology (e.g., through capital goods and foreign personnel) has provided an initial base for technological development, it hardly seems to have been assimilated and improved upon. As a result of unfavorable initial conditions (the near absence of local entrepreneurs and skilled technicians and engineers, concentration on the acquisition of skills and the lack of systematic monitoring of involvement of local personnel in major technological decisions), the degree of assimilation and foreign technology improvement has largely been low.

Macroeconomic Policy Improvement Is Necessary But Insufficient

The main reason why the expected relationship between policy improvement and industrial performance is not realized is basically because improvement in macroeconomic policy is a blunt tool for industrial development.

The definition of policy improvement is narrowly defined in terms of budget, exchange rate and inflation. These are elements of macroeconomic stabilization and in themselves are not necessarily sufficient to bring about structural adjustment.

It would, thus, appear that countries with policy improvements do better than those with policy deterioration, at least as far as manufacturing is concerned. However, this proves nothing about the impact of adjustment itself, since the groupings reflect only macroeconomic management. For instance, the elements of structural adjustment which have had far-reaching impact on industry such as import liberalization are not captured in the indicator of policy improvements.

There are other elements, however. Despite general agreement on the need for openness and greater reliance on market forces, the design, pace and content of the reform process remain controversial and unclear. These include the extent and speed of trade liberalization; the role of the government in resource allocation and industrial restructuring; the integration of incentive reforms with structural measures to improve the supply capacity of the economy; and the linking of reform to the level of development of the economy and to improvements in the capabilities of government to mount effective supportive policies.

It is important, however, to be clear about the essential features of structural adjustment.

First, stabilization is not structural adjustment. Macroeconomic stabilization generally precedes or accompanies adjustment, even if the dividing line between the two is blurred in practice. In principle, stabilization is a precondition for structural reforms.

Second, SAPs are based on strong assumption about market and government efficiency. The underlying assumption is that markets are essentially efficient in developing countries and that government interventions in resource allocation are essentially distorting and inefficient.

Third, it is extremely difficult to evaluate the effects of structural adjustment in practice. Actual SAPs differ greatly between countries in their design and content, and in the extent to which they are actually implemented by governments.

Firms were expected to respond to incentives but in practice they have not. The failure of firms to respond to incentives can be demonstrated by the failure to stimulate investment in general and investments in technology. This suggests that there is need to understand better the factors influencing firms level responses.

Many writers criticize the lack of perspective created by the disproportionately narrow focus (on trade policies) and call for wider policies on the ability of firms to respond to incentives and signals in an appropriate manner. The newly industrializing countries had benefited from import substituting industries, which built a base for technological and learning experience. In addition, selective protection played an important role in the larger "interventionism" package.

Industrialization Challenges Ahead

Changing Conceptualization of Industrialization

The guiding principle in considering the challenges ahead is that the changing circumstances in the world economy have an influence on the conceptualization of the industrialization agenda.

The context in which industrialization is and will continue to take place has changed. The imperatives of the globalization process and the opportunities and threats that are associated with it have altered the conditions of industrialization in significant ways.

Africa's position in the new global trade relations will largely be determined by what action is taken in two important directions: first, increasing the regional and international competitiveness of its production activities by changing the structure of exports towards more dynamic, non-traditional products (in terms of their demand prospects and their potential to effect technological change); and second, by tackling the structural bottlenecks inherent in the entire system of governance and those specifically addressing export promotion and industrial development. The need to effect change in the export structure will inevitably bring discussions of policy issues relating to export diversification, the transformation of production structures, and industrialization back on the development agenda.

One major consideration that will influence the way the industrialization problem is conceptualized relates to the changing character of innovations and their role in international trade and competitiveness. Industrialization, for the less industrialized countries in Africa, will have to take place under conditions of accelerating technical change and the pervasive application of new technologies. These developments support a kind of conceptualization of technological change, which emphasizes learning and the accumulation of technological capabilities. This is bound to have considerable implications for the conceptualization of the industrialization problem in Africa.

This approach to industrialization is consistent with the thrust of recent trade and growth models which have focused more explicitly on the microfoundations of innovations by addressing firm-level decision to invest in product or process innovations. The case studies of exporting firms in six countries have shown that ceaseless search for improvements in technology (especially product quality and cost-lowering process innovations) has been most instrumental in improving productivity (Wangwe 1995). Productivity growth, in turn, was a most important factor enabling exporting firms to succeed in the changing technological and market conditions. Exporting firms maintained and improved their market position by investing in technology and continuing to improve on it. Improvements were made not only in the firm's products but also in the process of production, in order to cope with pressure to keep costs at competitive levels or to improve product quality level (level and consistency). These responses were derived from signals given in export markets.

A major policy implication of the experience of industrial development in Africa is that, in conceptualizing the industrialization problem in Africa, fuller recognition will need to be given to the altering nature of technological change, with emphasis on learning and the accumulation of technological capabilities within firms, with the requisite support from the state in the form of various supportive infrastructural investments. This contrasts with the previous emphasis on the transfer of the capital and know-how required for an industrialization process which was primarily directed at import substitution.

Changing characterization of innovations, the nature of technological change and the centrality of technological change in trade and international competitiveness must mean that the conceptualization of the industrialization problem will need to change accordingly.

Import Substitution Industrialization Revisited

The challenge of industrialization in a more open and competitive environment will need to make import substitution more efficient and pay greater attention to export promotion and show greater appreciation of changing conditions in the world market. In doing so, however, many strengths and capabilities which were acquired in the import substitution phase will be deployed and be built upon.

The case studies of exporting firms have shown that most exporting firms started by serving the domestic market (Wangwe 1995). Import substitution firms grew up and built up various core capabilities by producing for the domestic market. The protection of the domestic market allowed them to accumulate resources, which were in turn invested in developing capabilities which enabled them to turn to exports at a later stage. In this sense, it can be argued that import substitution and export orientation are complementary. Import substitution has preceded exporting and has, under certain circumstances, formed an important basis for export orientation. For instance, export orientation programs such the Mauritian EPZs built on the capabilities which had been accumulated during the import substitution phase. The policy implication is that, if import substitution is effective in providing for the development of technological capabilities, it can establish the basis for building a competitive export sector. In the process of exporting, firms can develop efficient linkages and acquire technological capabilities. The challenge is to blend efficient import substitution and export orientation through a mix of policies which aim at maximizing the benefits from increased domestic demand and at stimulating both substantial (and efficient) import substitution and increased export orientation on the basis of growing technological capabilities.

Import Liberalization: Case for Selectivity

Import liberalization need not harm economies as the experience of some countries has shown. For instance, in a study of exporting firms in selected African countries three categories of trade liberalization regimes were found (Wangwe 1995).

The first regime is represented by countries like the Cote d' Ivoire and, to a lesser extent, Kenya, which were already fairly open. Import liberalization merely lowered and rationalized some tariffs but did not represent a major shock for industrial firms. The second regime is that of countries like Tanzania, where quantitative restrictions on imports had been quite pervasive, so that import liberalization came as a major policy shift. Competition from imports came suddenly, not giving much time for adjustment to the new competitive environment. In the case of Nigeria, the end of the oil boom around 1980 led to extensive use of tariffs and quantitative restrictions. The various foreign exchange conservation measures implemented in the period 1982-85 meant that several industries which were dependent on imported inputs had to operate considerably below capacity, hence reducing growth and worsening unemployment. The adoption of the SAP in 1986 represented a fundamental shift in the basic philosophy of economic management at the national level. The reforms included the adoption of a largely market-determined exchange rate and the removal or relaxation of quantitative restrictions on many tradable goods.

The third regime is represented by Zimbabwe and Mauritius, in which the implementation of trade liberalization was managed more selectively in a situation where the export sector was already quite diversified and firms had attained a reasonable degree of competitiveness. Trade liberalization had a constructive effect in that firms were given adequate time to make adjustments.

In Zimbabwe, for instance, import liberalization started with imported inputs through some form of an open general import licence system. The users of these inputs were made aware that the next phase of import liberalization would be applied to outputs. This message induced many firms to invest in technological improvements of various kinds in anticipation of a more competitive environment(Ndela and Robinson 1995). Managed import liberalization stands a better chance of providing an opportunity and incentive for firms to build up capabilities which can cope with a more competitive environment. LLDC countries that undertook selective trade liberalization performed better than those undertaking across-the-board trade liberalization (UNCTAD 1995).

According to UNCTAD (1995) Zambia's trade liberalization has been associated with deteriorating performance in many firms. The difficulties of deteriorating firms do not only arise from their own inefficiency but also from inefficiencies in the environment in which these firms are operating (e.g., distortions in financial market, high import protection of input providers, high energy costs). These factors were not dealt with prior to or in parallel with the introduction of liberalization measures.

Lessons for East Asia can be instructive in this context. Trade policy in Japan, Korea and Taiwan supported industrial deepening and the development of national firms, provided selective incentives to promote exports (e.g., tariff rebates, tax exemptions and export credits), selective protection (e.g., exchange control, tariffs and quotas) and liberalized specific industries as they attained international competitiveness (UNCTAD 1995).

What is at issue is the pace and content of the process by which liberalization is achieved; the role of the government in remedying market failures; and the tools of intervention that need to be retained or refashioned.

Economic reforms do not usually address any of the skills shortages that may be affecting the efficiency of African industry, yet many existing industries may become competitive if their human resources were improved. The design of SAPs should, therefore, include education and training as an integral part of the restructuring process.

It is important to note that a certain amount of capability development has already taken place in the industrial sector; and this is a valuable resource that should be conserved, rather than dissipated by shock therapy that leads to massive de-industrialization. Technological capabilities reside in groups of skilled and experienced persons, rather than in individuals; and the destruction of enterprises means that the stock of accumulated knowledge is effectively destroyed, even if the individuals concerned stay in the country. This important feature of industrialization is ignored by SAPs, but forms a vital part of the policy approach of the East Asian NIEs.

Education and Training

Technological learning does not occur in isolation. It involves interactions with other firms and institutions. Apart from physical inputs, it calls for various new skills from the education system and training institutes, technical information and services, contract research facilities, interactions with equipment suppliers and consultants and standards bodies. The setting up of this dense network of co-operation needs the development of special skills.

Science and Technology Infrastructure

There is also very little interaction between the industrial sector and the technology infrastructure that exists in many countries to provide R&D and technical support to enterprises. Many of the research institutes are poorly funded, and so have inadequate equipment and insufficiently motivated staff. They, thus, do not aggressively go out to search for, and offer, solutions to the technical problems of industry, preferring a more isolated existence. Much the same applies to technology information services to help local firms to locate and purchase foreign technologies.

The development of the science and technology (S&T) infrastructure and the provision of technical extension services to industry, especially to small and medium-sized enterprises (SMEs) is an important supply side measure that is capable of enabling firms to respond to incentives. The S&T infrastructure included such basic services as quality control and metrology; research and development; information on sources of technology; and assistance in the purchase of foreign technology. In this context, it is important to note that requirements of quality control have changed in the past two decades; and international trade in manufactured products increasingly requires stringent proofs of quality management.

Investment and Industrial Financing

In many countries in Africa economic reforms have been associated with uncertainty and high interest rates which can only be afforded by high return, short term and speculative/trading activities. In the case of Tanzania, for instance, the problems of financing industry are:
- Inadequate liquidity in the banking systems due to poor servicing of debts;
- Overcautious stance of external financing institutions and new foreign banks in the country due to past performance of many industrial firms;
- Inadequate policies to mobilize domestic savings into the financial system; and
- Lack of sound projects in industry compared with the more lucrative commercial activities especially when the state of supportive infrastructure and other aspects of the operating environment are taken into account.

Investment and financing of industrialization poses new challenges of domestic resource mobilization. The role of the state will be very instrumental in this process as the experience of some East Asian countries suggests. Fiscal support was provided by the governments of Japan, Korea and Taiwan to facilitate investments through reducing the effective tax rate on corporate income and allowing new firms to retain a higher share of profits and investment tax credits. Financial incentives were given in the form of low and stable interest rates, preferential policy loans and priority allocations of credit and foreign exchange. Competition policy was geared to productivity and capital accumulation sometimes restricting competition and in some cases promoting it. This suggests that Africa will need to face the challenges of mobilizing resources for financing industrialization in ways which do not rely on forced savings in agriculture.

Closer Government and Enterprise Sector Relations

Economic reforms in many countries in Africa have involved restructuring of public enterprises, a process which has been associated with dramatic changes in the ownership structure, capital structure, employment, technology investment, business support systems and marketing strategies. The shift in favor of privatization and private sector development is quite significant.

Change in management was also accompanied by greater autonomy of management (from interference by government) in the decision making of the enterprises. This makes it difficult to apportion the relative influences of the two changes i.e., different management team and increased autonomy.

Improvement in technology investments and modernization to cope with demands of market liberalization has followed privatization (e.g., Tanzania Breweries, TANELEC). In some cases, however, rehabilitation and investments in technology have not taken place and production has been delayed (e.g., Moshi Leather Tannery) in Tanzania.

The privatization process has been slower than anticipated partly because of political implications of the process and partly because like all other investments private investors have not found the policy environment sufficiently stable to warrant long term investments.

The political issue which has often been raised is that if nationalization was praised for transferring ownership of industry to the indigenous groups through state ownership then privatization to foreign investors or to ethnic minorities is interpreted as a retrogression. The financial and technological capability of a large part of would be indigenous investors limits their ability to engage in purchasing significant shares of the public enterprises which are being offered for divestiture.

The relationship between government and the enterprise section was found to have varied across countries in Africa. The relationship between government and the enterprise sector influences cooperation with the enterprise sector and effectiveness of government policy.

For instance, in the case of Mauritius the government and the enterprise sector cooperated in many ways and held consultations on matters affecting local entrepreneurs continuously gained control of industrial development. In the Cote d' Ivoire the government worked with and was supportive of enterprise sector development in a way which did not threaten the main actors in industry, even if they were non Ivorians.

In several countries (e.g., Tanzania, Kenya, and Nigeria and the post-independence Zimbabwe), the relationship between government and the enterprise sector (or significant parts of it) has been less cordial for historical reasons. Government intervention in industrial development was perceived as intending to address imbalances in society, as a result of which some leading actors in industrial development could be losers. In Tanzania the nationalization policy and the socialist policy were perceived as a threat to the private sector. In Kenya the way the Africanization policy was introduced and practiced was perceived as a threat to the Asian community, who were the leading local private-sector industrialist group. The indigenization policy in Nigeria posed a threat to some foreign investors. In post-independence Zimbabwe, too, the relationship between government and sections of the enterprise sector became less cordial as the government began to address some imbalances in society.

Local and Foreign Investment: Towards Complementarity
The relationship between foreign and local investment can be designed to be complementary. Foreign investment can augment local technological capabilities.

The positive role of foreign investment in building local technological capabilities has come out quite clearly in Mauritius, where local private capital has been progressively buying out foreign capital. This harmonious nationalization of investments has been facilitated by the existence of an entrepreneurial class which developed from the local plantocracy during the years when sugar production was dominant. The surpluses which were accumulated then were invested in industry. In addition, the macroeconomic environment and the climate for investment have been conducive for both local and foreign investment.

Outsiders (foreign firms in some form of partnership with local firms, non-indigenous entrepreneurs) have sometimes been instrumental in initiating the process of building up the capabilities that are necessary for improving competitiveness. This has occurred where these outsiders were incorporated into the national accumulation process and their capital and know-how were transferred to others (Wangwe 1995). The case studies in six African countries revealed an array of relationships between foreign capital and local capital. In some cases foreign investment preceded investment by local firms but the latter developed and gradually took over ownership from foreign-controlled firms. Foreign investment and other industrialization agents have a role in building technological capabilities. Foreign investment, in particular, could make a contribution to filling some important gaps in the capabilities of African firms. The role of government policy is important in influencing the outcome of those relationships.

Role of Regional Cooperation
Regional cooperation in R&D is particularly attractive when local conditions are important in determining the most appropriate technology. A study (Wangwe 1995) on exporting firms in selected African coun-

tries has shown that opportunities in the regional markets have been tapped on the basis of product quality and appropriateness to the specific conditions in the region. For instance, regional exports in agricultural machinery and other farm implements were found to be based on products which had been developed to suit agro-economic conditions in the region. Zimbabwean firms exporting agricultural machinery had developed products which suited the soil and climatic conditions in the region. Their competitiveness was a result of many years of continuous investment in searching and learning, as indicated by their R&D activities. The firms started by copying imported designs and made efforts, in response to demands from farmers, to make innovations to suit the specificities of the region. The development of agricultural machinery suited to the conditions of Southern Africa was encouraged by the domestic demand for such innovations from the large-scale farming community. The specificity of the technological adaptations gave the firms natural protection from international competition.

However, it is important to realize that even such natural protection is tenable only up to a point, beyond which there is danger of losing markets to competitors from other regions. Even if imported products are not as suitable to local conditions, competitors from outside the region have sometimes penetrated the regional market by supplying their products at lower prices or by supplying products whose quality of finish looks better. Thus specific local markets can be lost to others if continuous efforts are not made to develop competitiveness in terms of quality and price.

Regional cooperation could be useful in importing technology in all its forms. The collection of information on sources of equipment and technology is an expensive business; and institutions to serve groups of countries could marshal more resources than those confined to the smaller individual economies. Once imported, the technology could be adapted to local needs by regional technology institutes. This would be a highly skill-intensive task, where the regional sharing of the cost and benefit would clearly make economic sense.

Africa has demonstrated the slowest progress in developing regional integration and cooperation arrangements (UNCTAD 1993b). The challenge which emerges from this study is whether regional cooperation arrangements can be designed for Africa to facilitate (through investments, joint technological activities and trade) the process by which firms and other institutions in Africa build up technological capabilities. The African Economic Community and existing sub-regional economic cooperation arrangements should accord high priority to promoting trade expression, based on both exports and imports, by removing distortions, avoiding the duplication of large investments where national markets are small, reducing transaction costs (e.g., by trading arrangements which guarantee market access, regional marketing intelligence, improvements in the marketing infrastructure) and by redirecting trade flows.

Regional cooperation in setting up support systems for technology and training would greatly relieve the pressure on individual governments and allow for the sharing of experiences and knowledge. Some research and training centres can be viable only on a regional basis. While extension services necessarily have to be local, a network of such services over a region can reap various economies.

International Support

International support measures for industrial development include access and building technological capabilities.

Building of technological capabilities can be facilitated by international support measures such as facilitating access to technology, making full use of the transitional period accorded to full use of the transitional period accorded to LLDCs under WTO, participation in programs for developing human and physical infrastructure and technical assistance.

External financing such as official development assistance and international private sector finance can facilitate adjustment and industrial restructuring and improve supply response in the medium to long term.

FDI can be attracted by putting in place appropriate investment policies, promoting human and infrastructural capacity and creating a stable economic and political environment.

Greater access to external markets would facilitate reaping of economies of scale and foreign exchange earning.

Notes

1. Under these circumstances it follows that some of the industries which have been established are not foreign exchange users and may not have been justified at all in terms of their resource costs;
2. The main protection instruction were tariffs, import duty, drawbacks and rebates, foreign exchange quotas, licensing arrangements and other quantitative restrictions. This form of industrial trade policy encouraged the already established import-substitution industries.
3. The initial objectives of this policy were, temporarily to correct the balance of payments problem and to shield domestic industry form competition from

abroad. As a consequence the sector moved towards inward looking import substitution industrial development/structures.
4. Semboja H.H. 1995, "An Overview of the Performance of the Industrial Sector and Selected Industries during the 1980s-1990s," ESRF Technical Paper, Dar-es-Salaam.

Works Cited

Amsden, A., *Asia' s Next Giant*, New York: Oxford University Press (1989).
Browne M.B. and Tiffen P., *Short-Changed: Africa and World Trade*, London, Pluto Press (1997).
Grossman, G. and E. Helpman, "Trade Policy, Innovation and Growth." In: AEA Papers and Proceedings (1990).
Helleiner, G.K., "Trade Policy, Industrialization and Development." AEA Papers and Proceedings (1990).
Hussain, I. "Structural Adjustment and the Long-Term Development of Sub-Saharan Africa." In: R. Van Der Hoeven, and F. Van Der Kraaij, eds., *Structural Adjustment and Beyond in Sub-Saharan Africa*, The Hague, Netherlands, Ministry of Foreign Affairs, (1994a).
ITC, *Cocoa: Traders Guide*, Geneva International Trade Centre (1987).
Killick, T. Explaining Africa' s Post Independence Development Experience. Paper presented at the Second Biennial Conference on African Economic Issues, Abidjan, 13 - 15 October (1992).
Krugman, P., "Import Protection as Export Promotion." In: H. Kierzkowski, *Monopolistic Competition and International Trade*, New York: OUP (1990).
Krugman, P., "New Trade Theory and the less Developed countries." In: Calvo *et al.*, *World Development*, 20 (2) (1992).
Lucas, R., "On the Mechanics of Economics of Economic Development." In: *Journal of Monetary Economics*, 22 (1989).
Meier, G.M. and W.F. Steel, eds., *Industrial Adjustment in Sub-Saharan Africa*, World Bank (1989).
Ndela D and P Robinson, 'The Case of Zimbabwe' in Wangwe S. ed. Exporting Africa, Trade, Industrialisation and Technology in Africa. UNU-INTECH, Routledge Publications (1995).
Nurkse, R. "Patterns of Trade and Development", Wicksell Lectures Stockholm (1959).
Pack, H. and L. Westphal, "Industrial Strategy and Technological Change: Theory versus Reality." In: *Journal of Development Economics*, 22 (1989).
Prebisch, R., "Five Stages in My Thinking About Devlopment" in Bauer, P. Meier, G. and Seers D., eds., *Pioneers in Development*, New York, Oxford Univesity Press, (1954).
Riddell, R., *Manufacturing Africa*, ODI.S (1991).
Rodrik, D., "Closing the Productivity Gap: Does Trade Really Help?" In: Helleiner (1992).

Romer, P., "Increasing Returns and Long-Run Growth." In *Journal of Political Economy*, 98 (1990).

Steel, W.F. and L. Webster, "Small Enterprises under Adjustment in Ghana." In: *World Bank Technical Paper*, 138 (1991).

Stewart, F., Lall, S. and S. Wangwe, *Alternative Development Strategies in Sub-Saharan Africa*, Macmillan (1992).

Stewart, F., "Recent Theories of International Trade: Some Implications for the South." In: Kierzkowski (1984).

Stewart, F., "Are Adjustment Policies in Africa Consistent with Long-Run Development Needs?" In: *Development Policy Review*, 9 (1991).

UNCTAD, Developments and Issues in the Uruguayan Round of Particular Concern to Developing Countries, Note by UNCTAD secretariat, TD/B/39(2)CPR I, 15 March 1995.

UNCTAD, Follow up to the recommendations adopted by the conference at its eighth session: Evolution and consequences of economic spaces and regional integration processes, TD/B/40(1)7, 23 July (1993b).

UNIDO, *African Industry in Figures*, Vienna (1993).

Wade, R., *Governing the Market*, Princeton University Press (1990).

Wangwe, S. (ed.) *Exporting Africa: Trade, Industrialisation and Technology in Africa*. UNU-INTECH, Routledge Publications (1995).

World Bank *Accelerated Development in Sub-Saharan Africa: An Agenda for Action*. Washington, DC, The World Bank (1981).

World Bank *Adjustment in Africa: Reforms, Results, and the Road Ahead*. New York, Oxford Univesity Press (1994).

World Bank *Sub-Saharan Africa: From Crisis to Sustainable Growth, A Long-Term Perspective Study*. Washington, DC, The World Bank (1989).

Abbreviation Table

BOO:	Build Operate and Own
BOT:	Build Operate and Transfer
CBS:	Central Bureau of Statistics
CTI:	Confederation of Tanzania Industries
ESAP:	Economic and Social Adjustment Program
EU:	European Union
EPZ:	Export processing Zones
FDI:	Foreign Direct Investment
GDP:	Gross Domestic Product
ITC:	International Trade Center
ISI:	Import Substitution Industrialization
ISO:	International Standard Organization
LLDC:	Least Developed Countries
MVA:	Manufacturing Value Added
NIE:	Newly Industrialized Economies
NTE:	Non Traditional Exports
R&D:	Research and Development
SAP:	Structural Adjustment Program
SSA:	Sub-Saharan Africa
SME:	Small and Medium Enterprises
S&T:	Science and Technology
TNC:	Transnational Corporation
UDI:	Unilateral Declaration of Independence
UNIDO:	United Nations Industrial Development Organization
UNCTAD:	United Nations Conference on Trade and Development
USA:	United States of America
USAID:	United States Agency for International Development
VA:	Value Added
WTO:	World Trade Organization

Chapter 6

Structural Adjustment Programs and Poverty in Sub-Saharan Africa: 1985-1995

Ali Abdel Gadir Ali

Introduction

The impact of Structural Adjustment Programs (SAPs) on the performance of Sub-Saharan Africa (SSA) has been intensively investigated in the literature (see, for example, Elbadawi 1992; Lipumba 1995, 1994, and Mosley 1994). But, as early as 1987 the macroeconomic emphasis of SAPs was shown to have been inconsistent with the long-term development interests of the region. Special attention was drawn to the negative impact of these policy measures on the poor and vulnerable sections of the population (e.g., Cornia et al. 1987). On the basis of such criticism later versions of the policy package started to pay attention to these aspects, leading to a revival of interest in the analysis of poverty. Thus, for example, Squire (1991:177) identified the issue of the impact of SAPs on poverty as one of the most important debates of the 1980s.

Not surprisingly, the World Bank was a major contributor to the SAPs-poverty debate. Up to August 1995 it is reported that the Bank had published 24 relevant reviews on poverty issues and completed 22 poverty-assessment reports on SSA countries, with three poverty-assessment draft reports for Niger, Nigeria and Lesotho. The declared policy on poverty assessment reports is to have such reports completed for every country in which the Bank has an active lending program (see World Bank 1995). In addition, since 1993 three annual "status reports on poverty in sub-Saharan Africa" have been issued to "inform members of the Special Program of Assistance." A massive research program on poverty has developed within the World Bank which gave rise to numerous journal articles published in learned journals. This massive research output, which included theoretical contributions, model simulations and empirical studies, seems to have been "directed" at establishing a non-existent result that "adjustment programs" did not hurt the poor in Africa, and that "more adjustment rather than less adjustment" is needed for poverty reduction. A sample of such contributions will be reviewed in this chapter. At this juncture, however, we need to note that such research, notwithstanding its apparent rigor and quality, betrays an amazing ideological commitment to structural adjustment programs (see Taylor 1993:870).

Be the above as it may, and to focus attention, we will follow the World Bank (1990) and Ravallion (1994) in defining poverty as the "inability to attain a minimal standard of living." We will also follow convention in adopting the absolute poverty line as the most relevant for SSA. The most common approach in this respect is to define the absolute poverty line as the cost of basic food items deemed essential to attain some recommended food energy intake. Added to this cost of food, is a modest allowance for non-food items thought to be crucial for living in a social context "without feeling shame."

In the course of the chapter we shall also be adopting the direct approach to poverty analysis as suggested by Kanbur (1987). This approach requires that in investigating poverty we should start from a poverty measure summarizing the state of poverty in a given society. Such an approach contrasts with the indirect way of looking at poverty through social indicators such as expected life, child survival and education indicators (see, for example, Pio 1994 and Berg et al. (1994).

The most widely used poverty indicators in the literature belong to the family of additively separable measures. Such measures are required to satisfy a number of fairly reasonable restrictions which are summarized by, among others, Kakwani (1993b). For our purposes we note that the most important measures in use are the head-count ratio, (H), the poverty-gap ratio, $P(1)$, and the Foster-Greer-Thorbecke (FGT) measure, $P(a)$.

The head-count ratio is defined as the proportion of the population for whom income (or consumption expenditure) is less than the poverty line. It is understood in the literature that the head-count ratio is a measure of the spread of poverty. The poverty-gap ratio is defined as the aggregate poverty deficit of the poor relative to the poverty line. It measures the depth of poverty in a given society since it depends on the distance of the poor from the poverty line. The FGT measure of poverty weights the poverty gap for each individual by a poverty aversion parameter prevailing in society (Parameter A, which is non-negative). The interpretation of this measure, however, is a shade problematic. Its attraction, on the other hand, is that it gives more weight to the poorer in society. Moreover, for values zero and one of the poverty aversion parameter, the FGT measure gives rise to head-count ration and the poverty-gap ratio respectively. In what follows we shall have occasion to use the head-count and the poverty-gap measures.

Having noted the above, the remainder of this chapter falls into five sections. First, we deal with what we consider the most important issues in the study of poverty in SSA. According to our reading of the evidence, the important issues revolve around the fact that little is known about the magnitude and behavior of poverty in SSA and yet policies and programs are routinely being proposed for the region to get to a poverty-reducing growth path. Secondly, we review the most recent evidence on poverty in SSA, and point out that reported results only tell part of the story of the spread and depth of poverty in the region. An alternative methodology for the estimation of poverty is presented in the fourth. The proposed methodology takes into account the weak data base of the region and proposes a consistent way of studying the behavior of poverty. Thirdly, our alternative results are presented. Among other things, it is shown that poverty in SSA is a much deeper phenomenon than is commonly recognized and that the increase in poverty during the second half of the 1980s is much more dramatic than is commonly reported in the literature. Further, we also show that SAPs have had a negative impact on poverty. Finally, we discuss the policy challenge given our results and provide some concluding remarks.

Poverty Issues in SSA

At the core of the SAPs-poverty debate in SSA lies the issue of information about poverty itself: its magnitude and its behavior. The simple fact of life is that pretty little exists by way of information that could enable enlightened policy formulation (see, for example, Azam 1994). Moreover, the issue of causation from SAPs to poverty has proved to be a particularly hard nut to crack. In this section we review what we consider to be the most important aspects of this issue. Without attempting to provide a comprehensive and exhaustive listing, we deal with three such aspects namely: the behavior of poverty prior to the advent of SAPs, the effects of SAPs on poverty and growth and poverty reduction.

The Behavior of Poverty

As is probably well known the WDR 1990 was devoted to the issue of poverty in the context of development. Specifically, the behavior of poverty over time was dealt with by asking two questions: What happened to poverty up to the 1980? And what happened to poverty during the 1980s?

To answer the first question, data from eleven countries "which together account for 40% of the total population of the developing world and 50% of the poor" was utilized. Country specific poverty lines were employed to report the results and the head-count ratio and the income shortfall (the income-gap ratio) were used as measures of poverty. The results were reported as a comparison between a "first year" and a "last year" for each of the eleven countries. The years varied considerably between countries. Thus, for example, the "first year" spanned the period 1960 (for Brazil) and 1973 (for Malaysia); the "last year" spanned the period 1980 (for Brazil) and 1988 (for Colombia), (see World Bank 1990: Table 3.2). On the basis of these comparisons the World Bank (1990:40) concluded that the evidence "reveals considerable reduction in the incidence of poverty.... In sum, therefore, the evidence suggests that there has been considerable reduction in the number of the poor, and achievement of somewhat better standards of those who remained in poverty."

Despite this general conclusion, however, Africa was found to be different. We note in this respect that none of the eleven countries included in the sample was an SSA country. The reason for this neglect is cited as the absence of reliable intertemporal statistics on income distribution in most African countries. Despite the acknowledged lack of data, and using evidence on stagnant consumption per capita in the region, it was concluded that "progress in reducing poverty has probably been slowest in that region. Even assuming that the distribution of income did not worsen between 1965 and 1985, the number of Africans in poverty would have increased by 55 million" (World Bank 1990:41-42).

Thus, despite the lack of evidence the Bank was prepared to conclude that Africa was different as far as the behavior of poverty during the seventies was concerned. There is evidence, however, from at least one SSA country that such conclusion is not warranted. African countries, like many developing countries, were pursuing at that time internationally recommended development strategies which were very sensitive to social equity. The African state was very active in providing subsidized food, education and health and was providing employment. The needs of the poor were catered to in the context of the development strategy. Thus, for example, Ali (1992) has shown that in Sudan the headcount ratio increased only marginally during the seventies from 50% to 52.6% but at the same time the poverty-gap ratio has declined from 24.6% to 22.9% indicating an improvement in the lot of the poor.

The policy relevance of the above is rather obvious. In the limit, if one subscribes to the World Bank's view that Africa was different one would be prepared to write-off whatever policy initiatives followed during the seventies as irrelevant to the current poverty alleviation efforts.

SAPs and Poverty

The 1980s, it will be recalled, is the period during which SAPs were adopted by African countries like "secular gods"! The behavior of poverty during this period was addressed by the World Bank (1990:42) in an indirect way. Evidence pertaining to 13 countries (2 Europeans, 4 Latin Americans, 6 Asians and one African) was used. The results were reported for a first year (which varied between countries from 1977 for India to 1985 for China and Cote d' Ivoire) and a last year (which varied between countries from 1984 for Pakistan and 1988 for China and Colombia). On the basis of such evidence it is noted that "many observers have argued that the recession and adjustment were harmful to the poor...In the regions most severely affected by the recession poverty has increased.... In sub-Saharan Africa the only data available, those for Cote d' Ivoire, display a slight increase in the mid-1980s."

The WDR 1992, which dealt with the environment and development as its special topic, revisited the issue of the behavior of poverty during the 1980s. The method of analysis relied on aggregative estimates for the various regions. It is concluded that "new estimates prepared for this Report reveal negligible reduction in the incidence of poverty in the developing countries during the second half of the 1980s. The numbers of the poor have increased from slightly more than 1 billion to more than 1.1 billion in 1990. The experience of other developing regions has been markedly different from that of Asia. All poverty measures worsened in sub-Saharan Africa, the Middle East and North Africa and Latin America and the Caribbean" (World Bank 1992:42).

No attempt, however, has been made by WDR 1992 to establish a link between the increase in poverty and SAPs. Such a link could have been easily established by resorting to the information provided by Corbo and Rojas (1992:34) on the regional distribution of countries implementing SAPs. According to this information out of 25 "early intensive adjustment lending" countries, 16 were from SSA, 7 were from Latin

America and the Caribbean, one from North Africa and one from Asia. Thus, 24 out of 25 intensively adjusting countries belonged to the regions where "all poverty measures worsened" during the second half of 1980s. More important to note is that 64% of the intensively adjusting countries were from SSA. Similarly, another set of 25 "other adjustment lending" countries included 16 from SSA, 5 from Latin America and the Caribbean, 2 from Asia, and one each from North Africa and Europe. That is, out of 25 "other adjusting" countries, 22 belonged to the regions where "all poverty measures worsened" during the second half of the 1980s. The share of SSA in the "other adjusting countries" is, once again, 64%.

More direct evidence about the behavior of SSA rural poverty during the second half of the 1980s is provided by Ali (1993). Using information provided by IFAD, 10 SSA countries were classified according to World Bank conventions as "intensively adjusting" (Kenya, Malawi, Tanzania, Zambia, and Ghana), "other adjusting" (Gambia, Mali, and Gabon) and non-adjusting (Lesotho and Ethiopia). For each of the ten countries the head-count ratio as a measure of poverty is reported by IFAD for 1965 (or the closest year thereof) and for 1988 (see Jazairy et al. 1992).

For the intensively adjusting countries it was found that rural poverty has increased from 56.6% in 1965 to 62.4% in 1988. Similarly, for the "other adjusting" countries, the index of rural poverty increased from 45.1% in 1965 to 60.7% in 1988. The corresponding absolute number of the poor increased from 18.2 million to 36.2 million for the intensively adjusting group and from 2.3 million to 5.1 million for the "other adjusting" group. In contrast, the head-count ratio for the non-adjusting group decreased from 65.8% in 1965 to 43.6% in 1988 and the absolute number of the poor remained constant around 17 million. From these results, and without forcing the causation too hard, it can be concluded that such evidence suggests that SAPs must have had something to do with the behavior of poverty in SSA during the second half of the 1980s.

Once again the policy relevance of the effect of SAPs on poverty should be obvious. If one believes the World Bank (1990, 1994b), then one would be prepared to argue that the design of conventional SAPs need no surgical operations to graft a "human face" onto them.

Growth and Poverty Reduction

Information is equally lacking regarding the effect of economic growth (measured by the increase in per capita income) on poverty in SSA. In its recent report on adjustment in Africa the World Bank (1994b) addressed the issue in a more direct fashion. According to the Bank SAPs have "often been accused of hurting the poor. To address this issue, it is useful to focus on two questions: would the poor have benefited from less adjustment? And, to the extent that adjustment benefited the poor, could policy reforms have been designed differently to have benefited them more?" (World Bank 1994b:163). The method used to answer the questions was indirect where it is noted that adjustment "has contributed to faster GDP per capita growth in half of the countries examined, and there is every reason to think that it has helped the poor, based on the strong linkages between growth and poverty reduction elsewhere in the world" (World Bank 1994b:163). Thus, no evidence was provided on causality.

In view of the importance of the issue, it is perhaps instructive to spend more time on the type of growth experience studied by the Bank to support its conclusion. We note that the countries examined in the report are 29 SSA countries with population of half a million or more in 1991. The countries were chosen such that they had reasonable social stability and their adjustment programs were in place during the period 1987-1991. The growth experience was recorded as a comparison of the average growth rates for the periods 1981-86 and 1987-91, and the countries were classified into various groups depending on the purpose of the analysis. One such grouping is based on the observed improvement in macroeconomic policies. According to this convention, four groups have been identified (World Bank 1994b:138, Table 5.1):

A. Large improvement in policies: where there are six countries with a mean growth rate of -0.8% for the period 1981-86 (with half the countries recording negative growth rates and where Nigeria's growth rate was -4.6%) and a mean growth rate of 1.1% for the period 1987-91 (with no negative growth recorded for any country);
B. Small improvement in policies: where there are nine countries with mean growth rate of -1.1% for the period 1981-86 (with three countries recording positive growth) and a mean growth rate of -0.1 for the period 1987-91 (with four countries recording positive growth);
C. Deterioration in policies: where there are eleven countries with mean growth rate of -1.0% for the period 1981-86 (with four countries recording positive growth) and a mean growth rate of -2.6% for the period 1987-91 (with two countries with positive growth);
D. Unclassified: where there are three countries and where the growth rates are reported for two of them with no mean growth rates. We calculated mean growth rates for this group as 3.7% for the period 1981-86 and 2.05% for the period 1987-91.

Using the above information it is not difficult to show that the mean growth rate for the sample of 29 countries was 0.2% for the period

1981-86 and 0.11% for the period 1987-91. Excluding the "unclassified" group, we find that the mean growth for the remaining 26 countries was -0.96 % for the period 1981-86 and -0.5% for the period 1987-91. Using the Bank's causal relationship between "growth and poverty reduction," the above calculations do not support the Bank's contention that "adjustment-led growth has probably helped the poor." In this respect it is perhaps important to note that Mr. Jaycox, the vice-president of the Africa region in the Bank, provides a more realistic reading of the evidence presented by his staff. In his foreword to Husain and Faruqee (1994:viii) he notes that despite "these gains, per capita growth rates remain fairly low; even if sustained, they would be insufficient to make a dent in the rising incidence of poverty."

A further reasonable reading of the African evidence is provided by the Bank in its recent overview of revisiting Africa's development strategy. One of the questions posed by this report is "what has happened to the indicators of poverty in Africa over the last five years?" Lack of appropriate information, however, precluded a definitive answer to the question and on the basis of preliminary results from seven countries, the Bank concludes that the "analysis of poverty and social indicators over the last few years, suggests that while growth is important, the pattern of growth is important too" and called for decomposing the change in poverty into a growth component and a distribution component.

Recent Evidence on Poverty in SSA : A Review

World Bank

Despite the huge literature produced by the World Bank on poverty in Africa, this section will focus on the latest report produced by the Bank in this respect. The draft report carries the title "taking action for poverty reduction in sub-saharan Africa" and is authored by an Africa region task force. The report is structured in five chapters and nine annexes. Chapter 2 of the report embodies the latest World Bank estimates of poverty in SSA. Despite a commendable effort by the task force in producing a development sensitive report, we note that the poverty-information content of the report continues to leave quite a lot to be desired. The report is based largely on secondary information, highly selective use of data and a few preconceived conclusions. In what follows we review the most important findings.

A. The Depth of Poverty: the report notes that in SSA "at least 250 million people (about half of the population) are still surviving on less than the equivalent of $1 a day" (World Bank 1995c:5). Implied in this estimate is a head-count ratio of 50%, but the report does not explain the origin of the estimate nor the year of the estimate. This may have been derived from Chen, Datt and Ravallion (1994) (hereinafter CDR) as quoted by the "status report on poverty in SSA 1994: the many faces of poverty" (see World Bank 1994c:13). According to this second source the estimate of population living on less than $1 per person a day in 1990 in SSA is 52.89%. In Table 2.2 (p. 10) of the task force report, the percent of population living below the national poverty line in eight SSA countries is given (Gambia (64%), Guinea-Bissau (49%), Kenya (41%), Nigeria (33%), Cote d' Ivoire (37%), Cameroon (48%), Uganda (55%) and Zambia (68%). These head-count ratios were derived from household surveys and poverty assessment reports. Once again the report does not give the year of estimation and cautions against comparisons between poverty levels among countries due to the specific poverty lines used.

B. Growth Performance: the report notes that people in SSA, along with those in South East Asia "remain among the poorest in the world.... from 1970 to 1992 GDP per capita, in terms of purchasing power parity, grew by only $73. In terms of annual percentage change, the purchasing power of GDP per capita in the Africa region registered a 1.7 percent annual growth from 1972 to 1982 and an annual decline of 1.7 percent during the 1982-1992 period. On the other hand, in South Asia, real per capita GDP levels, which were at a much lower level than those in SSA in 1970, had increased by $420, or 2.3 percent per annum by 1992." These, no doubt, are revealing comparisons. However, the African Development Bank (1995:222-3) has recently shown that SSA was a high growth region where real GDP during the period 1965-1973 increased by an annual rate of 5.7% and where real per capita GDP increased by a rate of 3% per annum.

C. Income Inequality: the report notes that countries in SSA "reveal a much more unequal income (and expenditure) distribution than most other developing countries....The persistence of very high levels of inequality may cause a higher systematic demand for transfers, which may in turn reduce the rate of accumulation and hence, negatively influence economic growth and poverty reduction" (World Bank 1995c:7-8 and Table 2.1). Of five countries for which data is provided inequality has increased in three: Kenya (at an annual rate of 0.94% during the period 1981/82-1992), Nigeria (at a rate of 1.74% during 1985-1992) and Ethiopia (at a rate of 1.98% during the period 1989-1994), remained constant in Ghana (with a Gini coefficient of 41% for 1987/88 and 1991/

92) and declined in Cote d' Ivoire (with a Gini coefficient of 45% in 1985 and 35% in 1988). Combining these results with those about growth it becomes clear that in SSA not only is growth faltering but inequality is worsening as well. This makes efforts aiming at poverty reduction doubly demanding. The report recommends the need for further research on the explanation of the level and behavior of inequality in SSA which seems to be a sensible proposition.

D. Adjustment and Poverty: the report outlined conditions for successful poverty reduction. The required appropriate actions in this respect covered areas such as: political stability, good governance and sound institutions, sound macroeconomic policy and income growth, investment, resource mobilization and debt, sound natural resource management and the provision of social services for the poor (see World Bank 1995:14-29). Under sound economic management the report notes that there is no empirical evidence to show that the urban and rural poor suffered a decline in welfare as a result of adjustment programs. Most of the research on the basis of which this conclusion is based is conducted by the World Bank (quoted research work includes Easterly 1994, Demery and Squire 1995, and Bruno, Ravallion and Squire 1995) and by the celebrated Cornell Food and Nutrition Policy Program (e.g., Dorosh and Sahn 1993). While it may not be fruitful to dwell, at this stage, on the SAPs-poverty causation, such conclusion is contradictory to the report's own generalizations about growth and inequality in SSA. Moreover, the report also notes that there is growing evidence "that sound macroeconomic policy and growth will reduce poverty, and that there is no systematic evidence that the poor lose either directly or indirectly from policies which promote aggregate growth" (World Bank 1995:17). This, of course, is the argument of the World Bank (1993) all over again. We note, however, that nobody in his or her right senses would argue that "sound macroeconomic policics" and "sound economic growth" will not reduce poverty. But surely reasonable men and women can disagree on the definition of the adjective "sound" as it applies both to policies and growth. Otherwise, an unreasonable SSA development economist will be entitled to enquire as to why "reasonable" economists from outside the continent feel obliged to differentiate "poverty-reducing growth paths" from other growth paths (see e.g., World Bank 1995). Depending on the poverty nature of the growth path, it stands to reason that the poor could stand to lose directly or indirectly from "policies which promote aggregate growth."

Having noted the above, it remains to note that the task force report is a very sensitively written document compared to other reports which came out of the World Bank. Chapter 4 of the report provides an analysis of the Bank's operational work and its likely impact on poverty reduction. The chapter provides extremely valuable inside information about the impact of operations on poverty reduction and arrives at a number of sensible conclusions regarding the conventional design of SAPs. A careful reading of the chapter, however, would indicate that perhaps a more refined methodology is required to analyze the rich set of data provided on operations.

Despite it sensitivity to the SSA poverty problems the task force report, like many other research results, does not seem to have been taken into account in guiding operational work for the future. Evidence on this is to be found in the "Africa Region Strategy FY97-99" issued in October 1995. According to the strategy the Bank's central mission is stated as "to reduce poverty and improve the quality of people's lives." Estimates of poverty used in the strategy document put the number of the poor as 279 million with a head-count ratio of 50%, and it is noted that the number of the poor and their proportion are on the rise. At an aggregative level, SSA national income growth is projected at 3.8% per annum for the next 10 years, while population is projected to grow at 3% per annum implying an average per capita income growth of only 0.8% per annum. Such low rate of per capita income growth is expected to slowly reduce the proportion of the poor, but is expected to increase the number of the poor. It is calculated that a growth in national income of 5% per annum is needed to stabilize the number of the poor. For a significant progress in poverty reduction national income growth of 6%-7% will be required (see World Bank 1995d:2).

Given the estimates, and the declared "central mission," the World Bank (1995d:3 and 4) proceeds to note that higher "levels of economic performance will not be an easy "sell" in Africa. In effect, the strategy implies more and faster adjustment, not less. In the short run, this may mean more social pain and hardship, not less." Further, given the "central mission" and the SAPs implications of the need for higher economic performance, the Bank notes that implementing the strategy will require a changed agenda. Important among the changes required is "developing a new way of talking about structural adjustment, i.e., almost a new "vocabulary" for discussing macro-economic management that conveys more directly and accessibly the link between macro-economic management and poverty reduction."

The semantics of the strategy, as quoted above, imply that after all the new "central mission" is not that central; what remains central is the conventional SAPs policy design. If poverty reduction is the "overarching objective" of development in SSA, as the World Bank (1994a) declared, one would have expected that a new set of policy measures consistent with the new objective would be designed. At one extreme, the new set of policy measures could come out to be identical to that of the conventional SAPs design, but at another extreme it is possible that the new set would be totally different. The emphasis on "faster and more" adjustment, and the call for changing the details of the discourse from "adjustment" to "economic management" betrays an unfinished business of designing a new policy package. This stance, we suggest, is not much different from that of the World Bank (1994b).

Chen, Datt and Ravallion

The most recent estimates of poverty in SSA are those provided by CDR (1994). The concern of the authors was to investigate whether poverty was increasing in the developing world, but in the process they reported results for SSA based on a sample of 14 countries for which data was available. The investigation was conducted in terms of a comparison between two dates, 1985 and 1990. Due to the importance of these results we note the following about the methodology followed by the authors:

A. To achieve comparability across countries the authors used the same level of real consumption to define the poverty line. The same level of consumption is defined as the consumption level at 1985 purchasing power parity (where use has been made of the PPP exchange rates reported by Summers and Heston (1988, 1991).
B. The number of countries included in the sample is 44 (10 Asians, 13 Latin Americans, 14 SSA, 4 middle Eastern and North African, and 3 East Europeans). The total number of household surveys used in the analysis is 63, the data for 26 of which pertain to income rather than consumption expenditure. To adjust these, the authors estimated mean consumption by appropriately multiplying mean income from the survey by the ratio of private consumption to GNP derived from national accounts as estimated by the World Bank (1992).
C. The dates of the surveys used by the authors spanned the period 1981 to 1992. To estimate the 1985 and 1990 Lorenz curves the authors assumed that the Lorenz curve at the nearest survey date to be their best estimate of the Lorenz curve for 1985 or 1990; and if only one survey was available for the country then that survey is used for both dates.
D. Instead of computing poverty measures to conduct the comparison, the authors estimated poverty incidence curves (PIC). A PIC is generated by plotting the proportion of population (p) on the vertical axis consuming less than a given level z on the horizontal axis; each point on the PIC gives the head-count index of poverty. Poverty deficit curves $D(z)$ and

poverty sensitivity curves S(z) can be obtained from the PICs. Using these curves and assuming that the poverty line, z, is unknown, comparing poverty between two dates is conducted by invoking the idea of dominance.

E. First-order dominance says that "poverty cannot have increased between two dates if the PIC for the latter date lies everywhere above that for the former date up to a maximum poverty line." If the PIC for the two dates cross, then the ranking is ambiguous. The poverty deficit, D(z), and the poverty sensitivity S(z), curves, provide second-order and third-order dominance respectively. These dominance tests are found to be nested in the sense that first-order dominance implies second-order dominance which in turn implies third-order dominance.

F. To construct the PICs the authors estimated Lorenz curves from the survey data available to them where two different specifications of the Lorenz curve were tried: a general quadratic form and a beta form. Choosing between the two specifications was governed by the requirement to satisfy the conditions for a valid Lorenz curve and by a goodness of fit criterion.

G. The 14 SSA countries included in the analysis represented 36.62% of the 1990 population of all low and middle-income countries of the region. Only two countries of the SSA sub-sample had two surveys: Cote d' Ivoire (1985 and 1988) and Ghana (1987/88 and 1988/89). In all SSA countries consumption expenditure was used as the indicator of living standards; and in all of them, except Botswana and Ethiopia, expenditure per person was used as the ranking variable.

The results are reported in terms of PICs defined over five poverty lines. The poverty lines are defined as consumption levels per person per month in 1985 purchasing power parity dollars. The consumption levels used are $21, $30.42, $40, $50 and $60. PICs for regions are also reported. According to the results reported by the authors "poverty increased in sub-saharan Africa and the conclusion is robust for all poverty measures if one restricts the poverty line to $50 per month. The poverty deficit curves for this region show an increase in poverty for poverty lines up to a high level (above $60 per month)." A summary of the results is reported in Table 6.1.

Table 6.1: Poverty Incidence Curves for Sub-Saharan Africa (cumulative percent of population under the poverty line)

Poverty Line ($/person/month 1985 PPP)	1985	1990
21.00	31.65	33.44
30.42	51.40	52.89
40.00	64.98	65.55
50.00	74.09	74.11
60.00	80.15	80.00

Source: Chen, Datt, and Ravallion 1994:370, Table 3.

Thus, from table 6.1 we see that for all poverty lines below $60 per person per month poverty has increased during the second half of 1980s. The reported increase, however, is not dramatic: 1.79 percentage points (equivalent to an annual rate of increase of 1.11%) for the $21 poverty line; 1.49 percentage points (equivalent to an annual rate of increase of 0.57%) for the $30.42 poverty line; 0.57 percentage points (equivalent to an annual rate of increase of 0.18%) for the $40 poverty line; and a meager 0.02 percentage points (equivalent to an annual rate of increase of 0.01%) for the $50 poverty line. The recorded increase in mean income from $48.63 to $48.92 works out into an annual rate of growth of only 0.12%. Both in terms of the level of poverty for the various poverty lines and in terms of the magnitude of increase in poverty, these results seem to be underestimating poverty in SSA.

It is perhaps in realization of this underestimation that CDR proceeded to revise their estimates on account of a possible bias that may have been introduced by countries excluded from the sample. The revision is done on the assumption that the sample is representative for 1985 and that the Lorenz curve in the excluded countries did not shift between the two dates. The rate of growth in real mean consumption from national accounts was used to estimate poverty measures for 1990 in the countries not included in the sample, and the results were used to correct the 1990 aggregates for the region as a whole. The revised results gave rise to a sharper increase in poverty in SSA amounting to 4.64 percentage points for the $21 poverty line; an increase of 3.95 percentage points for the $30.42 poverty line, 2.65 percentage points for the $40 poverty line and an increase of 1.86 percentage points for the $50 poverty line. Mean consumption, however, is recorded to have declined from $48.63 to %46.38 with an annual rate of decline of 0.99%.

Demery and Squire

In a recent paper Demery and Squire (1996:40) set out to provide "the most convincing evidence to date that economic reform is consistent with a decline in overall poverty and that a failure to reform is associated with increased poverty." The authors describe their analysis as "solidly grounded" and is made as such by the recent availability of data from household sample surveys for six SSA countries at two points in time. Earlier contributions, which used to be quoted by the World Bank in support of its unrelenting enforcement of SAPs in SSA countries, are described by the authors as having been theoretical (e.g., Azam 1994), or having used indirect evidence based on modeling exercises (e.g., Bourguignon, de Melo, and Morrison 1991; Bourguignon, de Melo, and Suwa 1991; Dorosh and Sahn 1993), or having relied on anecdotal evidence (e.g., Watkins 1995).

In view of the strong claims made by the authors regarding their results we note the following about their methodology:

A. Six countries for which survey data is available are studied. These include Cote d' Ivoire (with national surveys 1985 and 1988), Ethiopia (rural surveys for 1989 and 1994), Ghana (national: 1988 and 1992), Kenya (rural: 1981 and 1991), Nigeria (rural: 1985 and 1992) and Tanzania (rural: 1983 and 1991). The authors concede (p.57, n. 2) that there are problems with the data.

B. The authors use the head-count index as their measure of poverty and note that "the poverty line is held constant in real terms through time." As we will show below if the poverty line is held as a constant proportion of mean expenditure, then observed changes in poverty could be attributed solely to changes in the distribution.

C. In decomposing the change in poverty into its growth and distribution components the authors observe in the "note" to Table 6.3 that their reported results "are the average of two decompositions, one using the initial period mean and Lorenz curve, and the other using the terminal mean and Lorenz curve. This procedure ensures an exact decomposition, that is, it eliminates any residual." Nothing is mentioned in the text about the decomposition procedure but it seems that the authors are adopting Datt and Ravallion (1989) decomposition.

D. Regarding the link between SAPs and poverty the authors use the recent World Bank (1994) macroeconomic policy stance index. The index is a weighted average score index on fiscal policy, monetary policy and exchange rate policy. The authors calculated an initial index corresponding to the first survey (as an average of three years) and a final index corresponding to the second year of the survey. The change in macroeconomic policy is taken as the difference between the two indices. The resulting change in policy is compared with the change in poverty it being noted that an increase in the policy index is taken as reflecting an improvement in policy performance.

Since the authors are using secondary information their results on incidence do not seem to merit any discussion. According to the results only Cote d' Ivoire recorded a deterioration in macroeconomic policy between 1985 and 1988 where the policy index recorded a decline of 1.65 points with poverty increasing by 5.3 percentage points. For the rest of the countries improvements in economic policy are reported with corresponding declines in the head-count ratios: Ethiopia (a policy improvement of 0.55 points and a decline in poverty of 3.6 percentage points), Ghana (1.35 and -1.95), Kenya (0.45 and -0.28), Nigeria (1.79 and -1.27) and Tanzania (2.76 and -1.83). Demery and Squire conclude that these results present "the most compelling evidence to date that improvements in the macroeconomic policy regime of the kind usually associated with World Bank and IMF-supported adjustment programs are consistent with a decline in the incidence of poverty overall. These results do not establish causality, but, at least in the six countries for which we have evidence, we can conclude that failure to implement an adjustment program has been doubly harmful to the poor." Needless to note that these are contestable results. Their importance, however, lies in the type of methodology used which admits the acceptability of drawing conclusions from association.

Alternative Estimates

Despite the sophisticated nature of the analysis of poverty based on the PIC, it remains true that the level of poverty in this type of analysis relies on the head-count ratio. Moreover, we note that the SSA results reported by CDR (1994), despite their claim to wider generality, are in essence dealing with the pure growth effect on poverty. The reason for this is to be found in the fact that only two of the countries in the SSA sample had two observations on the Lorenz curve. For the remaining countries the Lorenz curve is assumed to remain constant for the two dates. Of the two countries that had two observations, Cote d' Ivoire' s income distribution shows an amazing improvement as reflected by the decline in the Gini coefficient from 0.4463 in 1985 to 0.3455 in 1988, a decline of 0.1008 points in only three years! For Ghana, the second country which had two observations, the Gini coefficient increased from

0.359 in 1987/88 to 0.3674 in 1988/89, an increase of 0.0084 points in one year. Cote d' Ivoire's result is suspect, specially in view of the reported substantial decline in mean income from $ 98.67 in 1985 to $69.91 in 1988, unless one is prepared to accept a fairly strong positive relationship between growth and inequality! Ghana's result, on the other hand, is probably worthless given its very short-run nature. The above two observations, together with the assumption of constant Lorenz curve for the remaining 12 countries, render Chen, Datt and Ravallion claim to generality in dealing with the magnitude of poverty in SSA in 1990 dubious to say the least. In the remainder of this section we provide alternative estimates.

Data

Three aspects of our use of data are worth reporting about. These include per capita income and expenditure, Lorenz curves and poverty lines.

A. *Per Capita Expenditure*: to maintain maximum comparability with CDR (1994a) we will use their reported per capita consumption figures in 1985 PPP dollars for both 1985 and 1990. We note in passing that these reported figures can not be obtained from the original Summers and Heston (1991) data set presumably due to adjustments introduced, but not reported, by CDR.

B. *Lorenz Curves*: estimates of the Lorenz curve for the SSA sample of countries for 1985 as reported by CDR (1994b) will be used. Appendix Table A.1 provides the original reading of the Lorenz curves. These estimates provide distributional information for one point in time. Assuming that the Lorenz curves remain unchanged for 1990 would enable us to look at the pure growth effect on poverty. Using the Lorenz curves we estimated the average income for the various percentile groups of population. Such information is not provided in the original source of data, but is vital for estimating the average income of the poor which plays an important role in the estimation of all poverty measures.

C. *Gini Coefficients*: central to the analysis which will follow is the time behavior of Gini coefficients. Alternative reasonable values for the rate of change of the Gini coefficients will be required for estimating the distribution component of the change in poverty. A reasonable guide in this respect is the relevant literature on the Kuznets (1955) hypothesis on the relationship between inequality and development (see, for example, the competent review of this literature by Anand and Kanbur 1993a, 1993b, and the references cited therein). The Kuznets hypothesis has come to be known as the inverted-U hypothesis meaning that inequality is expected to increase during early phases of development, reaches a maximum and then declines as income per capita increases. From the World Bank classification of countries in the WDRs most of SSA countries are classified as low-income countries which implies that assuming an observed increase in the Gini coefficient will be consistent with Kuznets hypothesis. Moreover, as we indicated earlier the World Bank (1995c) reports that the Gini coefficient remained constant in Ghana during the period 1987-1992 at 41%, increased in Kenya from 51% in 1981/82 to 56% in 1992 (at an annual rate of increase of 0.94%), and in Nigeria from 39% in 1985 to 44% in 1992 (at an annual rate of 1.74%) and in Ethiopia from 41% in 1989 to 45% in 1994 (at an annual rate of 1.9%) while it declined in cote d' Ivoire from 45% in 1985 to 35% in 1988 (at an annual rate of 8%). We note that Cote d' Ivoire' s result is suspect. On the basis of these results we will assume alternative rates of increase for the Gini coefficient in SSA ranging from 0.8% to 2%.

D. *Poverty Lines*: in the absence of detailed country studies on poverty, country-specific poverty lines are hard to come by. An established convention in the literature is that of identifying a poverty line of a representative poor country eg. Indian poverty lines were frequently used as representative (see, for example, Kakwani 1980, World Bank 1990, and Ravallion, Datt, and van de Walle 1991). Another convention which is becoming popular in World Bank studies is to use a poverty line of "US$1 per person per day" at 1985 purchasing power parity. Using this latest convention as a guide to the reasonableness of a poverty line we calculated the ratio of the alternative poverty lines used to mean expenditure as reported by CDR for the SSA sample. For all poverty lines above $20 per person per month we found that the alternative poverty lines break down for four countries in the sense that the ratio of the poverty line to mean expenditure is found to be greater than unity. The relevant poverty line for these countries seems to be that of $21 with the following (z/u) ratios: Ethiopia (83.6% for 1985 and 134.5% for 1990), Guinea Bissau (121% for 1985 and 82.1% for 1990), Uganda (76.8% for 1985 and 75.7% for 1990) and Zambia (116.2% for 1985 and 105.4% for 1990).

For the remaining ten countries we calculated the following (z/u) ratios:

Botswana: for 1985 a minimum ratio of 38.6% corresponding to z = $30 and a maximum of 77% corresponding to z = $60, with a reasonable ratio 57.9% implying a poverty line falling between $40 and

$50. For 1990 a minimum ratio of 21.5% and a maximum of 43%, with a reasonable ratio of 43% corresponding to a poverty line of $60.

Cote d' Ivoire: for 1985 a minimum ratio of 30.4% and a maximum of 60.8%, with a reasonable ratio of 55.8% corresponding to a poverty line between $50 and $60. For 1990, a minimum ratio of 48% and a maximum ratio of 96%, with a reasonable ratio of 56%, implying a poverty line between $30 and $40.

Ghana: for 1985 a minimum ratio of 48.5% and a maximum ratio of 96.9%, and a reasonable ratio of 56.6% corresponding to a poverty line between $30 and $40. For 1990, a minimum ratio of 46.6% and a maximum ratio of 92.9%, with a reasonable ratio of 54.2% corresponding to a poverty line between $30 and $40.

Kenya: for 1985 a minimum ratio of 47.7% and a maximum ratio of 95.5%, with a reasonable ratio of 55.7% corresponding to a poverty line between $30 and $40. For 1990, a minimum ratio of 44.8% and a maximum ratio of 89.6%, with a reasonable ratio of 52.3% corresponding to a poverty line between $30 and $40.

Lesotho: for 1985 a minimum ratio of 47.6% and a maximum ratio of 95.2%, with a reasonable ratio of 55.6% corresponding to a poverty line between $30 and $40. For 1990, a minimum ratio of 43% and a maximum ratio of 86.1%, with a reasonable ratio of 50.2% corresponding to a poverty line between $30 and $40.

Mauritania: for 1985 a minimum ratio of 56.9% and a maximum ratio of 114%, with a reasonable ratio of 66.4% corresponding to a poverty line between $30 and $40. For 1990, a minimum ratio of 52.9% and a maximum ratio of 106%, with a reasonable ratio of 61.8% corresponding to a poverty line between $30 and $40.

Rwanda: for 1985 a minimum ratio of 61.7% corresponding to a poverty line of $20 and a maximum ratio of 176.5%, with 61.7% as the relevant reasonable ratio. For 1990, a minimum ratio of 81.9% corresponding to a poverty line of $20 and a maximum ratio of 234%.

Senegal: for 1985 a minimum ratio of 49.6% corresponding to a poverty line of $21 and a maximum ratio of 141.5%, with a reasonable ratio of 60.1% corresponding to a poverty line between $21 and $30. For 1990, a minimum ratio of 47.3% corresponding to a poverty line of $21 and a maximum ratio of 135%, with a reasonable ratio of 57.5% corresponding to a poverty line between $21 and $30.

Tanzania: for 1985 a minimum ratio of 41.4% corresponding to a povert line of $21 and a maximum ratio of 118.1%, with a reasonable ratio of 50.2% corresponding to a poverty line between $21 and

$30. For 1990, a minimum ratio of 43.1% and a maximum ratio of 86.2% with a reasonable ratio of 50.3% corresponding to a poverty line between $30 and $40.

Zimbabwe: for 1985, a minimum ratio of 39% and a maximum ratio of 77.9% with a reasonable ratio of 58.4% corresponding to a poverty line between $30 and $40. For 1990, a minimum ratio of 38.3% and a maximum ratio of 76.6% with a reasonable ratio of 58.4% corresponding to a poverty line between $40 and $50.

On the basis of these results we calculated an average reasonable (z/u) ratio of 58.7% for 1985 and 53.7% for 1990. The overall average reasonable (z/u) ratio for 1985 and 1990 is calculated as 56%. We hasten to note that this overall average (z/u) ratio is almost identical to that obtained from an absolute poverty line calculation for Sudan which was previously used for SSA by Ali (1995). In what follows we shall use a (z/u) ratio of 0.574 for all countries except for Ethiopia, Guinea Bissau, Rwanda, Uganda and Zambia where a ratio of 0.67 will be used following the World Bank convention for low-income countries. One advantage of this procedure is that it allows poverty lines to differ between different countries depending on their stage of development as reflected by per capita income. We also note that this reasonable (z/u) ratio is not much different from the average ratio computed for the sample of SSA countries reported in Tabatabai and Fouad (1993) where poverty lines are reported in US$.

A Decomposition Methodology

Following Ali (1992, 1994) and Kakwani (1993a), suppose that P is a poverty index which is a function of the poverty line, z, mean income of the population, u, and the Gini coefficient, m:

(1) $P = P(z, u, m)$

Logarithmic differentiation with respect to time will give the following result, where a * over a variable indicates the rate of change with respect to time (i.e. growth rate):

(2) $P^* = c\, z^* + e\, u^* + k\, m^*$

where c, e and k are elasticities of the poverty index with respect to the relevant arguments. For the FGT class of poverty measures it can be shown that the elasticity of the poverty index with respect to the poverty line is equal to the negative of its elasticity with respect to mean income (see, e.g., Ali 1992). In view of Kakwani's results regarding the sign of the elasticity of poverty with respect to mean income, the elasticity of the poverty index with respect to the poverty line is positive. Using this in equation (2), we have:

(3) $P^* = c[z^* - u^*] + k\,m^*$

Now, if the poverty line is assumed to be a function of mean income (an assumption which is usually invoked in the context of advanced countries, but also used for convenience in developing countries), and if the elasticity of the poverty line with respect to mean income is denoted by r, then (3) simplifies to:

(4) $P^* = c\{r - 1\}u^* + k\,m^*$

Equation (4) embodies a decomposition of the behavior of poverty into two components: the pure growth effect on poverty (represented by the first term on the right-hand-side) and the pure distribution effect (represented by the second term). Note that in the above decomposition, if the elasticity of the poverty line with respect to mean income is unity then the growth effect vanishes and the behavior of poverty would depend on the distribution effect. Further, note that if the poverty line is very sensitive to changes in mean income (i.e. r >1) then growth increases poverty.

Assuming that the Lorenz curves do not change between two dates amounts to assuming that m* is zero and we are left with the pure growth effect on poverty. To obtain the overall change in poverty over time one needs to capture the distribution component. This we will attempt to do with the SSA sample of Chen, Datt and Ravallion.

In our estimation of SSA poverty we shall be using two poverty measures which belong to the FGT class of measures, namely the head-count ratio and the poverty-gap measure. These are given as follows:

(5) $H = q/n$

where H is the head-count ratio, q is the number of the poor and n is the total number of population. The poverty-gap measure is given by:

(6) $P(1) = H \cdot [1 - v/z]$

where v is the mean income of the poor. Kakwani (1993-a) has shown that the elasticity of the poverty-gap measure with respect to mean income is given by:

(7) $e(1) = -v/[z - v]$

which implies that the elasticity with respect to the poverty line is:

(8) $c(1) = v/[z - v]$

Similarly, the elasticity with respect to the Gini coefficient is found to be equal to:

(9) $k(1) = [u - v]/[z - v]$

The above framework, with the associated parameterization, will be used to explore the magnitude of poverty in SSA. Before proceeding it should be noted that our proposed decomposition is similar to that of Datt and Ravallion (1992:277-78) who decompose the change in poverty into a growth component, a distribution component and a residual. They note that the "residual exists whenever the poverty measure is not additively separable between mean income and the (Lorenz parameters). In general, the residual does not vanish....If the mean income or the Lorenz curve remains unchanged over the decomposition period, then the residual vanishes." Ours is an exact decomposition applied in the case where the Lorenz curves are assumed to remain invariant between the two dates under study. Further note that the Datt-Ravallion decomposition is applicable where there are two observations on the level of poverty while ours can be applied when there is only one observation.

The Results

The 1985 Poverty

Table 2 reports our results for the CDR sample of 14 SSA countries for 1985. It should be noted that, except for our procedure regarding the determination of the poverty line, these results are comparable to those obtained by CDR. The table reports on the two poverty measures used, namely, the head-count and the poverty-gap ratios as well as the mean expenditure (income) of the poor, the poverty line and the mean expenditure of the population. As noted in the data section the mean expenditure of the population is taken from CDR and is denominated in PPP 1985 dollars. The poverty line and the mean income (expenditure) of the poor are derived from this figure and are as such denominated in PPP 1985 dollars. The table also reports the population weight used in the analysis, which is taken from ADB (1995). Weighted average results for the sample use these weights and the results are rounded to the second decimal place.

Table 6.2: Poverty in Sub-Saharan Africa in 1985

Country	Head ratio(%)	Mean income of the poor ($)	Poverty gap ratio (%)	Poverty line ($)	Mean expenditure ($)	Population weight
Botswana	50.26	24.34	22.89	44.70	77.83	0.0070
Côte d' Ivoire	40.31	36.36	14.44	56.65	98.70	0.0646
Ethiopia	34.36	17.28	8.34	22.82	34.06	0.2677
Ghana	29.00	25.19	8.47	35.57	61.96	0.0836
Guinea Bissau	53.86	5.19	29.82	11.63	17.36	0.0057
Kenya	53.17	19.66	24.20	36.08	62.86	0.1294
Lesotho	50.89	18.26	25.20	36.17	63.01	0.0102
Mauritania	32.17	15.02	16.20	30.26	52.72	0.0115

Rwanda	31.59	18.14	6.47	22.81	34.04	0.0394
Senegal	49.65	13.01	23.10	24.33	42.38	0.0415
Tanzania	53.53	13.84	28.12	29.15	50.79	0.1419
Uganda	37.10	13.70	9.34	18.31	27.33	0.0983
Zambia	48.53	7.61	18.03	12.11	18.07	0.0447
Zimbabwe	56.71	24.88	24.78	44.18	76.96	0.0546
Mean/Total	42.47	18.45	16.05	29.30	48.70	1.0000

Source: Author's calculations based on CDR 1994b.

On the above results we note the following:
A. The poverty lines: the weighted average poverty line for 1985 is calculated as US$29.3 per person per month. The highest poverty line is that of Cote d' Ivoire (US$ 56.7) while the lowest is that for Guinea Bissau (US$11.63 per person per month). For only two of the remaining 13 countries in the sample is the poverty line above US$ 40 per person per month, and for four of the remaining 11 countries the poverty line is above US$ 30 per person per month. Thus for only 50% of the CDR sample is the "global bench-mark poverty line" of US$ 30 per person per month used by the World Bank relevant. This implies that two values of the poverty line range used by CDR (US$40 and 50) are relevant for only three countries. Further, it is seen from the results that for 3 of the countries for which the poverty line is below the "global bench-mark," the poverty line is below US$ 21 per person per month, the lowest poverty line in the CDR range.
B. The average income of the poor: the weighted average expenditure of the poor for 1985 is calculated as US$18.45 per person per month, while the weighted average per capita expenditure is calculated as US$48.7 per person per month. Thus, the average income of the poor is 37.89% of per capita income and 62.97% of the poverty line. On both counts this indicates that poverty in SSA is a very deep phenomenon even when looked at from a PPP perspective. To further appreciate the depth of African poverty we note that the average expenditure of the poor in ten countries in the sample was below US$20 per person per month, and for only one country is this average above US$ 30 per person per month.
C. The head-count ratio: it will be recalled that the head-count ratio measures the spread of poverty among the population. For 1985 the weighted average head-count ratio is estimated as 42.47% meaning that 47% of the population in the sample were living below the poverty line of $29.3 per person per month. The lowest head-count ratio is estimated for Rwanda with only 31.59% of the population falling below the poverty line of

$22.81 per person per month, while the highest ratio of 56.7% is recorded for Zimbabwe implying that nearly 57% of the population were below the poverty line of $44. The distribution of the countries of the sample is such that four countries (with a population weight of 40.22%) have a head-count ratio of less than 35%, two countries (with a population weight of 16.29%) have a head count ratio in the range 35-45%, and the remaining countries have a head-count ratio in excess of 45%.

D. The Poverty-Gap Ratio: it will be recalled that the poverty-gap ratio is a measure of the depth of poverty. For 1985 the weighted average poverty-gap ratio is found to be 16.05%. The lowest ratios are calculated for Rwanda (6.47%), Ethiopia (8.34%), Ghana (8.47) and Uganda (9.34%). The highest ratios are calculated for Guinea Bissau (29.82%), Tanzania (28.12%), Lesotho (25.2%), and Zimbabwe (24.78%).

The Pure Growth Effect

Assuming that the Lorenz curves remain constant, poverty results for 1990 are reported in Table 6.3. Comparing these results with those for 1985 we get the pure effect of the growth in per capita income (expenditure) on poverty. The weighted average per capita expenditure for 1990 is calculated as US$49.26 implying a growth rate of 0.23% per annum. The weighted average head-count ratio for 1990 is calculated as 43.07% implying that 43% of the population of the sample were living below the poverty line of $29.49. Thus the growth of per capita income between 1985 and 1990 has caused poverty, as measured by the head-count ratio, to increase by an annual rate of 0.28% such that the head-count ratio increased from 42.47% in 1985 to 43.07% in 1990.

At the individual country level poverty as measured by the head-count ratio increased in six countries with a population weight of 48.58%, remained constant in one country (Mauritania, with a population weight of 1.11%), and declined in seven countries with a population weight of 50.31%. The countries which experienced increased poverty are Cote d' Ivoire (where per capita income declined at an annual rate of 8.7% while poverty increased by an annual rate of 0.59%), Ghana (where per capita income increased by an annual rate of 0.8% and poverty increased by a rate of 0.22% per annum), Rwanda (where per capita income declined by an annual rate of 5.5% while poverty increased by an annual rate of 0.37%), Tanzania (where per capita income increased by a rate of 6.5% per annum and poverty increased by a rate of 0.31% per annum), Uganda (where per capita income increased by an annual rate of 0.3% while poverty increased by an annual rate of 1.6%), and Zimbabwe (where per capita income increased by 0.34% per annum and poverty increased by 1.3% per annum).

The countries which experienced a reduction in poverty are Botswana (where per capita income increased by an annual rate of 12.3% while poverty declined by a rate of 0.03% per annum), Ethiopia (where income declined at a rate of 2.8% per annum and poverty declined at 0.23% per annum), Guinea Bissau (with income growth of 8.1% per annum and a poverty decline by 0.07% per annum), Kenya (with an income growth rate of 1.3% per annum and a decline in poverty at a rate of 0.14% per annum), Lesotho (with an income growth rate of 2.0% per annum and a decline in poverty at a rate of 0.09% per annum), Senegal (with an income growth rate of 0.9% per annum and a decline in poverty at an annual rate of 0.02%) and Zambia (with an income growth rate of 2% per annum and a decline in poverty at an annual rate of 0.28%).

Table 6.3: Growth induced poverty in Sub-Saharan Africa, 1990

Country	Head ratio(%)	Mean income of the poor ($)	Poverty gap ratio (%)	Poverty line ($)	Mean income of the pop. ($)	Population weight
Botswana	50.18	42.71	23.29	79.73	138.90	0.0071
Côte d' Ivoire	41.52	23.38	14.47	35.89	62.52	0.0665
Ethiopia	33.96	14.82	8.55	19.81	29.57	0.2635
Ghana	29.32	25.95	8.81	37.09	64.62	0.0835
Guinea Bissau	53.67	7.60	29.89	17.15	25.59	0.0054
Kenya	52.81	20.77	24.29	38.46	67.00	0.1312
Lesotho	50.64	19.76	25.64	40.02	69.72	0.0099
Mauritania	32.17	16.16	16.19	32.54	56.69	0.0111
Rwanda	23.19	13.18	6.29	17.17	25.63	0.0388
Senegal	49.60	13.65	23.01	25.46	44.36	0.0407
Tanzania	54.35	19.42	27.94	39.96	69.61	0.1423
Uganda	40.08	13.41	11.17	18.59	27.75	0.0997
Zambia	47.85	8.32	18.25	13.43	19.92	0.0453
Zimbabwe	60.43	26.56	24.71	44.94	78.30	0.0550
Average/Total	43.07	18.10	16.36	29.49	49.26	1.0000

Source: Author's calculations based on CDR (1994b).

(5) summarizes the results for the first two rates of growth in the Gini coefficient. The third higher rate of growth gave believable results.

Table 6.4: The True Values of Poverty Measures in SSA in 1990 (in percentages except for the elasticity)

Country	Elasticity	P (l; g= 0.8%)	P (l; g = 1.2%)	P (l; g = 1.8%)	H (g = 0.8%)	H (g =) 1.2%	H (g = 1.8%
Botswana	2.6	25.81	27.15	29.26	55.60	58.49	63.06
Côte d' Ivoire	3.1	16.01	17.01	18.59	45.93	48.80	53.33
Ethiopia	3.0	9.62	10.20	11.11	38.19	40.49	44.11
Ghana	3.5	10.06	10.81	11.93	33.49	35.99	39.71
Guinea Bissau	1.9	32.24	33.46	35.37	57.89	60.08	63.50
Kenya	2.6	26.92	28.31	30.52	58.53	61.87	66.70
Lesotho	2.5	28.30	29.71	31.93	55.91	58.69	63.08
Mauritania	2.5	17.88	18.77	20.67	35.52	37.29	41.06
Rwanda	3.5	7.23	7.74	8.55	37.94	39.55	43.69
Senegal	2.5	25.40	26.68	28.68	54.75	57.51	61.82
Tanzania	2.4	30.73	32.21	34.53	59.79	62.67	67.18
Uganda	2.9	12.45	13.15	14.25	44.69	47.20	51.15
Zambia	2.3	19.99	20.90	22.34	52.54	54.93	58.71
Zimbabwe	2.7	27.51	28.99	31.34	67.26	70.88	76.63
Average	2.8	18.16	19.13	20.55	48.02	50.66	54.46

Source: Author's calculations.

Poverty as measured by the poverty-gap ratio increased as a result of growth from 16.05% in 1985 to 16.36% in 1990, recording an annual rate of increase of 0.37% per annum. Nine countries experienced increased poverty while five countries experienced a decline in poverty. The countries where the depth of poverty increased are Botswana (at an annual rate of 0.35%), Cote d' Ivoire (0.04%), Ethiopia (0.5%), Ghana (0.79%), Guinea Bissau (0.05%), Kenya (0.07%), Lesotho (0.35%) Uganda (3.6%) and Zambia (0.24%). The countries where the depth of poverty declined are Mauritania (an annual rate of decline of 0.01%), Rwanda (0.56%), Senegal (0.08%), Tanzania (0.13%), Uganda (0.9%) and Zimbabwe (0.05%).

It is interesting to note that poverty increased on account of the two measures used for Cote d' Ivoire, Ghana and Uganda where poverty increased while it declined on account of the two measures only in Senegal. In six countries (Botswana, Ethiopia, Guinea Bissau, Kenya, Lesotho and Zambia) poverty declined on account of the head-count ratio but increased on account of the poverty-gap ratio. In all six countries such behavior can be explained by the fact that the rate of increase in the average income of the poor was less than that of the poverty line, implying a worsening of the lot of the poor. In three countries (Rwanda, Tanzania and Zimbabwe) poverty increased on account of the head-count ratio but declined in terms of the poverty-gap ratio. Once again the explanation is to be found in the fact that the average income of the poor increased by a rate higher than that of the poverty line. In Mauritania poverty remained the same in terms of the two measures it being noted that the average income of the poor increased at the same rate of the poverty line and the recorded decline of the poverty-gap at the rat of 0.01% could be due to rounding.

The 1990 Poverty

Using the framework specified in equation (4), together with the results of table 6.3, we are now in a position to explore the overall magnitude and change in poverty in SSA during the second half of the 1980s. To do this we need an estimate of the elasticity of the poverty-gap ratio with respect to the Gini coefficient, which we can obtain from equation (9), and a rate of growth of the Gini coefficient, for which we do not have any information. By using these two parameters we can estimate the distribution-related rate of growth of the poverty-gap ratio which will be added to the observed growth-related rate of growth. Starting from the poverty-gap calculated for 1985 and applying the overall growth rate we will be able to obtain the true poverty-gap ratio for 1990. Working from these estimates we can also obtain the true head-count ratio for 1990 which is consistent with the true poverty-gap ratio.

Table 6.4 reports what we call the true values of poverty in SSA for 1990. Three alternative rates of increase of the Gini coefficient are assumed based on the discussion of the recently reported behavior of inequality in SSA. The assumed rates are 0.8%, 1.2% and 1.8% per annum. The table also reports our calculated elasticities of the poverty-gap ratio with respect to the Gini coefficient. The elasticities in question range from a low of 1.9 for Guinea Bissau to a high of 3.5 for Ghana and Rwanda. The weighted average elasticity for the sample is given as 2.8.

It is perhaps clear from the table that poverty as measured by the head-count ratio has increased in SSA from 42.47% in 1985 to a minimum of 48.02% and a maximum of 54.46% in 1990 depending on the behavior of the Gini coefficient. The annual rate of increase of poverty as measured by the head-count ratio varies from 2.49% as a minimum to 5.1% as a maximum, with 3.58% being the annual rate of increase corresponding to an annual rate of increase in the Gini coefficient of 1.2%. Similarly, poverty as measured by the poverty-gap ratio has increased in SSA from 16.05% in 1985 to a minimum of 18.16% and a maximum of 20.55% in 1990.

At the country level poverty as measured by the poverty-gap ratio (at the minimum level corresponding to a rate of increase in the Gini coefficient of 0.8% per annum) recorded highest rates of annual increase for Uganda (5.9%), Tanzania (5.5%) and Ghana (3.5%) while the lowest rate of increase is recorded for Guinea Bissau (1.6%), Mauritania (2%) and Cote d' Ivoire, Zambia and Zimbabwe (2.1 % each). At the maximum level (corresponding to a Gini annual growth rat of 1.8%) the poverty-gap measure recorded the highest rates of annual increase for Uganda (8.8%), Ghana (7%), Rwanda (5.7%), Cote d' Ivoire (5.2%); the lowest rates are recorded for Guinea Bissau (3.5%) and Tanzania (4.2%). The head-count ratio, at the minimum level of estimated poverty, increased by the highest annual rates of 3.8% for Uganda, 3.7% for Rwanda and 3.5% for Zimbabwe., and at the lowest rates of 1.5% for Guinea Bissau, 1.6% for Zambia and 1.9% for Lesotho. At the highest level of estimated poverty, the headcount ratio increased by the highest annual rates of 6.7% for Rwanda, 6.6% for Uganda and 6.2% for Zimbabwe; and at the lowest rates of 3.4% for Guinea Bissau and 3.9% for Zambia.

Current Poverty Levels

Most of the discussion in the literature quotes poverty estimates for 1990 as representing current poverty levels. This we believe is compounding the problem of underestimating the extent of poverty in S SA. Using our decomposition procedure will allow us to estimate current poverty levels. The estimation starts from calculating the rate of growth of the poverty-gap ratio which is given by:

(10) $P^* \quad H^* + [v/(z - v)] \cdot [Z^* - v^*]$

where a * over a variable indicates the rate of growth over time and where v is the mean income of the poor and z the poverty line. The rates of change in the relevant variables are calculated from the results of Tables 2 and 4. Starting from the estimated poverty-gap measure for 1985 a projected poverty-gap measure for 1995 is obtained and using a projected income-gap ratio for 1995 a headcount measure is also obtained. Table 6.5 summarizes the results for the first two rates of growth in the Gini coefficient. The third higher rate of growth gave believable results for all countries except for Zimbabwe. The table reports the initial 1985 poverty-gap measure and the projected income-gap measure of 1995. The reported mean result for the sample of countries is a weighted average result where the 1994 population weights are used.

Table 6.5: Poverty in Sub-Saharan Africa, 1995

Country	P (1) 1985	P (1,g = 0.8%	P (1,g 1.2%	(1-v/z) 1995	H (g = 0.8%	H (g = 1.2%)
Botswana	22.89	29.10	32.20	0.4832	60.22	66.64
Côte d' Ivoire	14.14	18.14	20.45	0.3386	53.57	60.40
Ethiopia	8.34	11.10	12.47	0.2337	47.57	53.36
Ghana	8.47	11.95	13.78	0.3077	38.84	44.78
Guinea Bissau	29.47	34.44	37.10	0.5614	61.35	66.08
Kenya	24.20	29.94	33.13	0.4654	64.33	71.19
Lesotho	25.20	31.79	35.04	0.5165	61.55	67.84
Mauritania	16.20	19.73	21.75	0.5034	39.19	43.21
Rwanda	6.47	8.08	9.25	0.1878	43.02	49.25
Senegal	23.10	27.94	30.80	0.4620	60.48	66.67
Tanzania	28.12	33.58	36.88	0.5263	63.80	70.07
Uganda	9.34	16.60	18.51	0.3263	50.87	56.73
Zambia	18.03	22.15	24.23	0.3867	57.28	62.66
Zimbabwe	24.78	30.53	33.92	0.3813	80.07	88.96
Mean	16.05	20.57	22.85	0.3603	54.86	67.16

Source: Author's calculations.

Looking at the head-count results we note that at the minimum level of estimated poverty seven countries have a head-count ratio in excess of 60%, three countries with a head-count ratio in excess of 50% but less than 60%, two countries with a head-count ratio between 40%-50% and only two countries with a head-count ratio of less than 40%. The weighted average head-count ratio for the sample is 54.86%. The highest reported level of poverty is that for Zimbabwe where 80% of the population are projected to live below the poverty line; the lowest reported level of poverty is that for Ghana where 38.8% of the population are projected to live below the poverty line.

At the medium level of estimated poverty, corresponding to an annual rate of increase of the Gini coefficient of 1.2%, we note that the head-count ratio is above 60% in nine countries, between 50 and 60% in two countries and between 40 and 50% in countries. The weighted average head-count ratio for the sample is 67.16%. The highest reported level of poverty is that for Zimbabwe where nearly 89% of the population are projected to live iq/Poverty; the lowest reported level of poverty is that for Mauritania where 43% of the population are projected to live below the poverty line.

SAPs and Poverty

Following Demery and Squire 1996 it is possible to use our results to explore the association between SAPs and poverty. According to the method suggested by Demery and Squire we need to look at the change in poverty and the associated change in the macroeconomic policy stance index of the World Bank (1994b: Table B.1, 260-261). The macroeconomic policy stance index comprises three components, as follows:

A fiscal policy component: which includes fiscal balance excluding grants as a percentage of GDP and total revenue as a percentage of GDP. The overall change if the fiscal policy index is obtained as the addition of the score of the changes in the individual components. The scoring range for the fiscal balance is such that a score of -2 reflects a change in the fiscal balance of -9.9 to -5.0 percentage points while a score of 3 reflects a change of 5 percentage points or more. The scoring scale for the total revenue is such that a score of -1 reflects a change in total revenue of -4.0 percentage points or more, a score of 0 reflects a change of -3.9 to 3.0 percentage points and a score of 1 reflects a change of 3.1 percentage points or more;

A monetary policy component: which includes seigniorage and inflation. The overall change in the monetary policy index is calculated by averaging the scores for the change in the two components. The scoring range for the seigniorage is such that a score of 2 reflects a change of -3.0 to -2.1 percentage points while a score of -2 reflects a change of 2.0 to 3.9 percentage points. The scoring range for inflation goes from a score of 2 reflecting a change of -4.9 to -10.0 percentage points to a score of -3 reflecting a change of 31.0 percentage points or more;

An exchange rate policy component: which includes the real effective exchange rate and the parallel market exchange rate premium. The overall change in the exchange rate policy index is calculated by averaging the scores for the change in the two components. The scoring range for the real effective exchange rate (where the real effective exchange rate is defined such that an increase constitutes a depreciation) is such that a score of -2 reflects a change of -10.0 percentage points or more while a score of 3 reflects a change of 31.0 percentage points or more. The scoring range for the parallel market exchange rate premium is such that a score of 3 reflects a change in the premium of -100 percentage points while a score of -3 reflects a change of 51.0 percentage points.

The score for the overall change in macroeconomic policies is calculated by averaging the scores for the change in three components such that a score of 1.0 or more is taken to reflect a large improvement in policy, a score of 0-0.9 is taken to reflect a small improvement in policy and a score below zero is taken to mean a deterioration in policy.

The change in policy is calculated for averages of two periods 1981-86 (pre-reform period) and 1987-90 (adjustment period). Noting that we have poverty estimates for 1985 and 1990 the change in the policy index reported by the World Bank (1994b) can be applied directly. We note, however, that such information is available for ten of the countries in the CDR sample.

No attempt has been made to calculate the change in the policy index for the remaining four.

Table 6.6: SAPs and poverty in the SSA

Country	Head-Count ration 1985 (%)	Head-Count ration (g=0.8) 1999 (%)	Change in poverty (% points)	Change in macro policy (score)
Côte d' Ivoire	40.31	45.93	5.62	-1.3
Ghana	20.00	33.49	4.49	2.2
Kenya	53.17	58.83	5.66	0.5
Mauritania	32.17	35.52	3.35	0.5
Rwanda	31.59	37.94	6.35	-0.2
Senegal	49.65	54.75	5.10	0.5
Tanzania	53.53	59.79	6.26	1.5
Uganda	37.10	44.69	7.59	0.2
Zambia	48.53	52.54	4.01	-0.3
Zimbabwe	56.71	67.26	10.55	1.0

Source: Tables 2 and 4 and World Bank (1994b).

From table 6.6 we note that three countries recorded a large improvement in policies as reflected by a score of an overall change in microeconomic policies of greater than unity. In all three poverty has increased. In Ghana, the best performing country in terms of policy, poverty increased by 4.49 percentage points between 1985 and 1990, an increase of 15.48 per cent ranking as the fourth largest increase among the ten countries in the sample. Tanzania, the second best performing, country in terms of policy, saw poverty increasing by 6.26 percentage points between the two dates which works out as an increase of 11.69 per cent ranking as the fifth lowest percentage increase. In Zimbabwe, the third best performing country in terms of policy, poverty increased b-, the highest absolute number of 10.55 percentage point and the third highest percentage increase of 18.6.

At the other end of the scale we also have three countries recording deterioration in policy. The worst performing country in terms of policy is Cote d' Ivoire which saw poverty increasing by 5.62 percentage points ranking as the 5th highest percentage increase of 13.94. The second worst performing country in terms of policy is Zambia which saw the second lowest increase in poverty in absolute terms by 4.01 percentage points and the lowest increase in poverty of 8.26. Rwanda is the third worst performing country in terms of policy where poverty increased by 61,35 percentage point recording the second highest increase in poverty of 20.1%.

Given the above we recall that the association between the change in the policy index and the change in poverty enabled Demery and Squire (1996:40) to claim that they have provided "the most convincing evidence to date that economic reform is consistent with a decline in overall poverty and that a failure to reform is associated with increased poverty." In the same vein an African armed with the results of table 6.6 is entitled to claim that he has provided the most convincing evidence to date that economic reform as proxied by the policy index gives rise to increased poverty irrespective of the degree of success achieved in implementing the reform. Of course ordinary Africans going about their daily lives of eking a living have their own evidence about the damage that has been inflicted upon them by the ideological commitment of the World Bank to SAPS. But, alas, such evidence is seen as anecdotal and is not permissible in learned journals. However, it is the perception of poverty by ordinary Africans which frequently leads to IMF riots and policy reversals.

Concluding Remarks

In recent years a consensus has emerged among African economists of different persuasions that the order of the day in SSA continues to be "development" rather the management of economies in crisis. This position has recently been articulated by Lipumba (1994:68) who notes that the "main objective of economic reforms and adjustment policies in Africa must be to establish conditions that foster sustainable, poverty-eradicating economic growth. Attaining a sustainable balance of payments deficit, low inflation, and a competitive exchange rate are important goals if and only if they contribute to economic growth and improvement in the living standards of the majority of Africans who are in poverty. Eradicating poverty in Africa is a long-term goal, but it requires immediate action." Important in this position is the recognition that "the majority of Africans" are in poverty. This is the message we have attempted to convey by providing our alternative estimates. It is an important message that has a direct bearing on policies designed to alleviate poverty in Africa.

As is probably well known the current policy debate on alleviating poverty is dominated by the World Bank's approach to the subject. The approach is based on information generated to answer the question "who are the poor?" In the words of the 1990 WDR if "governments are to reduce poverty or to judge how their economic policies affect poverty, they need to know a lot about the poor.... Policies targeted directly to the poor can hardly succeed unless governments know who the poor are and how they respond to policies and to their environment" " (World Bank 1990:29).

This is, of course, the celebrated targeting approach to poverty alleviation which became famous in the context of SAPs (see, for example, Bardhan 1994. Besley and Kanbur (1993:67), who provided a rigorous treatment of the principles of targeting, also note that it "is in the wake of macroeconomic and structural adjustment that targeting seems to have attained a special significance in developing countries, as more and more governments have come under pressure to reduce expenditure. Indeed, targeting has come to be seen as a panacea in poverty alleviation."

Without getting involved in details, we note that Besley and Kanbur (1993) identified the ideal solution as that of "perfect targeting" but proceeded to note that, in the real world, problems associated with administrative costs, incentive effects and political economy militate against fine target in a scheme and recommend the advantages of more universalize schemes.

Be the above as it may, the "Who are the poor?" approach to designing policies to reduce poverty may be relevant when the head-count ratio is in the order of magnitude of 40% as the redistribution with growth literature believed in the early 1970s (see Chenery et al. 1974, Fishlow 1995, and Bardhan 1995. But when the head-count ratio goes to above 60% such that the "majority of Africans are in poverty," it seems reasonable to argue that designing poverty alleviation policies on the bases of identifying and targeting the poor is blatantly irrelevant. The reason for this is simple: the poor in this case are the majority of the population so that the ratio of the leakages (to the non poor) to the benefits (to the poor) is less than unity, which is a rough measure of the fineness of targeting. Alternatively, using Besley and Kanbur measure of fineness of targeting, given that it is increasing in the benefits to the poor, we find that as the head-count ratio increases the fraction also increases. This suggests that in poverty situations Like the ones prevailing in SSA, where the head-count ratio is above 60%, a strong case can be made for universal targeting.

Given the results reported in this chapter, a cynic would be prepared to argue that perhaps the relevant and cost-effective policy to alleviate poverty in SSA is one of reverse targeting. Under such a policy it is the rich who would be the subject of targeting. The objective of policy in this case would be to mobilize the required resources to alleviate or eradicate poverty. It can be shown, however, that the relative magnitude of resources to be generated in SSA for poverty eradication will be large. Using information from Tables 6.3 and 6.4 it is an easy matter to show that when the head-count ratio is 48.02% the required resources would work out as 11.05% of the total resources at the disposal of society. At the other extreme, when the head-count ratio is 54.46% the required resources are 12.57% of the resources of society. These are admittedly high rates of taxation, reflecting as they do the depth of poverty in the region.

In this chapter we have argued that the major challenge facing poverty alleviation efforts in SSA is the lack of knowledge about the behavior and magnitude of poverty. We have shown that estimates reported in the literature underestimate the incidence, depth and the behavior of poverty. By doing so, such results also underestimate the effort required on the policy front to alleviate poverty. Our alternative results suggest: that poverty in SSA is a very deep phenomenon as evidenced by the very low per capita income of the poor; that the majority of Africans are in poverty (the head-count ratio in the range 55-67%); that the lot of the poor has worsened (an increase in the poverty-gap ratio from 18% to 21-23%). Thus, during the second half of the 1980s poverty, as measured by the head-count ratio, has increased in SSA at annual rates in the range of 2.5%-5.1%.

Such poverty levels and trends, we conjectured, are incremental to the secular levels and trends established in the region in the 1970s. As such, therefore, we called for a medium-term poverty alleviation approach that would enable the region to regain its secular poverty levels and trends. Such a strategy, we noted, may or may not be consistent with the currently dominant policy orientation of macroeconomic stabilization and adjustment. In particular, we argued that the current targeting approach to poverty alleviation may be a shade irrelevant.

In this respect it is perhaps important to note that in revisiting Africa's development strategy, the World Bank (1994a) identified "two clear legs to the strategic" agenda of sustainable poverty reduction. The first leg emphasizes the importance of encouraging labor-intensive growth, labor being the factor owned by the poor. This is to be undertaken in the context of a "stable and undistorted macroeconomic framework" that encourages private enterprise and removes biases against agriculture. The second leg requires the provision' of broad-based health, education and infrastructural services. Both propositions, we suggest, fundamentally signal the crucial importance of the equity dimension in the process of development, and are generally acceptable. Disagreement, however, may arise over how, to go about doing them. This naturally raises the political economy question of the role of the African state in asset redistribution.

In some countries of the region, for example, the issue of land redistribution continues to be a major political and social issue. Bardhan (1995:11) identifies land reform as one of the contested issues in the context of poverty alleviation. After reviewing the various arguments, lie notes that redistributive "land reform, by changing the local political structure in the village, gives more 'voice' to the poor and induces them to get involved in local self-governing institutions and management of public goods. Even local markets (say, in farm output or credit or water) function somewhat more efficiently when the leveling effects of land reform improve competition and make it more difficult for the local oligarchy to comer the market." In the context of the changing African political landscape towards more democratic governance regimes, the above political economy consideration of the dynamic effect of land reform can not be dismissed as irrelevant to poverty alleviation (see also Lipton and van der Gaag 1993 and Fishlow 1995).

Given the depth and trends of poverty in the region, political economy considerations relating to international cooperation will also have implications for poverty alleviation. Elbadawi (1995) argues the case for substantial debt relief to the region on account of the external shocks and the negative impact of debt on growth. The current dominant approach of both bilateral and multilateral donors to this cooperation has emphasized the creation of "social funds" to tidy the poor during the transitional period of adjustment. Stewart and van der Geest (1995) have recently evaluated the experience of "social funds" in selected countries. Their conclusion, which is rather obvious, is that "add-on temporary institutions, depending heavily on external funds, have been poorly targeted and have not been able to provide for effective poverty reduction during adjustment"; local institutions, with no funding from outside, were round to be more effective. One reason for this is the small size of the "social funds" compared to the "incremental poverty" being created in the process of adjustment (see also Emmerij 1995).

Works Cited

Ali, A.A.G. "Poverty and Growth: A Comment on Kakwani." Unpublished memo, (1994).

Ali, A.A.G. "Adjustment Programs and the Environment in Africa: Some exploratory Results." *Eastern Africa Social Science Research Review*, 9, 2 (1993).

Ali, A.A.G. "Structural Adjustment Programs and Poverty Creation in Africa: Evidence from Sudan." Eastern Africa Social Research Review, 8, 1 (1993).

Anand, S. and R. Kanbur. "Equality and Development: A Critique." *Journal of Development Economics* 41 (1993a).

Anand, S. and R. Kanbur. "The Kuznets Process and Inequality-Development Relationship." *Journal of Development Economics* 40 (1993b).

Azam, J-P. "The Uncertain Distributional Impact of Structural Adjustment in Sub-Saharan Africa." In: R. van der Hoeven and F. van der Kraaij, (eds.), Structural Adjustment and Beyond in Sub-Saharan Africa; London, James Currey (1994).

Bardhaii, P. "Research on Poverty and Development: Twenty Years Aller Redistribution with Growth." Paper presented to the Annual World Bank Conference on Development Economics, World Bank, (1995).

Berg, E., Hunter, G., Lenaghan, T. and M. Riley. "Poverty and Structural Adjustment in the 1980s: Trends in Welfare Indicators in Latin America and Africa." CAER discussion paper no. 27, Development Alternatives Incorporated, USA, Maryland (1994).

Besley, T. and R. Kanbur. "Principles of Targeting." In: Lipton and van der Gaag, (eds.) (1993).

Bourguignon, F., de Mello, J. and C. Morrisson. "Poverty and Income Distribution during Adjustment: Issues and Evidence from the OECD Project." World Development, 19. No. 11, pp. 1485-508 (1991).

Bourguignon, F., de Mello, J. and A. Suwa. "Distributional Effects of Adjustment Policies: Simulations for Archetype Economies in Africa and Latin America." World Bank Economic Review, 5. No. 2, pp. 339-66 (1991).

Bruno, M., Ravallion, M. and L. Squire. "Equity and Growth in Developing Countries: Old and New Perspectives on Policy Issues." Paper presented at the Conference on Income Distribution and Sustainable Development, Washington DC., IMF (1995).

Chen, S., Datt, G. and M. Ravallion. "Is Poverty Increasing in the Developing World?" *Review of Income and Wealth*, Series 40, 4 pp. 59-76 (1994a).

Chen, S., Datt, G. and M. Ravallion. "Statistical Addendum to: Is Poverty Increasing in the Developing World?" Policy Research Department, World Bank (1994b).

Chenery, H., Ahluwalia, M., Bell, C., Duloy, J. and R. Jolly. "Redistribution with Growth." Oxford, Oxford University Press (1994).

Corbo, V. and P. Rojas. "World Bank-supported Adjustment Programs: Country Performance and Effectiveness." In: V. Corbo, S. Fischer and S. Webb, (eds.), *Adjustment Lending Revisited: Policies to Restore Growth*. World Bank (1992).

Cornia, G., Jolly, R. and F. Stewart, (eds.). "Adjustment with a Human Face." 1, *Protecting the Vulnerable and Promoting Growth*. Oxford, Clarendon Press (1987).

Datt, G. and M. Ravallion. "Growth and Redistribution Components of Changes in Poverty Measures: A Decomposition with Applications to Brazil and India in the 1980s." *Journal of Development Economics* 38 no 2 pp. 275-95 (1992).

Demery, L. and L. Squire. "Macroeconomic Adjustment and Poverty in Africa: An Emerging Picture." *World Bank Research Observer*, 11, 1 pp. 35-39 (1996).

Dorosh, P. and D. Sahn. "A General Equilibrium Analysis of the Effect of Macroeconomic Adjustment and Poverty in Africa." Cornell Food and Nutrition Policy Program (CFNPP) Working Paper no. 39, Cornell University 1991).

Easterly, W. "Why is Africa Marginal in the World Economy?" Washington DC., World Bank (1994).

Elbadawi, I. "Consolidating Macroeconomic Stabilization and Restoring Growth in Sub-Saharan Africa." Paper presented to the ODC conference on "Africa's Economic Future." Nairobi, AERC (1995).

Elbadawi, I. "Have World Bank-supported Adjustment Programs Improved Economic Performance in Sub-Saharan Africa." World Bank, WPS 1002 (1992).

Elbadawi, I. and B. Ndulu. "Long-Term Development and Sustainable Growth in Sub-Saharan Africa." Paper presented to SAREC International Colloquium on New Directions in Development Economics; Stockholm (1994).

Emmerij, L. "A Critical Review of the World Bank's Approach to Social-sector Lending and Poverty Alleviation." International Monetary and Financial Issues for the 1990s; 5, Geneva, UNCTAD (1995).

Fishlow, A. "Inequality, Poverty and Growth: Where Do We Stand?" Paper presented to the Annual World Bank Conference on Development Economics, World Bank (1995).

Horton, S., R. Kanbur and D. Mazumdar. "Openness and Inequality". An invited paper for IEA World Congress, Tunis (1995).

Husain, I. and R. Faruqee, (eds.). "Adjustment in Africa: Lessons from Country Case Studies." World Bank (1994).

Jazairy, I, Alamgir, M. and T. Panuccio. *The State of World Rural Poverty: An Inquiry into Its Causes and Consequences*. New York, New York University Press (1992).

Kakwani, N. "Poverty and Economic Growth with Application to Côte d'Ivoire." *Review of Income and Wealth*, Series 39, 2 (1993a).

Kakwani, N. "Measuring Poverty: Definitions and Significance Tests with Applications to Côte d'Ivoire." In Lipton and van der Gaag, (eds.), (1993b).

Kanbur, R. "Measurement and Alleviation of Poverty." IMF Staff Papers, 36 (1987a).

Kanbur, R. "Structural Adjustment, Macroeconomic Adjustment and Poverty: A Methodology for Analyzes." World Development, 15 (1987b).

Lipton, M. "Successes in Anti-poverty." Issues in Development discussion paper no. 8, Geneva, ILO (1996).

Lipton, M. and J. van der Gaag, (eds.). "Including the Poor: Proceedings of a Symposium Organized by the World Bank and the International Food and Policy Research Institute." World Bank (1993).

Lipton, M. and J. van der Gaag. "Poverty: A Research and Policy Framework." In: Lipton and van der Gaag (eds.), (1993).

Lipumba, N.H.I. "Structural Adjustment Policies and Economic Performance of African Countries." International Monetary and Financial Issues for the 1990s, 5, Geneva, UNCTAD (1995).

Lipumba, N.H.I. "Africa Beyond Adjustment." Policy Essay no. 15, Washington DC., ODC (1994).

Mosley, P. "Decomposing the Effects of Structural Adjustment: The Case of Sub-Saharan Africa." In: R. van der Hoeven and F. van der Kraaii, (eds.). *Structural Adjustment and Beyond in Sub-Saharan Africa*. London, James Curry (1994).

Pio, A. "The Social Impact of Adjustment in Africa." In: G. Cornia and G. Helleiner, (eds.). From Adjustment to Development in Africa. New York, St. Martins Press (1994).

Ravallion, M. "Poverty Comparisons." Fundamentals in Pure and Applied Economics, Switzerland, Hardwood Academic Press, Chur (1994).

Ravallion, M. and 0. Datt. "Growth and Poverty in Rural India." Policy Research Working Paper no. 1405, World Bank (1995).

Squire, L. "Introduction: Poverty and Adjustment in the 1980s." *World Bank Economic Review*, 5, 2 (1991).

Stewart, F. and W. van der Geest. "Adjustment and Social Funds: Political Panacea or Effective Poverty Reduction?" Employment Papers no. 2, Geneva, ILO (1995).

Summers, R. and A. Heston. "The Penn World Table (mark 5): An Expanded Set of International Comparisons, 1950-1988." *Quarterly Journal of Economics* Vol. 106, no 5 pp. 327-68 (1991).

Summers, R. and A. Heston. "A New Set of International Comparisons of Real Product and Price Levels Estimates for 130 Countries, 1950-1985." *Review of Income and Wealth* vol. 34 no. 1 pp. 1-25 (1988).

Tabatabai, H. and M. Fouad. "The Incidence of Poverty in Developing Countries: An ILO Compendium of Data." World Employment Programme Study; Geneva, ILO (1993).

Taylor, L. "The World Bank and the Environment: The World Development Report 1992." World Development, 21 no. 5, pp. 869-81 (1992).

Watkins, K. "The Oxfam Poverty Report." Oxford, Oxfam, U.K (1995).

World Bank. "Africa Region Strategy FY97-99." Washington DC., World Bank (1995a).

World Bank. "Status Report on Poverty in Sub-Saharan Africa 1995: Incidence and Trends of Poverty." Africa Technical Department, Washington DC., World Bank (1995b).

World Bank. "Taking Action for Poverty Reduction in Sub-Saharan Africa: Report of an Africa Region Task Force." Technical Department, Africa Region, Washington DC., World Bank (1995c).

World Bank. "Africa's Development Strategy Revisited: An Overview." Africa Region, World Bank (1994a).
World Bank "Adjustment in Africa: Reforms, Results and the Road Ahead." Oxford, Oxford University Press (1994b).
World Bank. "Status Report on Poverty in Sub-Saharan Africa 1994: The Many Faces of Poverty." Africa Technical Department, World Bank (1994c).
World Bank. "World Development Report." World Bank (1992).
World Bank. "World Development Report." World Bank (1990).

Appendix 1: Poverty Relevant Publications of the World Bank

Year publication	Title of Report	Comments
1982	A Report on the Poverty Task Force	Reviewed the Bank's work during the 1970s.
1983	Focus on Poverty: A Report by a Task Task Force of the World Bank	Called for strengthening the Bank's approach to poverty.
1986	Poverty and Hunger: Issues and Options for Food Security in Developing Countries	Discussed the importance of poverty alleviation in addressing food security problems.
1987	Protecting the Poor During Periods of Adjustment	Discussed the social costs of adjustment and explored measures to protect the poor.
1988	(i) Status Report on the Bank's Support for Poverty Alleviation. (ii) Report of the Task Force on Poverty Alleviation. (iii) Report of the Task Force on Food Security in Africa (iv) The Challenge of Hunger in Africa: A Call to Action	(i) Called for addressing poverty issues in adjustment lending. (ii) Advocated preparation of poverty profiles. (iii) Recommended a more vigorous approach to growth and adjustment. (iv) Drew a comprehensive food security action plan.
1990	World Development Report 1990: Poverty	Proposed a two-pronged strategy for poverty reduction.
1991	(i) Assistance Strategy to Reduce Poverty (ii) Food Aid in Africa: An Agenda for the 1990s (iii) Food Security and Disasters in Africa: A Framework for ACTION	(i) Operationalized the WDR90 appoach (ii) Food aid to be used to combat poverty. (iii) Discussed the impact of natural Disasters on food security. (iv) Raiseds awareness that 20% of world population remain in poverty.
1992	(i) The Poverty Reduction Handbook.	(i) Provided a framework for preparing

		poverty assessments.
	(ii) World Development Report 1992: Development and the Environment	(iii) Emphasized that poverty reduction and environmental conservation are mutually reinforcing objectives.
1993	(i) The Social Dimension of Adjustment Program: A General Assessment	(i) Called for improvements in operational areas relating adjustment and poverty.
	(ii) A Review of the World Bank's Efforts to Assist African Governments in Reducing Poverty	(ii) Noted that increasing the poverty focus of SAPs and sector lending is a most effective means of poverty reduction.
	(iii) World Development Report 1993: Investing in Health	(iii) Stressed that cost-effective provision of health to the poor is an effective way of reducing poverty.
	(iv) Implementing World Bank's Effort to Reduce Poverty	Assessed Bank's efforts through its sectoral lending.
1994	(i) Poverty Reduction and the World Bank: Progress in Fiscal 1993	(i) Emphasized Bank's commitment to reduce poverty.

Source: Adapted from World Bank 1995:90-93.

Appendix 2: SSA Lorenz Curves: Share % of Total Household Consumption

Country %	Lowest 10%	Lowest 20%	Second quintile	Third quintile	Fourth quintile	Highest 20%	Highest 10%
Botswana (85/86)	1.43	3.60	6.88	11.43	19.29	58.89	42.93
Côte d'Ivoire (85)	1.96	5.14	9.53	13.79	21.01	50.53	34.60
Ethiopia (81/82)	3.68	8.56	12.67	16.36	21.10	41.31	27.52
Ghana (87/88)	2.81	6.94	11.64	16.12	22.07	43.23	28.09
Guinea Bissau (91)	0.50	2.06	6.47	12.02	20.59	58.86	42.35
Kenya (92)	1.24	3.39	6.72	10.73	17.32	61.84	47.87
Lesotho (86/87)	1.04	2.87	6.40	11.25	19.49	59.99	43.57
Mauritania (87/88)	0.38	3.53	10.69	16.21	23.25	46.32	30.22
Rwanda (83/85)	4.41	9.70	13.09	16.65	21.64	38.92	24.58
Senegal (91/92)	1.36	3.50	6.97	11.60	19.31	58.62	42.82
Tanzania (91)	0.86	2.44	5.73	10.43	18.70	62.70	46.45
Uganda (89/90)	3.80	8.52	12.09	15.97	21.49	41.93	27.20
Zambia (91)	20.29	5.57	9.58	14.16	20.98	49.71	34.16
Zimbabwe (90)	1.75	3.98	6.29	10.01	17.38	62.34	46.94

Source: Chen. S., Dat, G. and M. Ravallion, Statistical addendum to "Is Poverty Increasing in the Developing World?" World Bank: Policy Research Department, (1994).

Chapter 7

The Elusive Prince of Denmark:
Structural Adjustment and the Crisis of Governance in Africa

Adebayo O. Olukoshi

Introduction

There has emerged a growing concern among scholars and policy makers, including bilateral and multilateral donors, to establish and elaborate a connection between structural adjustment and governance in Africa, Latin America, and Asia. This concern has been driven primarily by the desire for more tangible results from the on-going quest for market-driven economic reform in the adjusting countries of the developing World, especially those of Africa. It has resulted in the devotion of increasing intellectual and material resources to the study of the ways in which the framework for political, legal, and administrative governance can be better made to complement the reform objectives represented by the structural adjustment programs (SAPs) championed by the International Monetary Fund (IMF) and the World Bank.

Although there is not one universally accepted definition of governance, even among the bilateral and multilateral donors who have built it into their policy packages, there is now no doubt that many scholars and Western policy makers see an organic inter-connection between the politico-administrative and legal framework of the adjusting countries of Africa and their prospects for "successful" and "sustainable" market-based economic reform and generalized recovery. This inter-connection has been variously expressed in the literature: for some (Carter Centre 1989), attempting to reform African economies without addressing issues of governance is like seeking to undertake a program of *perestroika* without a simultaneous policy of *glasnost*. For others, like the World Bank (1994b), armed with the benefit of hindsight, it is now claimed to be quite clear that a quest for structural adjustment without a governance component would be tantamount to having "*Hamlet* without the Prince of Denmark."

This chapter is concerned to assess the implications for governance in Africa of the design and implementation of IMF/World Bank structural adjustment as part of a wider quest for establishing the case for closer attention to be paid to democratic practices and the democratic aspiration of the peoples of Africa in the formulation and operationalization of efforts at economic reform. It is a key assumption of the study that politics is central to the design and implementation of any economic reform project, particularly one as ambitious and all-embracing as the structural adjustment programs of the IMF and the World Bank. On the face of things, this assumption will appear to coincide with the increasing concern to factor politics into the implementation of economic reforms associated with the World Bank's recent "political economy" (World Bank, 1994b) and the public choice approach that has grown in popularity since the late 1980s. However, we differ from both the Bank and the public choice theorists insofar as ours is not a concern to "save" the neo-liberal adjustment project by seeking ways, including those that are clearly Machiavellian (Waterbury 1989, for example), of subordinating politics and political actors to the demands of the orthodox structural adjustment model in the belief that it is a model that is essentially coherent, settled and inevitable, the only challenge left being to (re-)orient politics and public administration in its support.

Underlying our own assumption on the centrality of politics to the process of economic development is the view that the structural adjustment framework for economic reform should, itself, be open to problematization and be seen as a legitimate target for contestation and reformulation by the social forces that must bear the costs of its implementation under donor pressure. There is nothing sacrosanct or settled about the neo-liberal structural adjustment model; persistence with it in the framework of popular disaffection may, in fact, compound Africa's crisis of governance. We therefore question the underlying assumption that informs the World Bank's "political economy" and the thinking of the public choice theorists that the problem in Africa today is not so much with an orthodox economic reform model, which is both "rational" and aims to restore "rationality" to African economies, but with the framework for politics and governance which is essentially "irrational" and "dysfunctional" on account of all-pervading "(neo-)patrimonial" structures and processes built into the post-colonial state form.

We take the view that precisely because of problems that inhere in the design, initiation, and implementation of IMF/World Bank structural adjustment, the quest for the Prince of Denmark within its boundaries will remain elusive. What Africa needs is not so much "good" governance defined in narrow technocratic, functionalist terms that are meant to further the goals of an adjustment model that is as controversial as it is contested but a system of *democratic governance* in which political actors have the space to freely and openly debate, negotiate, and design an economic reform package that is integral to the construction of a new social contract on the basis of which Africa might be ushered into the 21st century.

Background Context to the Governance Debate

When structural adjustment first made its entry into the economic policy environment of developing countries, there was very little direct, explicit, or formal concern within the World Bank with political (or "political economy") questions generally and issues of governance particularly. This is so in spite of the fact that the neo-liberal adjustment model carried its own specific political-ideological load and the process of securing its adoption by African governments involved explicit political maneuvering not only in terms of the deployment of a host of conditionality and cross-conditionality clauses, but also through the attempts by officials of the multilateral financial institutions to forge links with state officials who were thought to be most receptive to the aims of the program.

There were, of course, policy intellectuals (Lal 1983 and 1987, for example) who quite early in the implementation of the adjustment program were very clear about the regime types that they thought were required to ride roughshod over domestic interest group resistance to the implementation of the IMF/World Bank reform package. It was, however, only in 1989, almost a decade after the implementation of structural adjustment began in earnest in Africa, that the Bank first used the term "governance" in its discussion of Africa's developmental problems (World Bank 1989). It marked the beginning of an open and explicit entanglement by the Bank with politico-legal and administrative issues even if protestations continued from within the institution that its mandate precluded intervention in the internal political affairs of its members (World Bank 1994a). It is a protestation which always has been and remains essentially formal.

There were two broad and inter-connected developments— one intellectual, the other linked to the end of the Cold War— which provide the context for the open embrace by the World Bank and the rest of the donor community of issues of governance. Regarding the intellectual context, there emerged, in the period from the second half of the 1980s, a growing concern to move beyond the narrow *macro-economic* terms within which the tone for much of the early debate on structural adjustment was set to address broader *macro-political* and *macro-social* issues which, in the view of sympathetic critics of the Bank, operating mostly within the public choice theoretical approach, had been neglected. Growing international concern about the social costs of adjustment implementation, epitomized by the publication in 1987 of UNICEF's plea for "adjustment with a human face" (Cornia *et al.* 1987), combined with the increasing awareness of the lackluster performance of the adjustment policies in tackling Africa's deepening economic crisis to propel interest in the analysis of the political and administrative context for economic reform and how they might be managed. This interest fed into the Bank's own quest for credible explanations for the limited achievements of the adjustment model in stemming Africa's economic decline even after the overwhelming majority of the countries on the continent had embraced the reform package.

Initially, Bank officials, responding to criticisms about the failure of the "magical" forces of the market to stem Africa's economic crisis, argued that it was too early to pronounce on the efficacy of policies that were aimed at undoing some 30 years of accumulated distortions. In any case, the African "predicament," bad as it was, would have been even worse had structural adjustment not been introduced, this in spite of the fact that many countries left it very late before taking "painful" but "necessary" steps towards reform. Later, the explanatory framework shifted to the view that the persistence of the African economic crisis had to do mainly with widespread "slippage" and the lack of "commitment" to the reform package by the local policy and political elite. Most African governments were accused of adopting a "stop-go-stop" approach to adjustment implementation which worked to the detriment of a speedy overall economic recovery.

Attempts were also made by the Bank to demonstrate that those countries which showed more "will," "courage," or "commitment" in the course of reform implementation (the so-called "strong" adjusters) generally recorded a better economic performance than those, the so-called "weak" adjusters, which did not show an appreciable level of "commitment." Side by side with this, exercises were carried out aimed at comparing the situation in African countries before and after adjustment implementation with the conclusion pointing, predictably, to the improvements that occurred after the embrace of market-based reforms. Countries with adjustment programs were also compared with those without adjustment programs with the conclusion suggesting, again predictably, that the former did better on the whole than the latter. However, these efforts at explaining the impact of adjustment implementation neither failed to dent growing criticism of the adjustment package on account of its adverse social costs nor succeeded in concealing evidence of its limited economic achievements. It was against this background that some scholars began to raise doubts about the efficacy of efforts at adjustment implementation without proper cognizance being taken of issues of "political governance." Thus it was that concern with the "political economy" of structural adjustment began to emerge as an industry in its own right.

The questions which were raised within the framework of this "political economy" were many and varied: Why do governments find it difficult to embrace programs of economic reform and why do they leave it so late before introducing reform measures? Can economic adjustment occur without a simultaneous program of political and administrative reforms? What opportunities for coalition politics exist in the promotion of "necessary" economic reforms? Which sections of the state elite can be expected to be reliable allies in the quest for market-led reforms? How might technocrats be "insulated" from undesireable interest group pressures that might compromise the integrity of the adjustment package? What capacities exist locally for initiating or grasping orthodox market reforms? What lessons can be learnt about the timing, phasing, and sequencing of reform policy implementation? Which regime types are best suited for structural adjustment implementation? How might the "winners" from structural adjustment be supported to constitute a local resource for the program and how might "losers" be compensated, out-maneuvered, or side-stepped in order to prevent them from obstructing the implementation of the program? (Lal 1983, 1987; Callaghy 1989, 1990; Nelson 1989, 1990; Waterbury 1989; Grindle and Thomas 1991; Haggard and Kaufman 1989, 1992; Widner 1992; Bates and Krueger 1993; Haggard and Webb 1994).

The question of how "winners" from the reform process could be constituted into a politically viable and durable coalition for market-based economic policies was soon to flower into a full-scale discussion about the necessity for "local ownership" of the adjustment programs and how this might be bolstered. As we shall see later, this concern with the "political economy" of adjustment, enthusiastically embraced by the Bank in what has been described as an internal "paradigmatic shift" (World Bank 1994b), dovetailed into the main issues which were being debated within the public choice theoretical approach. The hallmark of this "political economy" is its spirited attempt to transpose some of the key assumptions and categories of neo-classical economics to the arena of politics. That is why, although some of the public choice theorists would opt for a slightly broader perspective on governance than the Bank, this can only be little more than a difference of emphasis within the club, united as they are by a common set of assumptions about the nature of politics.

The growth of interest in the governance issues associated with structural adjustment implementation was strengthened by the international political changes largely triggered off by Mikhail Gorbachev's twin program of *perestroika* and *glasnost* and the eventual termination of the East-West Cold War as we once knew it (Beckman 1992; Gibbon 1992; Mkandawire 1994). The wave of "democratization" that swept through the former Soviet bloc, including the defunct Soviet Union itself, towards the end of the 1980s and in the early 1990s and the evident resurgence of consciousness, on a global scale, with regard to questions of "democracy," "human rights," and "popular participation," however defined, fed into and reinforced the intellectual debates that were developing about economic reform and governance in Africa.

So too did the earlier experience of the countries of Latin America with transitions from military to civilian rule through elections serve to underline the urgent necessity to address the possible linkages between economic reform and the framework for political and administrative governance in Africa, especially as countries such as Argentina, Brazil, and Chile where political reforms had been undertaken began to enjoy some modest economic growth. Within Africa itself, the end of the Cold War exposed many regimes for the first time to the full force of domestic political pressures for reform from which they were previously shielded by one or the other of the Cold War rivals in their quest for the maximization of strategic/geo-political advantages. As several authoritarian African regimes and life presidencies buckled under sustained popular domestic pressure for political reforms and many others were compelled to abandon, formally at least, their political monopoly and concede multiparty elections, interest in the question of governance became even more central.

Although, on the face of things, it would seem that much of the current interest in the subject of economic reform and governance emanates from the international donor agencies active in Africa, especially the World Bank, it is important to emphasize at this stage that long before the donor community turned its attention to this question, numerous African groups and social forces had been involved in struggles for the expansion of the political space on the continent as well as for the creation of structures of governance that would permit the will of the majority of the people to prevail. This is evident from the entire history of the anti-colonial struggle which was as much about political reforms as about economic and social change with a view to enhancing individual liberties and popular participation.

The early post-colonial period, characterized by the abandonment, in many cases, of the African anti-colonial nationalist project, also witnessed spirited struggles against the imposition of one party and military rule, the institution of personal rule with all of the clientelist networks woven around it, the proliferation of corruption and bureaucratic red-tape, and the spirited efforts at various levels aimed at the de-politicization of the people. These struggles continued with varying levels of intensity until the mid-1980s onwards when, with the rapidly changing international environment to which we drew attention earlier, the domestic social forces for change in the way Africa had been governed politically and economically were emboldened and re-asserted themselves to unleash popular pressures for political reform towards the end of the 1980s (Anyang' Nyongo 1987; Mamdani *et al.* 1987; Mamdani and Wamba-dia-Wamba 1995).

So strong were the domestic pressures for political and economic reforms in Africa towards the end of the 1980s, and so massive was the level of popular participation in the struggle for change that some commentators (Legum 1992, for example) were to remark, rather hastily, that Africa was on the threshold of a "second liberation." Whereas the first liberation resulted in the historic defeat of the forces of colonialism, the "second liberation" was leading to the defeat of personal, autocratic rule within the framework of a system of political monopoly either by a single party or by the military. It was expected that this "second liberation" would result in the emergence of an era of "democratic" governments that promote "rational" economic policies. Yet many of those who popularized this view, fascinated as they were by the images of tens of thousands of African men, women and children (hitherto considered as politically docile), actively demanding political reform, hardly bothered to examine the democratic content of the demands and the sustainability of democratic change in the context of deepening economic crises, prolonged structural adjustment and the resurgence of competing ethnicities (Olukoshi 1995).

Be that as it may, there is no doubt that the confluence of domestic and external pressures for political and economic reform in Africa reinforced the impetus in the scholarly and donor communities to develop a keen interest in the linkage between adjustment and governance on the continent. As we shall see, the World Bank was to immerse itself fully into the governance issue, attempting as it did, to bring its peculiar bureaucratic/technocratic interpretation to bear on the operationalization of the term. A category of donors, namely Western governments and their official aid agencies, even went on to embrace a new "political conditionality" under which economic aid was tied to the progress of African governments in introducing political reform and respect for human rights (Mkandawire 1994; Olukoshi and Wohlgemuth 1995; Stokke 1995).

Yet, the notions of political and economic reform which the donors have generally attempted to promote in Africa run counter to those held by the main bearers within the continent of the struggle for democratization and popular participation. Whereas, to cite one example, the donors see structural adjustment and what some World Bank officials describe as "good governance" as being compatible, many of the social forces in the vanguard of the struggle for democratization not only reject the structural adjustment programs but also insist that they are incompatible with popular participation and responsive, democratic governance (Beckman 1988, 1990, 1992; Bangura 1992, see, also, Box 7). At the root of these conflicting positions between the democratic forces pushing for change in Africa and the donors championing reforms are sharply differing perceptions of the African problem. Let us now examine the theoretical sources of these differences as they relate to the governance debate.

Structural Adjustment and Governance: The Theoretical Context and Contestation

Structural adjustment programs were introduced into Africa on a massive scale from the early 1980s onward at a time when most African economies were already caught in deep crises of accumulation. These crises manifested themselves not only in terms of rapidly declining output and productivity in the industrial and agricultural sectors but also in terms of worsening payments and budget deficits, acute shortages of inputs and soaring inflation, growing domestic debt and a major problem of external debt management, decaying infrastructure, a massive flight of capital and declining *per capita* real income, among others. The reform programs which were introduced under donor pressure and implemented under the supervision of the IMF and the World Bank were, ostensibly, aimed at stabilizing the African economies, re-structuring the basis for accumulation, and permitting the resumption of growth (Tarp 1993; Gibbon and Olukoshi 1996). What the medium- to long-term effects of the adjustment programs would be not only on the economy but also on society, the practice of politics and the processes of administration became the subject of a major theoretical debate involving two broad schools. The differing positions articulated by both schools, namely, the neo-liberal "political economy" /public choice and the radical political economy schools, is, in many respects, a function of their understanding of the sources of the African economic crises and the role of the post-colonial state in the developmental process.

The Neo-liberal "Political Economy" /Public Choice School

Championed largely by Africanists based in North American universities and soon embraced by the World Bank as it developed its "political economy" of African development, the neo-liberal "political economy" / public choice school rests on the assumption, whether explicitly stated or not, that "democracy" and economic liberalization are two sides of the same coin. It takes as its starting point, the view that the post-colonial African state, by its very nature and, therefore, by definition, is at the heart of the economic and governance crises pervading the continent. This state, stripped of the most basic checks and balances of the (late) colonial period, has failed signally in its developmental mission on account of various inter-related factors: its "excessive" and "counterproductive" intervention in domestic economic processes to the detriment of market forces and the private sector; its over-bureaucratization and bloated size; the domination of its apparatuses by clientelist networks and an "urban coalition" that orients it against the rural (productive) sector and "rational" macro-economic policies; its submission to "rampant/macro populism" as it panders to a vociferous "urban coalition" ; its monopolization of the main economic levers in society with the resultant proliferation of rent-generating/-seeking niches/activities; and its over-centralization of development which has discouraged local (private) initiative.

Underpinning the failure of the post-colonial African state, making it an almost inevitable outcome, is its essentially "(neo-)patrimonial" nature and the "rent-generating/-seeking" motivation of African policy makers. These have been central to the adoption by the state of policies that "distort" markets through protectionist tariff and non-tariff barriers, misguided import substitution industrial development programs, overvalued exchange rates, artificial price-fixing/price controls, a host of subsidies, and the preference for state monopolies. In extending its reach as part of its goal of achieving short-term political order, the state has encouraged the proliferation of patronage institutions and networks which help to consolidate the position of a legitimacy-hungry elite by enabling it, in part, to "buy" the support and/or acquiescence and silence of other social forces while it dips its snout deeply and uninterruptedly into the public trough. Given the domination of the economy by the "(neo-)patrimonial" state or by a "(neo-)patrimonialist" state logic, it is not surprising that the failure of the state easily translates into the failure of the economy. This is made especially so as the fragility of the patronage networks and structures that underpin the state means that they must constantly be re-constituted, producing, in the process, acute regime and/or policy instabilty and resource misallocation on a stupendous scale to the detriment of long-term national development (Bates 1981; Sandbrook 1985, 1986, 1991; Joseph 1989; World Bank 1989; Barkan

1992; Hyden and Bratton 1992; Landell-Mills 1992; Widner 1992).

So central to the discourse of the neo-liberal "political economy" /public choice theorists was their thesis of the "neo-patrimonial" / "rent-seeking" sources of the failure of the post-colonial state that much of their intellectual output was devoted to producing a host of adjectival appellations that were thought to best capture its nature and *modus operandi*. Thus, in the course of the 1980s and early 1990s, depending on the taste and preferences of particular authors within the school, the post-colonial state was variously characterized as "prebendal," "parasitic," "personalistic," "clientelist," "kleptocratic," "unsteady," "over-extended," "predatory," "crony," "soft," "weak," "lame," "rentier," "sultanist," and, finally, "neo-patrimonial." The appellations served essentially to underline the perceived negative role of the state in the economy and society. In so doing, they reinforced the World Bank's own spirited efforts at stereotyping the state as both inherently ineffective and illegitimate.

The adverse perception of the state and its role in the works of the neo-liberal "political economists" /public choice theorists often contrasted sharply with the approving terms in which they discussed the private/voluntary sectors. This perception also fed into the state-retrenching logic of the IMF/ world Bank structural adjustment model, a retrenchment process which, in practice, not only involved the curbing of state expenditure, the retrenchment of many public sector employees, and the privatization/commercialization/ outright liquidation of public enterprises but also spirited efforts at reforming the civil service (ostensibly to make it "leaner" and "fitter"), curbing the interventionist ambience of the state, and limiting it to the provision of an "enabling" environment for private sector-led development.

In the view of the neo-liberal "political economy" /public choice theorists, structural adjustment which is aimed, *inter alia*, at encouraging the emergence of economic "rationality" through the unfettered rule of the impersonal forces of the market and the promotion of the growth of the private sector should be beneficial for the emergence and sustenance of democracy and a better system of governance than the "neo-patrimonial" structures that pervade the continent and which underlie public policy. Democratization, signaled by the dispersion of power away from the one-party state and state monopolies, should, in turn, help to break the hold of "minority" urban-based interests on economic policy to the benefit of the rural poor and the army of informals (Herbst 1990).

Elaborating this position, Larry Diamond argues that part of the reason for the failure of democracy to sink roots in Africa has had to do with the fact that the bourgeois class which could have championed it was either non-existent (because of state domination of the economy) or too heavily dependent on the state and immersed in compradorial, rent-seeking activities. Structural adjustment, in altering the basis for economic activity, will encourage the formation of a bourgeoisie more grounded in production and, therefore, much more autonomous of the state. It is this bourgeoisie that will, out of self-interest, seek to promote democratization and a more open system of government not based on clientelism, neo-patrimonialism, or prebendalism. For Diamond (1988:27), therefore" ...the increasing movement away from statist economic policies is among the most significant boosts to the democratic prospects in Africa."

In addition to the expectation that structural adjustment will "sanitize" African economies and lead to the creation of a non-parasitic bourgeoisie that will then bear the flag of democracy and transparent governance, the "new" "political economy" /public choice theorists have also argued that the process of the retrenchment of the state should result in the strengthening of civil society and associational life which, in turn, should enhance Africa's democratic prospects. The adjustment program, by emphasizing the role of the private sector and encouraging the channeling of resources to private, non-state groups will not only help to "thicken" civil society, that bastion of liberty and democracy, but will also generate interest at the level of society in how the state is governed. Not a few studies have been published within this framework celebrating the rise and activities of voluntary associations and non-governmental groups in civil society (Azarya and Chazan 1987; Bratton 1989, 1990; Chazan 1988; Rothchild and Chazan 1988).

Attempts have been made by several scholars working within the neo-liberal "political economy" /public choice approach to show that the vacation by the state of certain economic and social spheres as part of the process of structural adjustment has helped to stimulate the rise of voluntary and non-governmental groups and thus promoted the pluralism essential for democracy and transparent governance. Closely related to the flowering of associational life is the growth in informalization as various individuals and groups, unable, for a variety of reasons, to gain or maintain access to the state and the resources it controls, "disengage" from it and move into the parallel economy and, in so doing, extending the process of informalization and the "liberterian" and "democratic" impetuses which it (inherently) carries.

Viewed historically, it is difficult not to be astonished by the one-sided anti-statism that underlies much of the neo-liberal "political economy" /public choice approach. The international environment at the time when most African countries attained independence was decidedly in favor of state interventionism in the development process (Killick 1989; Chaudry 1993; Mkandawire 1995; Olukoshi 1996; Havenevik and van Arkadie 1996). This, essentially, was as true for the centrally-planned economies of the East bloc as for their capitalist rivals in the West. It was also true for the developing as for the developed countries. As it pertained to the developing countries, including those of Africa, a variety of theories, ranging from the "big push" approach to the "Gerschenkron thesis," was developed and popularized in support of an interventionist role for the state in the struggle against underdevelopment. From the late 1970s, however, as the neo-liberal ideology gained in ascendancy, the interventionist role of the state in the development process came under severe attack. From being the cornerstone of development, the state now came to be seen as the millstone holding back a system of market-led development.

Yet, the impression created by the neo-liberal "political economy" / public choice school that the role of the post-colonial African state in the development process was wholly "dysfunctional" and that only unremitting stagnation characterized African economies under the regime of state intervention flies in the face of the evidence available on the growth levels which were achieved by African countries in the 1960s and during part of the 1970s (Mkandawire 1995; Gibbon and Olukoshi 1996; Havnevik and van Arkadie 1996). In fact, record growth rates, in some cases as high as 9 per cent, were recorded by many African countries during the period to the early 1970s, growth rates which far overshadow the rare 3-4 percent average growth rates which have been celebrated as "quite good" during the structural adjustment years of the 1980s and 1990s.

In selectively painting a picture of failure with which to adversely stereotype the post-colonial state, the neo-liberal "political economists" attempt to make the case for their own alternative, largely idealized vision of the role which African states should play. They push for a minimalist state whose role is to produce an "enabling environment" for the functioning of an essentially self-regulating market economy based on free competition—and the "thickening" of civil society. This is a vision of the economy and of the role of the state in it which corresponds to no known actual experience in the recent history of the world (UNRISD 1995). But it is one which many of the neo-liberal "political economists" felt able to advocate as they sought to transpose the assumptions and categories of neo-classical economics into political science.

Precisely because of the attempt to transpose the categories of neo-classical economics to the political arena and to use these to designate social institutions and actors, the neo-liberal "political economy" /public choice school produces a set of rigid dichotomizations opposing the state to the market, the rural to the urban, the formal to the informal, agriculture to industry, and civil society to the state. Yet, as Bangura and Gibbon point out, this approach to seeking to grasp the African reality overlooks the fact that most of the relations designated by these categories systematically interpenetrate and overlap one another (Bangura and Gibbon 1992). For, a key characteristic of African economies and societies is the prevalence of "grey" areas which blur and, sometimes, blend the dichotomizations that are central to the arguments of the neo-liberal "political economy" /public choice approach. This suggests that, contrary to the assumptions of the neo-liberal "political economy" /public choice school, a correct reading of the politics of reform in Africa can not be obtained from deductions deriving from the kinds of rigid dichotomizations that are integral to their analyses. The reductionism that pervades this approach perhaps explains why the World Bank and its sympathetic critics have repeatedly and systematically misjudged and misread the effects of structural adjustment in Africa (Gibbon *et al.* 1992).

Furthermore, for many of the neo-liberal "political economy" /public choice theorists, civil society and informalization are treated too uncritically as the arena of democracy and democracy itself is defined in largely functionalist, managerial terms. Additionally, they embrace the structural adjustment model uncritically, accepting at face value the objectives which its authors attribute to it and ignoring the vigorous contestation that has been going on concerning its basic assumptions and the consequences of its implementation. Also problematic is their elevation of "neo-patrimonialism" and "rent-seeking" to the status of an explanatory *deus ex machina* which does not allow for the validity of other motivations for the actions of social actors, including the state. Moreover, their identification of power and exploitation exclusively with the state has been criticized for failing to acknowledge that power relations and exploitation can also and are, indeed, found in civil society.

In other words, like the state, civil society also embodies contradictory tendencies and processes which its uncritical equation with democracy conceals. Thus, civil society is not, exclusively, a domain of "liberty" and "democracy" and the tendency to oppose it to the state in a one-sided manner hardly helps to deepen our insights into the ways in which the two inter-penetrate. Finally, there is nothing self-evident that the creation/existence of a private capitalist class "autonomous" of the state will necessarily be supportive of democracy in Africa. Numerous empirically-based studies carried out on the politics of the private capitalist class and its organizations in various African countries suggest, in fact, that they could, and do have strong anti-democratic proclivities (Mamdani and Wamba-dia-Wamba 1995; Mkandawire 1996).

The Radical Political Economy School

Against the neo-liberal "political economy" /public choice school, most radical political economists contend that structural adjustment and democratic governance are not necessarily or even essentially compatible. This position is arrived at and argued at different levels by different authors working within the broad radical political economy school (Gibbon et al. 1992). At one level, it is premised on the essentially repressive thrust of the adjustment package itself in an analytic frame in which authoritarianism is seen and treated as a property that inheres in the neo-liberal reform model. At another level, the view is argued that structural adjustment, given its unpopularity in many African countries and its failure to deliver quick and tangible benefits, intensifies the authoritarianism that has always been a property of the state in colonial and post-colonial Africa.

Furthermore, there are those who establish the structural adjustment-authoritarianism linkage by focusing on the political strategies of the IMF and the World Bank in the early years of adjustment implementation in Africa, strategies which, according to them, exposed an unmistakable initial/persistent preference by the international financial institutions for authoritarian regimes which were thought to be more capable of taming resistance to the market-based reform measures. In some cases, these authoritarian regimes used the additional resources that adjustment implementation provided to resist or thwart domestic pressures for democratization. Of course, there are many instances in which the different levels of analyses overlap and the synthesis which follows does not, for the purposes of this chpater, separate them.

According to many of the radical political economists, the neo-liberal doctrine, the goals which it seeks to achieve, and instruments by which they are to be achieved all carry a repressive load that is directed against a range of local social forces opposed to it, including especially the middle class of professionals and the working/unemployed poor. These forces are often defined in the neo-liberal model as the selfish and self-serving "urban parasites" on whose behalf an inefficient economic order that is biased against the rural poor and the denizens of the informal sector was erected in much of Africa. This way, the repression of opposition, mostly urban-based, to the adjustment project is justified ideologically.

Yet, opposition to structural adjustment is inevitable because all over Africa, its implementation has entailed the imposition of additional economic burdens on the working people and the poor through the devaluation of currencies, the freezing of wages and salaries, massive public sector retrenchments, the imposition of so-called cost recovery measures in the educational and health sectors, the elimination of subsidies (real and invented), the curtailment of the welfare and social expenditure of the state and the high inflationary consequences of price deregulation and rapid devaluation, among others. These are measures which are unpopular in their own right; they are made even more so by the fact that they come as an external imposition implemented by regimes whose legitimacy is, more often than not, in tatters.

The implementation of adjustment measures has often been accompanied by attempts at undermining the organizational capacity and autonomy of the opponents of the adjustment package. This is all the more so as opposition to the adjustment process gathers a nationalist momentum and the state is unable to show tangible results from the "pains" of adjustment either in the short- or long-term. Since a majority of the peoples of the adjusting countries are "losers" from the adjustment process, no significant constituency is able to emerge to make the case for the reform package against its opponents. It is left to state officials, faced with donor conditionality, to attempt to force through the program and, in doing so, silence the critics administratively.

Resistance to structural adjustment is, however, not limited just to the working poor; it includes significant sections of the middle class, especially the professionals, and the many members of the manufacturing class whose interests are adversely affected by the program. Attempts at silencing the opposition therefore translate into widespread repression and high-handedness executed by officials representing a contested state. In the view of the radical political economists therefore, authoritarianism rather than democracy is the flipside of structural adjustment implementation in Africa. Being economically repressive, the program requires an equally politically repressive framework for its implementation. Regimes implementing orthodox structural adjustment not only resort to undemocratic methods of pushing it through, they also make spirited efforts to prevent autonomous organization for alternatives to the neo-liberal project. The authoritarian import of the adjustment model and the repressiveness associated with its implementation are reinforced by the undemocratic logic and practice of donor conditionality (Bangura 1986, 1989a, 1989b; Hutchful 1987; Mustapha 1988, 1992; Campbell 1989; Ibrahim 1989, 1990; Beckman 1990, 1992; Bangura and Beckman 1991; Gibbon *et al.* 1992; Olukoshi 1991, 1992; Mamdani 1991; Mkandawire, 1991, 1996; Nyang' oro and Shaw 1992; Mkandawire and Olukoshi 1995).

According to some of the radical political economists, it is the process of organization of resistance to the authoritarianism and repression associated with structural adjustment implementation that begins to open up (new) democratic possibilities based on the self-organization of groups opposed to the program and in spite of state repressiveness. If, therefore, the period of implementation of structural adjustment has witnessed the growth of democratic pressures in many African countries as evidenced by the public demonstrations for political change in all the four corners of the continent, it is not because of structural adjustment *qua* structural adjustment but in spite of it. In their bid to protect themselves against the repressive economic content and political repercussions of structural adjustment, various groups adversely affected by the program, and whose organized resistance at the trade union, student union, and professional association level the state attempted to prevent, had no option other than to bear the flag of democratization and the freedom of association.

The emergence of open resistance to authoritarian rule during the adjustment years was aided by the wave of "democratization" that swept through the former Soviet bloc and the extremely limited achievement of the market-based economic reform policies themselves. Yet, those countries of Africa where transitions have recently been made to elected forms of government, compelled as they have been under the prevailing international regime to stick to orthodox structural adjustment, are finding that their fragile "democracies" are in peril precisely because of their persistence with an adjustment process that continues to be marked by poor economic results and huge social costs. In many of these countries, the forces in the forefront of the democracy campaign were also quite vociferous in their opposition to IMF/World Bank structural adjustment. These "choiceless democracies," as Mkandawire (1996) describes them, are imperiled by the fact of their being trapped in an unpopular neo-liberal net that appears to be leading nowhere as far as economic growth and development are concerned even as they carry big social costs.

On balance, the insights that emanate from the perspectives of the radical political economy school appear to be far closer to reality in much of Africa than the positions conveyed by the neo-liberal "political economy" /public choice theorists. However, in criticism of the radical political economists, it has been argued that their approach more accurately captures the authoritarian outcomes of adjustment implementation in countries with strong civil societies but is less helpful in understanding the political dynamics in countries with weak or non-existent civil societies. In the latter case, exemplified by Tanzania and Mozambique, a process of "democratization from above," which is tied to attempts at legitimating a new development strategy favoring private accumulation with state resources and discarding all developmentalist and welfarist pretences, is one possible outcome which many of the radical political economists would seem to rule out *ab initio* (Gibbon 1992).

It has also been argued, in the light of the contemporary experience in Africa whereby several elected governments are sticking to othodox adjustment policies, that a more interesting issue to focus on is not whether adjustment and democracy can occur simultaneously and/or co-exist but "what the dual pursuit of adjustment and democratization implies for the consolidation of each other, and more specifically for the immediate future of the democratization process" (Mkandawire 1996: 28). This, of course, does not rule out the possibility, as indeed several scholars have concluded, that continued implementation of orthodox adjustment might adversely affect the consolidation of efforts at political reform in African countries where multi-party elections have been held (Olukoshi 1995; Mkandawire 1996).

Donor Parameters and Prescriptions for "Good" Governance in Africa

Diagnosing the African Crisis of Governance

We have already noted that much of the perspective which the donor community, especially the World Bank, brought to bear on its governance work was heavily informed by the output of the public choice school. In many respects, the World Bank, given the resources at its disposal and its wide reach, came to set the pace for other donors, including many of the bilaterals, on the governance question. That being the case, our discussion here of the parameters and prescriptions for the Africa governance programs of the donors will draw primarily, though not entirely exclusively, on the perspectives and policies which the World Bank developed and has tried to apply in relation to the countries of the continent. This is done without prejudice to the slight differences in definition and emphasis which exist among the donors. For the Bank itself, governance is defined as "the manner in which power is exercised in the management of a country's economic and social resources for development" (World Bank 1992: 52).

Once it formally joined the governance bandwagon in 1989, the World Bank was to quickly immerse itself into the task of operationalizing the term for its own objectives. For this purpose, the 1989 report of the Bank entitled *Sub-Saharan Africa: From Crisis to Sustainable Growth* served as a launching pad. Its diagnosis of the governance – development matrix in Africa more or less replicated the main positions of the public choice approach. According to the Bank, at the heart of the litany of Africa's development problems is a fundamental crisis of governance. This has manifested itself not only in terms of a relentless decline in the quality of government, growing bureaucratic obstruction, and weakened judicial systems, but also in increasing political and administrative arbitrariness, the collapse of the rule of law, and an all-pervasive culture of corruption and rent-seeking. In most African countries, "a deep political malaise" stymies all developmental action. The situation is complicated by the increasing lack of local "capacity" both in the public and private sectors (World Bank 1989, 1990, 1992, 1994a, 1994b; Nunberg and Nellis 1990; Landell-Mills 1992; Serageldin and Landell-Mills 1991; Wai 1991, 1994; Dia 1993, 1996; Adamolekun and Bryant 1994).

According to the Bank, soon after independence in Africa, the civil service in most countries not only became overbloated but was also politicized and came tobe staffed by inexperienced people brought in through patron-clientelist networks. In time, as the agencies and organizations of government proliferated, the civil service wage bill ballooned, claiming an increasingly disproportionate share of governmental revenues and the national GDP until many countries came to find it difficult even to pay wages and salaries regularly, if at all. The rapid expansion of the public sector wage bill not only created serious fiscal problems but also had the effect of "crowding out" other critical items of current expenditure, especially the maintenance of physical infrastructure and the procurement of essential supplies and equipment. Thus it was that the widespread situation was created, among others, where doctors and nurses lacked basic medicines and equipment with which to treat patients, teachers lacked chalk and books for teaching, and postal workers had no stamps and money orders to sell. The decline in motivation that resulted from this situation only reinforced the diminishing productivity of the civil service, a development not helped by the uncompetitive remuneration of those in the management cadre (World Bank 1989, 1992; Nunberg 1989; Dia 1993; Adamolekun and Bryant 1994; Lindauer and Nunberg 1996).

Given the host of problems that bedeviled it, it is not surprising that the post-colonial African bureaucracy found it hard to cope with the stresses of "rapid modernization," a task made more daunting by the embrace by Africa's independence political leaders of "inappropriate" and "ill-adapted" developmental models that were built on Northern values, institutions, and technology. These developmental models, like the state itself, were poorly rooted in African societies, including their history and culture. The state took on the dominant role in the economy partly because of its distrust for private, especially foreign, business and partly to enable it gain access to resources with which to pamper an "urban coalition" of interests from which its officials were largely recruited. Whatever existed by way of private enterprise was rendered uncompetitive by the heavy costs imposed on it by the degradation of state-maintained physical infrastructure, the breakdown of other basic public functions, red tape, and corruption. The entire political environment was not one which inspired confidence in private investors (World Bank

1989, 1990, 1992, 1994a, 1994b; Dia 1993, 1996; Lindauer and Nunberg 1996).

Elaborating on the Bank's perspective on the crisis of governance in Africa and the ways in which it has stifled development generally, its then President, Barber Conable (1991:3), noted that:

> All too often, there is a lack of government accountability to the governed, a lack of encouragement that would liberate entrepreneurial instincts, and a general lack of fair competition between farmers and firm.

He added that in much of Africa,

> Open political participation has been restricted and even condemned, and those brave enough to speak their minds have too frequently taken grave personal risks. I fear that many of Africa's leaders have been more concerned about retaining power than about the long-term interests of their people…

The political and other costs of Africa's record of poor governance are, according to the World Bank, legion. Conable, in taking stock of some of those costs, stated that

> The political uncertainty and arbitrariness evident in so many parts of sub-Saharan Africa are major constraints on the region's development. Investors will not take risks, entrepreneurs will not be creative, people will not participate if they feel they are facing a capricious, unjust or hostile political environment.

Elaborating further, Conable argued that

> Patronage and negotiation have thwarted the formation of professional cadres. Investment in human resource development has lacked direction and commitment. Such practices are a direct cause of Africa's economic growth rate failing in the 1980s to keep pace with population growth, of the debilitating brain drain from the region, and of the extra-ordinary fact that there are more expatriate advisers in Africa today than there were at the end of the colonial period.

The Principles and Parameters for a Solution to Africa's Governance Crisis

Based on its diagnosis of the sources and nature of Africa's crisis of governance, the World Bank proceeded to outline general prescriptions for a solution. The central focus of the prescriptions is a quest for the restructuring of the post-colonial African state in order to make it more supportive of the Bank's long-term strategy for liberating the forces of the market and promoting private enterprise (Beckman 1992). The Bank had no doubt that there is an organic linkage between the institution of a system of "good" governance and the prospects for the successful implementation of structural adjustment. By encouraging the rule of the impersonal forces of the market and instituting economic "rationality" into the process of resource allocation, a system of open and accountable government would be encouraged. The nurturing of open and transparent governance will, in turn, make it difficult to justify "irrational" economic decisions. For this purpose, the trimming down of the state and its re-orientation away from being an entrepreneur to being a promoter of the private sector remained a central objective. The achievement of this objective was expected to have a "liberating" effect on civil society. It would also result in the "empowerment" of the people. Furthermore, Africa should follow, rather than resist, "the world-wide trend towards privatization" (World Bank 1989:55).

In specifying the elements that were thought to be essential for the attainment of "good" governance in Africa, Bank staff identified the following key principles and parameters: greater accountability (financial and political) by public officials, including politicians and civil servants; transparency in governmental procedures and processes; a concerted attack on corruption; predictability in governmental behavior and in the political system; rationality in governmental decisions; competent auditing of governmental transactions; the drastic curbing of bureaucratic redtape; elimination of "unnecessary" administrative controls in order to plug avenues for rent-seeking; the promotion of the free flow of information; the encouragement of a culture of public debate; the institution of a system of checks and balances within the governmental structure; the decentralization of government; respect for human rights; judicial autonomy and the rule of law; the establishment of a reliable legal framework; and the protection of property and enforcement of contracts. Bank officials also add the issue of capacity-building to enable African technocrats to initiate and implement market-based economic reforms as an essential element of the quest for "good" governance in Africa.

Other donors, particularly the main Western governments and their aid agencies, explicitly advocated and attempted to implement a "new" political conditionality linking aid and other official resource flows to Africa to respect for human rights and the implementation of "democratic" reforms by African governments. As stated by Douglas Hurd (British High Commission, Lagos 1990:1), British Foreign Secretary until 1995, "the relief of poverty, hunger, and disease is one of the main tasks of overseas aid. Aid must go where it can clearly do good." He stated further that

> Countries tending towards pluralism, public accountability, respect for the rule of law, human rights, and market principles

should be encouraged. Those who persist with repressive policies, with corrupt management, or with wasteful and discredited economic systems should not expect us to support them with scarce aid resources which could be better used elsewhere.

In essence, for the leading donor countries, democracy, defined in terms of multi-partyism, elections and public accountability, had become the flip side of the neo-liberal market reform project in Africa. Where the Bank couched its political intervention in the affairs of African countries in governance terms that enabled it to claim not to have preferences for particular regime types, the bilateral donors felt no such inhibition and through political conditionality attempted to specify political forms for African countries.

Operationalizing the Principles and Parameters for "Good" Governance

In attempting to give operational content to its vision of a system of governance in Africa that is supportive of its reform project on the continent, the World Bank embarked on the articulation and implementation of a host of policy measures either on its own or in collaboration with other (mostly smaller, bilateral donors) and multilateral agencies of the United Nations. At one level, this involved the intensification of efforts at securing the implementation of existing Bank and Fund policies that were thought to be relevant to the governance agenda. In this regard, public enterprise privatization and the reform of the civil/public service were pursued with even more vigor. With regard to the latter, the Bank's focus was primarily on cost containment, staff retrenchment, elimination of "ghost workers" from the payroll, equipment provisioning, revision of the civil service code, public financial management reform, decentralization of government, and efforts at "professionalization" (World Bank 1994a, 1994b; Adamolekun and Bryant 1994; Dia 1993, 1996; Lindauer and Nunberg 1996).

As of the end of 1994, the Bank was supporting various civil service reform programs in 29 African countries, namely, Angola, Benin, Burkina Faso, Cameroon, Central African Republic, Cape Verde, Comoros, Congo, Gabon, The Gambia, Ghana, Guinea, Guinea Bissau, Madagascar, Malawi, Mauritania, Mozambique, Niger, Nigeria, Rwanda, Sao Tomé and Principe, Senegal, Sierra Leone, Sudan, Tanzania, Togo, Uganda, Zaire, and Zambia. The quest for public enterprise and civil service reform formed the kernel of the Bank's public sector management program in Africa during the 1980s and 1990s. At the heart of the program was a commitment to cost containment which came to be seen as central to the prospects for the (eventual) creation of an efficient and effective public service as stabilization is supposed to be to the restoration of economic growth through a package of adjustment measures (Lindauer and Nunberg 1996).

Beyond the attempts at pushing the privatization of public enterprises and the reform of the civil service, the Bank also invested in several countries in projects in the area of accounting and auditing as part of its stated goal of improving financial accountability in the public sector, including support, as in the case of Zambia, to the Public Accounts Committee of the Parliament. Public expenditure reviews became a regular feature of discussions between Bank officials and African governments, with the former aiming to directly influence the expenditure outlays of the latter within and among sectors. Open competitive tendering for contracts and the organization of competition in service delivery were also undertaken in a number of countries. In this regard, the Bank' s country procurement assessment reviews were used to try to influence the practices of various governments as were the financial accountability assessments undertaken in Ghana and South Africa.

Furthermore, attempts were made to set up public works agencies outside the governmental structure and to introduce what is described as "beneficiary participation" in the design of projects. Private sector groups were surveyed with a view to identifying aspects of the civil service and/or its operations that might be reformed to better serve the goal of creating an environment conducive to investors. One outcome of this was the introduction by many countries of "one-stop" investment advisory, vetting, and approval centres that were expected to cut the bureaucratic redtape—and corruption associated with the procurement of governmental approval for proposed investment projects (World Bank 1994a, 1994b; Adamolekun and Bryant 1994; Dia 1996).

In the area of enhancing the rule of law, the Bank concentrated most of its attention on ways by which legal institutions in Africa could be strengthened and "outdated" laws reformed. One of the underlying assumptions informing the Bank' s work in this regard is the view that an "appropriate" legal system is necessary for stability and predictability. As part of its strategy, the Bank pushed for specific reforms in Angola, Cape Verde, Cote d' Ivoire, Ghana, Guinea, Mali, and Uganda aimed at re-orienting the legal regime more explicitly and "efficiently" in support of property rights and contracts as they affect private sector loans and credits.

Projects focusing on legal training (targeted mainly at legal draftsmen) and the renovation/development of court infrastructure were also earmarked for execution in Burkina Faso, Mozambique, Tanzania, and

Zambia. Furthermore, the Bank's Africa region, through its Women in Development (WID) unit, sponsored several studies and workshops on legal constraints affecting the "economic empowerment" of women. Within this framework, it established links with national legal associations and law-related research groups in various countries as strategic entry points for the Bank's legal reform work, both generally and in relation to women (World Bank 1994b).

In addition to measures such as the promotion of open, competitive tendering for the supply of goods and services to the government by the private sector, including NGOs, the Bank sought to improve transparency in the governmental process by encouraging several African countries to publish official gazettes advertising public tenders and announcing their award. Burkina Faso and Mauritania are just two of the several countries where this strategy was pushed through. Governments were also encouraged, as in Kenya, to publish a summary version of their annual budget plans for circulation locally. This practice was expected to stimulate public debate on the economy and the public expenditure pattern adopted by the incumbent regime. In a bid to encourage "informed" reporting and public discussion of economic reform issues, the Bank's Economic Development Institute (EDI) developed training programs for journalists invited from time to time from various African countries. Publications from the Bank were also routinely targeted at the media in all of the adjusting countries of Africa and briefing sessions regularly scheduled to "explain" the objectives of the economic reform process, the successes recorded and the problems that persist (World Bank 1994a).

The Bank's work in the area of institution-building was extended, in principle and in practice, to support for non-governmental/ "grassroots" organizations. These organizations are seen as viable replacements for the state in several spheres; they are also central to the Bank's strategy for "empowering" the people and "thickening" civil society as a counterweight to the "(neo-)patrimonial" state. Not only was the establishment of NGOs "grassroots" organizations explicitly encouraged in various African countries, attempts were also made to make them beneficiaries of procurement contracts awarded by the state and project funds supplied by donors, including the Bank. Indeed, in several countries, the Bank inspired the establishment of NGOs for the execution of public works projects. The experiences of the Hometown Associations (HTAs) of West Africa were also studied with a view to replicating them in other parts of the continent as part of the quest for the expansion of NGO involvement in community development (World Bank

1989, 1992, 1994a; Landell-Mills 1992).

Also central to the governance program of the Bank in Africa is "capacity building" in the public and private sectors. The Bank's work in this area was premised on the assumption that a key element in the African crisis is the absence or collapse of "effective" policy-making and managerial capacity that is both up to date and relevant to the changing international economic environment in general and the promotion of market-based reform in particular. Indeed, for the Bank, capacity-building came to be regarded as yet another "missing link" in Africa's quest for development. That being so, the donor community, led by the Bank, invested resources in programs and projects aimed at building capacity on the continent. In 1991, several donors came together to establish an Africa Capacity Building Foundation (ACBF) in Harare, Zimbabwe.

Various forms of "technical" and "institutional" assistance were offered to different African governments to upgrade skills, improve procedures, strengthen organization, and encourage the more effective utilization of existing skills and assets, especially in the management of market-based reforms. NGOs and private sector agencies were important beneficiaries of the capacity building efforts of the donors, with the ACBF playing an important role in this regard. The goal of building capacity was complemented by another project, championed by the United Nations Development Program (UNDP), ostensibly to retain talent in Africa and stem/reverse the brain drain from the continent (World Bank 1991; Dia 1996). An associated objective of capacity building, partly also serving as a justification for it, was the promotion of local "ownership" of the donor adjustment model.

Furthermore, as part and parcel of the governance-adjustment linkage of the Bank, school curricula, especially at the tertiary level, were brought under scrutiny and recommendations made for course/departmental rationalization. As with the reforms of the civil service, the claim was that the educational sector reforms proposed by the Bank will make African tertiary education more "relevant" and result-oriented, not to speak of cost effective. The staff-student ratio in many countries was brought under scrutiny and proposals for cost recovery on many of the services provided by schools were presented. The administrative structure of higher educational institutions also came under close scrutiny and reform proposals outlined. Loans were made available to governments for the re-equipping of laboratories and libraries and the rehabilitation of physical infrastructure within the overall framework of the rationalization plans that the Bank advocated and the implementation targets which governments accepted.

Also, there was the claim that the reduction of the number of universities/polytechnics will help to channel resources to develop human resources to an intermediae technical level lacking in much of Africa. Such rationalization might also enable resources to be freed for support to the primary education sector. The re-organization of university curricula should also enable governments to de-emphasize the liberal arts and strengthen the technical and engineering services. A closer linkage between the educational sector and the private sector was also advocated by the Bank as part of the reform agenda. This way, not only would universities be able to attract some private sector funding, the training of human resources would also be more closely linked to the demands of the market generally and the labor market in particular.

Apart from the specific ideological-political load which the donor brand of governance carries, it is worth remarking that during the 1980s and early 1990s, there was a clear preference expressed for the insertion of technocrats into the political and administrative structures and processes of African countries, and their elevation to a high profile as part of the "good" governance – efficient adjustment implementation linkage. Where it was assumed that Africa's "old guard" patrimonial rulers were less inclined to faithfully implement structural adjustment because of its potential for undermining the clientelist networks on which their personal rule rested, the technocrats were thought to be driven purely by considerations of competence and professionalism required for effective reform implementation.

Thus, in various African countries, most notably Benin, Cote d' Ivoire, Nigeria, Senegal, and Togo, technocrats either took over some of the highest political offices or were placed in strategic ministries at the heart of government. In less evident or celebrated cases, technocrats, including nationals of some of the African countries concerned, were seconded to the national bureaucracy, especially the economic ministries, whilst keeping their international pay/perks. Questions of how these technocrats might be "insulated" from untoward advances and pressures from local interests formed a key element in the discourse of the public choice theorists. Such then, was the thrust of the donor governance program. It is a program that has been subjected to extensive criticism as much over its content as over its ideological-political fabric and context and its coupling to the neo-liberal economic adjustment project. Let us now proceed to elaborate on some of the criticisms.

Structural Adjustment and Governance in Africa: The Elusive Prince of Denmark

The point should be stressed at this stage that at the heart of the concern in the donor community generally, and the World Bank in particular, about questions of governance in Africa is the desire to promote the emergence of a conducive, and in their view, more legitimate political context, backed by the requisite administrative capacity, for the successful implementation of orthodox structural adjustment. In this thinking, the very adjustment model which is being pushed is hardly problematized. The possibility that the implementation of that model may, in fact, feed into and exacerbate Africa's governance problems is, therefore, ignored. All of the Bank's investment in political theory is, thus, primarily designed to find political supporting blocks for orthodox adjustment to be more "effectively" implemented.

Yet, policy and its implementation are always the objects of contestation among different forces in every political system. It is never clear, however, why in the case of Africa, the Bank and its governance theorists consider this not to be applicable, and where it is acknowledged, to treat it as *solely*, or even *primarily*, the result of the "selfish" and "illegitimate" machinations of "vested interests" that are steeped in a variety of "(neo-) patrimonial" relations. In this fundamental sense, the donor governance program amounts to little more than an attempt to "save" an adjustment model that has been a source of intense controversy and contestation. In doing this, the Bank has adopted an approach which, at one level, pretends that opposition to adjustment is either absent or is not fundamental. At another level, it has sought to selectively co-opt the language and some aspects of the platform of the forces which have articulated and led opposition to structural adjustment and the authoritarian state in Africa. This way, a decisive ideological-political attempt is made to neutralize the opposition.

Although the language of the Bank's governance discourse ("civil society," "accountability," "empowerment," "rule of law," "popular participation," etc.) is one which, on the face of things, fits into the renewed global interest in issues of democracy, the governance program is, in fact, reduced to a managerial/technocratic affair tailored to the goals of an adjustment program that, in the view of many, has, at the very least, contributed to the reproduction/intensification of authoritarianism in Africa. The question of *democratic governance* in Africa is, therefore, one which is still unresolved. This is without prejudice to the fact that there has been some association in the Bank literature, and in the writings of some its intellectual supporters, between governance and "democracy" (World Bank 1994b).

The question of democratic governance also goes beyond the inauguration of an elected government dedicated to the pursuit of unpopular economic reforms that are the product of an external imposition and which exact huge costs without showing tangible results. In the end, the challenge of democratic governance is the reality that some of the main bearers of the struggle for democracy in Africa are also in the frontline of the resistance to structural adjustment implementation. Many of them have sharpened their strategies for the democratization of their societies from their experience of resisting the authoritarian practices pre-dating, but reinforced by the implementation of structural adjustment. The concerns and interests of the various forces in society, not least those of the opposition to structural adjustment, will have to be taken much more seriously in a political governance framework which does not foreclose discussion on the economic reform model that will be adopted.

Taking stock of the governance agenda of the Bank, Beckman, in one of the most powerful published critiques of its 1989 report, persuasively argued that the institution's intervention in that arena seeks to boost state capacity for orthodox adjustment implementation not by addressing the objections of those opposed to the program but by seeking to undercut their political and ideological legitimacy. This is done partly by feigning a consensus that does not exist and partly by the promise of a better, rosier tomorrow that may never come but which helps to shift focus away from the current practice and consequences of structural adjustment. But much more fundamentally, the Bank attempts to blunt and discredit the extensive and powerful reservoir of nationalist opposition to structural adjustment as a foreign imposition by dismissing the forces of nationalism as being, historically, the peddlers in post-independence Africa of "inappropriate" ideas of modernization borrowed uncritically from abroad.

In developing this line of attack, not only does the Bank resort to a systematic distortion of the post-colonial developmental experience (for example, the claim that most governments drew up "comprehensive five-year plans" and invested in "large, state-run *core* industries" which is contrary to the empirical evidence), it also attempts to distance itself from developmental objectives and strategies (for example, support for state development finance companies set up to act as "trustees" for the emerging private sector) with which it was ideologically and financially involved. Furthermore, the Bank's resort to labeling the opponents of its adjustment project as "selfish," "narrow" and "urban-based" sidesteps the reality that the allegedly "narrow" interests which organized interest groups in much of Africa convey appeal to wider popular aspirations that strike a chord with various groups in urban as well as rural areas (Beckman 1992).

For Beckman, as for other critics, the governance program of the

Bank is also aimed at establishing an alternative basis of popular legitimacy for structural adjustment. That is why notions like "grassroots empowerment," "mobilization," "civil society," "equity," and "participation" have been central to its governance discourse. But at the hands of the Bank, the political-democratic side of these concepts are downplayed and their technocratic-managerial ones played up. Thus, for example, in the Bank's "political economy," "empowerment" refers, in the main, to freedom for local private entrepreneurs and not to the institutionalization of popular participation in collective decision-making. Seen from this angle, the donor concern is, therefore, less with democracy and more with "development." "Empowerment," for example, is seen as something to be encouraged insofar as it is considered to be good for "development." The "rule of law" is discussed and operationalized in terms which focus on the creation of a legal framework within which private, mainly foreign, business confidence can grow. But such notions of empowerment and the rule of law do not address the concerns of the opponents of the adjustment process who consider themselves as losers in one sense or another. Thus, in the end, the Bank's governance agenda is part of a broader project of managerial populism which does not signal the dawn of a new era of "adjustment with a democratic face" which is able to dovetail into what Mkandawire (1996) has described as "democracy without tears." For, stripped of its rhetorics, the governance program of the Bank fails to address the question of why the politics of adjustment has been so repressive in Africa (Beckman 1992).

When the governance positions espoused by the Bank are taken on their face value and squared up with the practice of structural adjustment initiation and implementation, it emerges clearly that there are strong grounds for questioning their compatibility. At one level, attention has been drawn to the fact that the relationship between the donors and African governments, both generally and in the process of adjustment initiation and implementation particularly, is hardly a democratic one. As Mkandawire (1996:35) notes, donor resource flows to Africa have mainly been disbursed "within essentially authoritarian structures."
This institutional reality has been reinforced during the adjustment years through the tightening and intensification of donor conditionality and cross-conditionality, a process aided by the World Bank's heavy investment in the strategy of donor consultation and co-ordination. It has also been reinforced by the *TINA* ("there is no alternative") ideology with which African governments have been confronted at every turn. The logic of conditionality, donor co-ordination and *TINA* compel adjusting governments to embrace a reform project which they may not

necessarily believe in and strive to implement it in the face of domestic opposition, popular disaffection, and limited results. It also sidesteps the domestic policy process, further erodes national sovereignty over basic economic policy decision-making, and undermines local policy-making capacity (Beckman 1992; Stokke 1995; Mkandawire 1996; Engberg-Pedersen *et al.* 1996; Engberg-Pedersen 1996; Gibbon and Olukoshi 1996; Olukoshi 1996).

At another level, the logic and demands of donor conditionality and cross-conditionality have meant that, for all intents and purposes, African governments have increasingly had to devote a good proportion of their time and resources to accounting to donors. Officials are constantly working on reports for the Bank, the Fund, a host of bilateral donors, and the Paris and London clubs. They are also required to spend time with a variety of visiting evaluation/monitoring missions. They too undertake missions of their own to the Bretton Woods institutions, the Paris and London Clubs, and other donors to defend their record and make the case for financing. The net effect of all this is that in many countries, governmental effectiveness has been further impaired even as key economic ministries are brought directly under donor influence. As Gibbon (Gibbon and Olukoshi 1996) notes, "the increasing 'donorization' of many branches of government activity...(added) to ministerial administrative burdens."

The undermining of governmental effectiveness in Africa also reinforced the erosion of the state's political capacity, defined as the "the ability to construct and maintain a working political coalition capable of sustaining the implementation of state policy" (Beckman, 1992: 95). In the midst of all this, the Bretton Woods twins have taken on the status of "offshore governments" which, for all intents and purposes, control the content and direction of economic and social policy in Africa, exact accountability from African officials to their head offices in Washington, D.C. but strive to avoid taking on a commensurate amount of responsibility for the consequences of the policies whose implementation they enforce (Olukoshi 1996).

Furthermore, the practice of adjustment implementation in most African countries has hardly conformed with the most elementary norms of "transparency" and "accountability" by rulers to the ruled. From the time discussions about structural adjustment initiation are started, they are shrouded in mystery which officials (both on the donor side and local state side) insist is necessary in view of the "delicate" nature of the negotiations. The local media is suffused with rumors based on information that leaks from the secret negotiations. Newspaper cartoonists attempt to capture the atmosphere of secrecy with images of IMF and World Bank officials wearing dark spectacles arriving in African capitals in the dead of the night and taking the first flight out when dawn breaks. Once the adjustment deal is sealed, another round of secrecy and evasion surrounds its content and the timing and sequencing of its implementation.

As the adjustment policies begin to be sprung on the people, leaders embark on "news management" mostly aimed at limiting the flow of information on the reform program. Exercises in "data massaging" are undertaken in order to paint a rosier picture of economic performance than is really the case. The language of economic discourse becomes more and more obscure such that the ordinary people tend to be intimidated into silence. As we noted earlier, the officials charged with implementing the unpopular reform program come to see themselves as being more accountable to the donors who exercise direct leverage over them and not to the populace who bear the brunt of the market reform policies. The entire process is one which does not allow much room for consultation with the people or internal and open policy debates within the governmental system. Indeed, only a tiny elite in the central bank, the office of the president/prime minister and the ministry of finance have a full picture of the entire adjustment package. Others have, at best, only a selective picture. Inter-ministerial co-ordination therefore becomes difficult as does planning on a national scale. Morale among those in the policy apparatus who are excluded becomes low.

Yet, it is not as if, for all of the costs which the market reform project exacts, the structural adjustment years have been marked by a taming of corruption. In fact, the problem of corruption has been deepened in many cases, fuelled by the decline in real income and living standards which has encouraged the intensification of petty graft among some categories of workers as a survival strategy. Among ruling class elements, new forms of "market-driven" corruption have emerged/intensified linked to the entire environment/process of liberalization generally, and, more specifically, to the privatization/commercialization/liquidation of public enterprises, the auctioning of foreign exchange, the floatation of enterprises on local stock exchanges, and the deregulation of interest rates. In Nigeria, the probe of the banking and financial sector which began in 1994 has produced startling evidence of "market-driven" corruption in the period since 1986 when the country's adjustment program started. Yet, the financial sector deregulation policies of the Nigerian state were once hailed as a hallmark of the "successes" of adjustment in Nigeria. In Kenya, the spectacular fortune illicitly amassed by Nicholas Biwot, once a pillar of the ruling Kenyan African National Union and close confidant of President Daniel Arap Moi, and his cronies was built on a systematic manipulation of the structures and processes of adjustment-related deregulation. Land-related scandals deriving from attempts at privatizing what is by far Africa's most important resource have been reported in all corners of the continent. The examples are endless and they suggest that the market is not exactly the anti-corruption antidote that it has been presented to be. Both the public and private sectors continued to be wracked by corruption in much of Africa.

The crisis and adjustment years in Africa have had far-reaching, adverse consequences for the administrative structure and capacity of many countries on the continent which the governance program of the donors has hardly addressed in a manner that can offer an enduring basis for meaningful renewal. Many civil service organizations have been severely weakened not only by staff retrenchments and the effects of the exercises on the morale of those retained but also by the collapse of the income (and purchasing power) of those employed in the public sector in the face of currency devaluation and massive inflationary pressures. The sharp decline in the real value of civil service pay, even where it was already underway during the pre-adjustment period, was reinforced by the devaluation measures that became the defining feature of the quest for adjustment and the inflationary pressures associated with them (Robinson 1990). In many countries, wage and salary freezes were also imposed; several countries even carried out cuts in the nominal pay of their civil servants. From being largely competitive with the remuneration available to employees in the local private sector, civil service pay levels in most African countries were, during the adjustment years, to fall way behind what was on offer in the local private sector; they became even less competitive internationally. One consequence of this development is that experienced and qualified personnel have, increasingly, been drained out of the public sector partly into the local private sector but overwhelmingly into the international labor market.

It is important to underline the fact of the collapse of public sector wages and salaries because this is one issue that has been central to the collapse of morale and effectiveness in the public sector. In a sense, the wholesale, one-sided anti-statist ideology on the platform of which the neo-liberal reform project for Africa was inaugurated was one which cast the civil service in bad light and meant that the overall emphasis of the World Bank's reform effort was on cutting it down to size. The centrality of the goal of restoring internal fiscal balances to the donor stabilization-adjustment program also meant that there was little room for any serious consideration of the ways in which the pay and purchasing power of civil servants could be enhanced. Thus, the anti-statism of the adjustment package reinforced, and was reinforced by, the demand management, deflationary thrust of the economic reform program. The focus of attention was, therefore, on cost containment/reduction which, almost always, translated into the retrenchment of workers and the freezing of pay.

Of course, suggestions were made that the savings from the reduction of civil service sizes might be used to enhance the pay of those who remain in public sector employment. But this was more an ideological proposition for legitimating retrenchment rather than a serious strategy for pay review with a view to making the civil service an attractive place within which to make a career. Not surprisingly, the proposition never gathered steam any where in Africa. Little wonder then, that the goal of making the civil service more professional, efficient and effective has been undermined by the very adjustment package on the basis of which reform is being pursued.

Faced with diminishing real incomes (and purchasing power) and ever-rising costs of living, many public sector employees have resorted to moonlighting activities in their bid to supplement their wages and salaries. These activities are almost always undertaken using office hours; offices have also been used as informal market outlets for selling a variety of consumer items at a discount. In addition, state resources (personnel, vehicles etc) are sometimes mobilized in support of the multiple livelihood strategies that have become prevalent. The consequences on the effectiveness of the civil service have been telling. It is not surprising that the civil service all over Africa has found it increasingly difficult to attract and retain high caliber local personnel with the requisite experience and expertise. The resultant "capacity gap" has largely been tackled through a resort to the employment of consultants, mostly from abroad.

The use of highly paid "independent" expatriate consultants who are remunerated in foreign currency and at internationally competitive rates to execute specific adjustment-related tasks has not only meant that the issue of the level of pay of local staff is effectively side-stepped but has also bred resentment which has deepened demoralization and, occasionally, inspired acts of sabotage against the consultants and the adjustment process itself. In the context of all of the foregoing, it is not difficult to understand why there are presently more expatriate "experts" in crisis-ridden Africa than there were at the dawn of independence in the 1960s (Beckman 1992; Mustapha 1992; Bangura 1994a; Mkandawire 1996; Olukoshi 1996).

The explicit preference which was to be shown for technocrats, the basic reasons that informed that preference and proposals on how the technocrats might be "insulated" from local influences represent an extension of the concern within the Bank and among some in the public choice school to seek ways of circumventing politics and getting orthodox adjustment policies implemented. In this sense, the preference for technocrats and for their insulation directly contradicts the donor concern with "transparency" and "accountability." It is remarkable that this preference gained in momentum even as the civil services of most African countries were suffering serious problems of de-professionalization and the collapse of morale. Many of the high profile technocrats who were elevated into senior governmental positions saw themselves as owing little responsibility to be accountable locally. This tended to be particularly so where politicians and the public were aware that particular technocrats had the full backing of the international financial institutions, the creditors, the bilaterals, and the "market." In such situations, technocratic accountability is swung decisively in favor of donors and local scrutiny is dismissed, a trend which "insulation," the quest to make technocrats "autonomous" of domestic social forces, tends to reinforce and accentuate.

The adjustment years have also been marked by a further erosion of the legitimacy of the post-colonial African state, with implications for its political capacity to implement policies. The efforts at retrenching the state not only helped to curb its social reach but also further undermined the post-colonial social contract on the basis of which it sought to construct ideological legitimization, build political alliances, relate with the opposition, and secure the co-operation/support of autonomous centres of power (Beckman 1992; Olukoshi and Laakso 1996). Furthermore, the adjustment years have been associated with the collapse of a pattern of expectations, concretized in specific group and community demands, focused on what the role of the state is understood as being. This collapse of expectations is reinforced by the widespread awareness that structural adjustment has come to Africa as an external imposition. In the search for alternatives, individuals and groups are driven into ethno-political and religious organizational frameworks that pose direct challenges to the post-colonial secular, national-territorial nation-state project. That is why some commentators have argued that the crisis of governance in Africa is also, in essence, a crisis of structural adjustment (Beckman 1992; Olukoshi and Laakso 1996).

In order for democracy and democratic governance to prevail and become consolidated, there must be social groups within various countries that have an interest in them as specific political projects. Yet, ironically, those very social groups that have, historically, been the main bearers of the struggle for the democratization of the African state, economy, governmental structure, and society are not only among those most adversely affected by structural adjustment but are also rejected in the adjustment model, and by the most influential donors, as "parasites" and "vested interests" to be undermined politically as part of the bid to dislodge the influence of the "urban coalition" which allegedly makes "rational" economic policy-making and implementation impossible. These social forces—workers, students, professionals, academics, etc and their organizations have been specifically targeted for disorganization by the ideologues and executioners of the neo-liberal adjustment project. Trade unions, students' organizations, professional associations and other organized groups opposed to structural adjustment have generally been smashed all over Africa or their organizational capacity severely weakened through constant official harassment, proscription, arrests and imprisonment, and staged-managed divisions (Beckman 1992; Bangura and Beckman 1991; Mustapha 1988; Mkandawire and Olukoshi 1996; Olukoshi 1996).

The notions of democracy and democratization which the students, workers, professionals and other groups with an active interest in the reform of African politics carry run counter to those which the donors have tried to push. Quite apart from the instrumentalist approach which the donors adopt to the democracy question in Africa, their overwhelmingly managerial and technicist operationalization of the issue of governance does not strike a chord with the explicitly political perspectives developed by the forces in the vanguard of the campaign for the democratization of Africa. Many of the activists emphasize, in their articulation of the African democratic project, political as well as economic and social elements. Democracy for them is not just a question of multi-party politics and electioneering even if the right of the people to freely elect their leaders is recognized as non-negotiable; it includes a vast array of social and economic reforms whose adoption are widely perceived as being necessary for the establishment of a more just social order. It is a definition of democracy which necessarily calls for an interventionist, "developmentalist" state, not for the unbridled retrenchment of the state. It calls for the thorough reforming of the state and its broad-ranging restructuring in order to tackle the problem of state failure but it also firmly rejects the World Bank/IMF program for the re-definition of the role of the state. For these groups, there is a fundamental incompatibility between structural adjustment and democratization and this is brought out by the experiences at the various national conferences in parts of Francophone Africa.

From Benin Republic to the Congo, Niger and Mali to Chad, and even the then Zaire, the main issues which dominated the agenda of the national conferences not only relate to administrative and political reforms, the limiting of the powers of the executive, the subordination of the military to the authority of elected politicians, the strengthening of the judiciary and its independence, and the re-organization of the military under civilian governmental authority but also to far-reaching economic reforms based on a "developmentalist" state. Stinging criticisms against IMF/World Bank structural adjustment programs have been commonplace even as many of the conferences acknowledge the necessity for far-reaching economic reforms in order to stem the tide of African economic decline. There was a recognition too of the need to strengthen state capacity even as the democratization of state structures and procedures are undertaken. Quite clearly, popular perception within Africa on the reform of African economies and the democratization of the state and society run counter to the views held and pushed by the donors, especially the World Bank.

Turning to the question of capacity-building which has featured in the discussion on governance and structural adjustment, and in support of which the World Bank launched an African Capacity Building Foundation, it has been argued and demonstrated by critics that donor-driven market reforms have contributed substantially to undermining indigenous capacity in Africa. While few will doubt that the availability in abundance of professionally competent economists, policy analysts, managers, auditors, jurists and other professionals is essential to the promotion of "good" governance, the experience of the last decade under the regime of structural adjustment and its associated authoritarianism has, as we noted earlier, been an exacerbation of the brain drain from Africa. Hordes of highly qualified personnel, trained at great expense to their countries, unable to cope with or accept the social, economic, and political costs of adjustment, have sought greener pastures in Europe, North America, and the Middle East. For some, frustration, arising out of the lack of basic equipment with which to perform their tasks and an anxiety to keep abreast of changes in their professional fields have been the factors motivating the decision to leave their countries. Those professionals who, for whatever reason, have stayed behind have either had to engage in moonlighting in order to earn extra money to supplement their diminishing real income or have opted for political appointments which are often unrelated to their professional training.

Without diminishing the importance of human resource development to the transformation of Africa, the politics of the World Bank's capacity building initiative for the continent ought to be exposed for what it is, namely that it is an initiative aimed at generating a ready intellectual and professional constituency for the Bank's adjustment programs. In this regard, it would be correct to argue that the initiative continues the World Bank's recent practice of defining the competence of the African professional in terms of the professional's willingness to imbibe and follow the institution's current approach to macro-economic and other policy changes in Africa. Many an African professional who have refused to see the world through the Bank's lenses have been dismissed as lacking in skill or denigrated as "rent seekers."

Given that a majority of African intellectuals are either hostile to or skeptical about the Bank's neo-liberal economic reform project in the countries of the continent, it is little wonder that the institution has strongly sought to push an exaggerated version of the view that Africa lacks a competent professional class. Clearly then, the Bank's capacity-building program is an attempt to produce professionals who will support structural adjustment and extend its logic to their spheres of competence. This way, it should be possible to claim that adjustment measures are home-grown or that the adjusting governments are really the owners of the programs. The issue of capacity building for strengthening democracy is not one that concerns the Bank. Indeed, in emphasizing the need for Africa's technocratic/professional class to be insulated from society, the Bank seeks to diminish accountability to the people by their governors. An insulated technocracy, unresponsive to democratic pressures and accounting only to donors is bound to a recipe for continued authoritarianism.

The disorientation and paralysis which many an African civil service has suffered has also been the fate of the educational sector where, in the face of Bank-inspired reforms, academics, students, and university administrative staff have felt compelled to embark on struggles to protect their rights and interests. These struggles have covered a broad range of issues, including pay, the content of the curriculum, the courses that are taught, the attempts at importing and imposing expatriate "experts," receiving internationally competitive salaries paid in foreign exchange, on the universities even as their local colleagues continue to suffer collapses in their real income, the imposition of cost recovery measures, and the continuing decay of facilities and infrastructure, including libraries. The authoritarian responses of the state to these struggles and the unwillingness of the Bank, for a variety of reasons, to yield ground has meant that many African institutions of higher learning have been crippled by a cycle of protest strikes with costly consequences for the economies of various countries. It is ironic that in an era when the Bank claimed to have integrated governance into its repertoire of policies, academic freedom and access to education have never seemed more threatened in the African post-colonial experience.

Beyond Structural Adjustment and Towards Democratic Governance in Africa

Fifteen years after structural adjustment made its grand entry into the African economic crisis management environment, debates have continued regarding its efficacy both as a macro-economic model and as a framework for tackling the continent's crisis of governance. As part of these debates, attempts have been made to argue the case for tempering the social costs of adjustment and prescribe alternatives to the neo-liberal project. Perhaps the best known of these is UNICEF's plea, made in 1987, for adjustment with a human face and the Economic Commission for Africa (ECA)'s blueprint entitled *African Alternative Framework to Structural Adjustment Program for Socio-Economic Recovery and Transformation (AFF-SAP)* published in 1991. But both the UNICEF and ECA critique of the adjustment experience in Africa have been overwhelmingly concerned with the macro-economic and macro-social aspects of structural adjustment and although they inevitably enter into a discussion of the market-state dynamic, governance questions remain underdeveloped in their discourses.

Thus, although it is clear that the donor approach to governance which we have focused on in this study leaves a lot to be desired, the task of fashioning out an alternative remains to be undertaken. We offer here broad principles that could inform such an endeavor. In doing so, our concern is for a system of *democratic governance* in Africa within which the search for economic recovery can be undertaken. In our view, there can be no trade off between democratic governance and economic recovery/growth in Africa; both are necessary and desirable and should be seen as going together beyond the managerial populism that serves as the framework for the donor endeavor.

The starting point in the quest for an alternative framework for governance which, by definition and in practice, is democratic is the need to recognize from the outset that the project must not be subordinated to the goals and exigencies of structural adjustment or, for that matter, any other economic model discussion of which is foreclosed because it is seen as something inherently right or inevitable or both. The tragedy of the donor approach to governance is that, *ab initio*, it was constructed with a view to facilitating/accommodating an essentially unreconstructed economic reform model that was itself already the object of much contestation.

Within the framework of the donor approach, the question that was posed was: How can the governance program be employed to facilitate the implementation of "painful" but "necessary" economic reforms? As

Mkandawire (1996) has noted, this way of posing the governance question is premised on a certain perverse approach to politics which has dominated the literature in recent times. Under this approach, "politics is reduced to servicing a technocratically defined "welfare function" instead of technocrats devising the instruments necessary to meet a *democratically specified* "social welfare function" (Mkandawire 1996:40). This, inevitably, limits the scope for democratic governance by foreclosing debates on a possible economic reform model. It undermines the political capacity of the state to strike the compromises and consensus necessary for policy implementation generally. It also prevents discussions that are necessary within the local policy bureaucracy. For Mkandawire, as for us and many others, the central question to be posed is: how do you carry out economic reform projects without undermining democratic governance and its consolidation? (Mkandawire 1996; Olukoshi 1996; Beckman 1992; Amadeo and Banuri 1991).

In view of the extremely limited results that have flowed from structural adjustment implementation and the huge costs which the program has exacted, it is clear that the quest for an alternative framework for democratic governance will entail revisiting the question of the kind of policy package that is required for reforming the economies of the continent and restoring them to the path of growth. In doing this, several points of principle will need to be taken cognizance of. At one level, it will be necessary to recognize that in Africa, there can be no question of a trade-off between the state and the market; the state does have a decisive role to play in the developmental process and its reinstatement, without the encumbrances which previously undermined its efficacy, must be a central concern of post-adjustment Africa (Mkandawire 1995; Olukoshi 1996).

At another level, the creation of a political and institutional framework through which democratic demands can be made on the developmental state is a task which must also be seen as central to this process. What this suggests is that policies generally and economic reform policies in particular ought to be underpinned by a clear social consensus if they are ultimately to play a part in reinforcing the legitimacy of the state and consolidating democratic governance. The establishment of a social consensus is the surest path to a sound foundation for domestic "ownership" of economic reform policy. Furthermore, the tendency to reify the market "into a neutral, apolitical and ahistorical institution" or to fetishize it with human attributes such as "anger," "disappointment," "displeasure," "nervousness" (Mkandawire 1996:36) which has been a by-product of global neo-liberalism will have to be discarded in favor of strategies for making the market accountable to the state and the state to the people (UNRISD 1995; Olukoshi 1996).

The question of the centrality of domestic social consensus to the quest for economic reform and the prospects for democratic governance has been brought to the fore by the experiences of those African countries which, during the late 1980s and early 1990s, made a transition from military or single party rule to multi-party forms of politics. Without exception, the elected governments of those countries, several of which consist of a new coalition of forces, have had to stick to the donor economic adjustment project whether or not there is domestic popular support for the program and without opening up an internal debate on the issue. On account of a number of factors, including unrefined donor conditionality, Africa's elected governments have been confronted with a choicelessness as far as economic policy is concerned, a bizarre state of affairs since democracy is, partly at least, supposed to be about choice. In sticking to the neo-liberal reform project, several of the elected governments have presided over the undermining/dissolution of the coalition of mostly anti-adjustment forces that propelled them to power in the first place. They have also found it difficult to submit the adjustment packages to full and open parliamentary scrutiny, preferring instead to muscle parliament or resort to executive fiat.

As their popular support base has shrunk, many of Africa's elected governments, those very same regimes whose inauguration was supposed to signal the dawn of a new day in Africa, have increasingly become more authoritarian—and corrupt to a point where many have found it difficult to distinguish between them and the *ancien regime* which they replaced (Beckman 1992; Olukoshi 1995; Mkandawire 1996). This recourse to authoritarianism, and the flowering of corrupt practices associated with it, has repeatedly threatened the parliamentary coalition of the government of Bakili Muluzi in Malawi, severely discredited the Chiluba regime in Zambia, permitted an opportunistic military *coup d' etat* in Niger, paved the way for the defeat of the Soglo regime in Benin and the return of Mathieu Kerekou to power, resulted in an increasing public exhibition of intolerance in Ghana by an increasingly intemperate Jerry Rawlings. It is because of this that several scholars have argued that the manner the orthodox adjustment packages which the elected governments of Africa have been compelled to stick to already packaged and sealed "pose serious problems to the consolidation of democracy in Africa" because their "rigid prerequisites, inflexible built-in positions and the proliferation of cross-conditionalities...force decision-makers into a take-it-or-leave-it corner, ruling out dialogue or creative political compromises within society at large" (Mkandawire 1996:34).

Africa's post-adjustment strategy for economic reform and democratic governance, must, if it is to overcome the failings and shortcomings of the pre- and post-political reform, orthodox adjustment years, aim at the adoption of economic policy measures which are the product of a domestic social contract and which strengthen, rather than weaken and/or undermine fragile democratic institutions and processes that are in need of consolidation. The necessity for the policy measures to be sensitive to the social/welfare aspirations of the populace cannot be overemphasized. Few doubt that drastic measures need to be undertaken to reform African economies and polities. However, such drastic reform efforts stand a better chance of being achieved where they are built into a negotiated social contract in whose making various interest groups have a stake. The contract will also provide a framework within which state and governmental legitimacy can be reconstructed for the formulation and implementation of policies. Groups are not inherently averse to making sacrifices, including taking cuts in their consumption levels, where they are satisfied that these are temporary measures carried out within the boundaries of a negotiated social bargain in which they are stake holders.

The construction of a social contract at this time will require a concerted attack on policy making "traditions" and structures which are part and parcel of authoritarianism on the continent. At one level, this will involve efforts at dismantling the authoritarian logic that permeates the existing patterns of donor-recipient relations. At another level, it will entail the recovery by domestic policy and political forces of the initiative for reform design and implementation as sanctioned through democratic institutions and processes. Furthermore, it will entail closer attention to the impact of economic reform instruments and policies and a recognition of the limits of state's political capacity as dictated by the balance of social forces. It is precisely because of failure to come to terms with these factors or the attempt to "manage" them out of existence that has ensured that the quest for democratic governance in Africa has proved elusive. But it need not be so.

Works Cited

Adamolekun, L. and C. Bryant, *Governance Progress Report: The African Regional Experience*. Washington, D.C.: World Bank, (1994).

Ajayi, S. I., "State of the Macro-Economic Effectiveness of Structural Adjustment Programs in Sub-Saharan Africa," in R. van der Hoeven and F. van der Kraaij (eds), *Structural Adjustment and Beyond in Sub-Saharan Africa*. London: James Currey, (1994).

Amadeo, E. and T. Banuri, "Policy Governance and Management of Conflict," in T. Banuri, ed., *Economic Liberalization: No Panacea*. Oxford: Clarendon Press, (1991).

Anyang' Nyongo P. (ed.), *Popular Struggles for Democracy in Africa*. London: Zed

Books, (1987).
Azarya, V. and N. Chazan, "Disengagement from the State in Africa: Reflections on the Experience of Ghana and Guinea," *Comparative Politics in Society and History.* 20, 1 (1987).
Bangura, Y., "Structural Adjustment and the Political Question," *Review of African Political Economy.* 37 (1986).
Bangura, Y., "Crisis, Adjustment and Politics in Nigeria," *AKUT* 38, Uppsala, (1989a).
Bangura, Y., "Crisis and Adjustment: The Experience of Nigerian Workers," in B. Onimode (ed.) *The IMF, the World Bank and the African Debt, Vol. 2, Social and Economic Impact,* London: Zed Books/IFAA, (1989b).
Bangura, Y. and B. Beckman, "African Workers and Structural adjustment, With a Nigerian Case-Study." In: D. Ghai (ed.) (1991).
Bangura, Y. and P. Gibbon, "Adjustment, Authoritarianism and Democracy in Sub-Saharan Africa: An Introduction to Some Conceptual and Empirical Issues." In: P. Gibbon, Y. Bangura and A. Ofstad (eds.) (1992).
Bangura, Y., "Economic Restructuring, Coping Strategies and Social Change: Implications for Institutional Development in Africa." *Development and Change,* 25, 4 (1994a).
Bangura, Y., *Intellectuals, Economic Reform and Social Change: Constraints and Opportunities in the Formation of a Nigerian Technocracy.* Dakar, CODESRIA Monograph Series, (1994b).
Barkan, J., "The Rise and Fall of a Governance Realm in Kenya." In: G. Hyden and M. Bratton (eds.) (1992).
Bates, R., *Markets and States in Tropical Africa.* Berkeley: University of California Press, (1981).
Bates, R. and A. Krueger (eds.), *Political and Economic Interactions in Economic Policy Reform: Evidence from Eight Countries.* Oxford: Basil Blackwell, (1993).
Beckman, B., "The Post-Colonial State: Crisis and Reconstruction." *IDS Bulletin,* 19, 4 (1988a).
Beckman, B., "Comments on Goran Hyden's State and Nation Under Stress," in Swedish Foreign Ministry, *Recovery in Africa: A Challenge for Development Co-ooperation.* Stockholm: Foreign Ministry, (1988b).
Beckman, B., "Whose Democracy? Bourgeois Versus Popular Democracy in Africa," *Review of African Political Economy.* 45/46, (1989).
Beckman, B., "Structural Adjustment and Democracy: Interest Group Resistance to Structural Adjustment and the Development of the Democracy Movement in Africa." Stockholm, Mimeo, (1990).
Beckman, B., "Empowerment or Repression?: The World Bank and the Politics of Adjustment." In: P. Gibbon, Y. Bangura and A, Ofstad (eds.) (1992).
Bratton, M., "Beyond the State: Civil Society and Associational Life in Africa," *World Politics.* XLI, 3, (1989).
Bratton, M., "Non-Governmental Organizations in Africa." *Development and Change,* 21, 1, (1990).
British High Commission, Lagos, "Hurd Highlights Need for Good Government

in Africa and Elsewhere." Press Release, 11 June, (1990).

Callaghy, T., "Towards State Capability and Embedded Liberalism in the Third World: Lessons for Adjustment." In: J. Nelson (ed.) 1990, (1989).

Callaghy, T., "Lost Between State and Market: The Politics of Economic Adjustment in Ghana, Zambia and Nigeria." In: J. Nelson (ed.) (1990).

Campbell, B., "Structural Adjustment and Recession in Africa: Implications for Democratic Process and Participation." Atlanta, Mimeo, (1989).

Carter Centre, *Perestroika Without Glasnost*. Atlanta: Africa Governance Program of the Carter Centre of Emory University.

Chazan, N., "Ghana: Problems of Governance and the Emergence of Civil Society." In: L. Diamond, J. Linz and S. Lipset (eds.) (1988).

Chaudry, K., "The Myths of the Markets and the Common History of Late Developers." *Politics and Society.* 21, 3, (1993).

Conable, B., "Reflections on Africa: The Priority of Sub-Saharan Africain Economic Development." World Bank, Washington, D.C., (1991).

Cornia, G., R. Jolly and F. Stewart (eds.), *Adjustment with a Human Face* (2 Vols.) Oxford: Clarendon Press, (1987).

Dia, M., *A Governance Approach to Civil Service Reform in Sub-Saharan Africa*. Washington, D.C., World Bank Technical Papers No. 225, (1993).

Dia, M., *Africa's Management in the 1990s and Beyond: Reconciling Indigenous and Transplanted Institutions*. Washington, D.C.: World Bank, 1996).

Diamond, L., "Roots of Failure, Seeds of Hope." In: L. Diamond, J. Linz and S. Lipset (eds.) *Democracy in Developing Countries Vol. 2 Africa*. Boulder, Co.: Lynne Reinner, (1988).

Diop, M. and M. Diouf, *Le Senegal Sous Abdou Diouf.* Paris: Karthala.

ECA, , *African Alternative Framework to Structural Adjustment Programs for Socio-Economic Recovery and Transformation*. Addis Ababa: ECA, (1989).

Elbadawi, I., D. Ghura and G. Uwajaren, *World Bank Adjustment Lending and Economic Performance in Sub-Saharan Africa.* Washington, D.C., Policy Research Working Paper No. 1001.

Engberg-Pedersen, P., "The Politics of Good Development Aid: Behind the Clash of Aid Rationales." In: K. Havnevik and B. van Arkadie (eds.) (1996).

Engberg-Pedersen, P., P. Gibbon, P. Raikes and L. Udsholt (eds.) (1996, forthcoming), *Limits to Adjustment in Africa: The Effects of Economic Liberalization 1986-1994.* London: James Currey.

Fukuyama, F., *The End of History and the Last Man*. London: Penguin, (1992).

Ghai, D., *IMF and the South: Social Impact of Crisis and Adjustment.* London: Zed Books, (1991).

Gibbon, P., Y. Bangura and A. Ofstad (eds.) *Authoritarianism, Democracy and Adjustment: The Politics of Economic Reform in Africa.* Uppsala: SIAS, (1992),.

Gibbon, P., "Structural Adjustment and Pressures Toward Multipartyism in Sub-Saharan Africa." In: P. Gibbon, Y. Bangura and A. Ofstad (eds.) (1992), *Ibid.*

Gibbon, P. and A. Olukoshi, *Structural Adjustment and Socio-Economic Change in Sub-Saharan Africa: Some Conceptual, Methodological and Empirical Issues.* Uppsala: NAI), forthcoming, (1996).

Grindle, M. and J. Thomas (eds.), *Public Choice and Policy Change: The Political Economy of Reform in Developing Countries.* Baltimore: Johns Hopkins University Press, (1991).

Haggard, S. and R. Kaufman, "Economic Adjustment in New Democracies." In: J. Nelson (ed.) (1989).

Haggard, S. and R. Kaufman (eds.), *The Politics of Economic Adjustment: International Constraints, Distributive Conflicts and the State.* Princeton, N.J.: Princeton University Press, (1992).

Haggard, S. and S. Webb (eds.), *Voting for Reform: Democracy, Political Liberalization and Economic Adjustment.* Oxford: Oxford University Press, (1994).

Havnevik, K. and B. van Arkadie (eds.), *Domination or Dialogue: Experiences and Prospects for African Development Co-operation.* Uppsala: NAI, (1996).

Herbst, J., "The Structural Adjustment of Politics in Africa." *World Development,* 18, 7, (1990).

Huntington, S. P., *The Third Wave: Democratization in the Late Twentieth Century.* Norman: University of Oklahoma Press, (1991).

Hutchful, E., "The Crisis of the New International Division of Labor: Authoritarianism and the Transition to Free Market Economies in Africa." *Africa Development,* 12, 2, (1987).

Hyden, G. and M. Bratton (eds.), *Governance and Politics in Africa.* Boulder, Co: Lynne Reinner, (1992).

Ibrahim, J., "The State, Accumulation, and Democratic Forces in Nigeria." Uppsala: Mimeo, (1989).

Ibrahim, J., "Expanding the Nigerian Democratic Space." Bordeaux: Mimeo, (1990).

Ibrahim, J., "The Weakness of "Strong States" : The Case of Niger Republic." In: A. Olukoshi and L. Laakso (eds.) (1996).

Joseph, R., "Governance in Africa." In: Carter Centre, (1989).

Ka, S. and N. van de Walle, "Senegal: Stalled Reform in a Dominant Party System." In: S. Haggard and S. webb (eds.) (1994), *op cit.*

Killick, T., *A Reaction Too Far: Economic Theory and the Role of the State in Developing Countries.* London: Overseas Development Institute, (1989).

Lal, D., *The Poverty of Development Economics.* London: Institute of Economic Affairs, Hobarth Paperback 16, (1983).

Lal, D., "The Political economy of Economic Liberalization." *World Bank Economic Review,* 1, 2, (1987).

Landell-Mills, P., "Governance, Cultural Change and Empowerment." *Journal of Modern African Studies,* 30, 4, (1992).

Landell-Mills, P. and I. Serageldin, "Governance and the External Factor." In: *Proceedings of the World Bank Annual Conference on Development Economics,* Washington, D.C.: World Bank, (1991).

Legum, C., "The Postcommunist Third world: Focus on Africa." *Problems of Communism,* 41, 1-2, (1992).

Lindberg, D. and B. Nunberg (eds.), *Rehabilitating Government: Pay and Employment Reform in Africa.* Washington, D.C./Aldershort: World Bank and Avebury, (1996).

Mamdani, M., T. Mkandawire and E. Wamba-dia-Wamba, *Social Movements, Social Transformations and the Struggle for Democracy in Africa.* Dakar: CODESRIA, (1988).

Mamdani, M., "Uganda: Contradictions in the IMF Program and Perspective." In: D. Ghai (ed.) (1991), *op cit.*

Mamdani, M. and E. Wamba-dia-Wamba (eds.), *African Studies in Social Movements and Democracy.* Dakar: CODESRIA Books, (1995).

Mkandawire, T., "Crisis and Adjustment in Sub-Saharan Africa." In: D. Ghai (ed.) (1991), *op cit.*

Mkandawire, T., "Adjustment with a Democratic Face." In: G. Cornia, T. Mkandawire and R. van der Hoeven (eds.) *Africa's Recovery in the 1990s: From Stagnation and Adjustment to Human Development.* London: Macmillan, (1992).

Mkandawire, T., "Adjustment, Political Conditionality and Democratization in Africa." In: G. Cornia and G. Helleiner, *From Adjustment to Development in Africa: Conflict, Controversy, Convergence, Consensus?* (London: Macmillan, (1994).

Mkandawire, T., "Beyond Crisis: Towards Democratic Developmental States in Africa." Dakar, Mimeo, (1995).

Mkandawire, T. and A. Olukoshi (eds.), *Between Liberalization and Repression: The Politics of Structural Adjustment in Africa.* Dakar: CODESRIA Books, (1995).

Mkandawire, T., "Economic Policy-Making and the Consolidation of Democratic Institutions in Africa." In: K. Havnevik and B. van Arkadie (eds.) (1996).

Mosley, P. and J. Weeks, "Has Recovery Begun? Africa's Adjustment in the 1980s Re-visited." *World Development.* 20, 10, (1993).

Mustapha, A. R., "Ever-Decreasing Circles: Democratic Rights in Nigeria, 1978-1988." Oxford, Mimeo, (1988).

Mustapha, A. R., "Structural Adjustment and Multiple Modes of Livelihood in Nigeria." In: P. Gibbon, Y. Bangura and A, Ofstad (eds.) (1992), *op. cit.*

Nelson, J. (ed.), *Fragile Coalitions: The Politics of Economic Adjustment.* New Brunswick: Transaction Books, (1989).

Nelson, J. (ed.), *Economic Crisis and Policy Choice: The Politics of Economic Adjustment in the Third World.* Princeton, NJ: Princeton University Press, (1990).

Nunberg, B., *Public Sector Pay and Employment Reform: A Review of the World Bank Experience.* Discussion Paper 68, World Bank, Washington, D.C., (1989).

Nunberg, B. and J. Nellis, "Civil Service Reform and the World Bank." PRE Working Paper 422, World Bank, Washington, D.C.

Nyang'oro, J. and T. Shaw (eds.), *Beyond Structural Adjustment in Africa: The Political Economy of Sustainable and Democratic Development.* New York: Praeger, (1992).

Olukoshi, A., "The Politics of Structural Adjustment in Nigeria." Uppsala, Mimeo, (1991).

Olukoshi, A. (ed.), *The Politics of Structural Adjustment in Nigeria.* London: James Currey, (1992).

Olukoshi, A. and L. Wohlgemuth (eds.), *A Road to Development: Africa in the 21st Century.* Uppsala: NAI, (1995).

Olukoshi, A., "Africa: Democratizing Under conditions of Economic Stagnation." Dakar, Mimeo, (1995).

Olukoshi, A., "The Impact of Recent Economic Reform Efforts on the state in Africa." in K. Havenevik and B. van arkadie (eds.) (1996), *op. cit.*

Olukoshi, A. and L. Laakso (eds.), *Challenges to the Nation-State in Africa.* Uppsala: NAI, (1996).

Robinson, D., *Civil Service Pay in Africa.* Geneva: ILO, (1990).

Rothchild, D. and N. Chazan (eds.), *The Precarious Balance: State and Society in Africa.* Boulder, Co.: Lynne Reinner, (1988).

Sandbrook, R., *The Politics of Africa's Economic Stagnation.* Cambridge: Cambridge University Press, (1985).

Sandbrook, R., "The State and Economic Stagnation in Tropical Africa." *World Development,* 14, 3, (1986).

Sandbrook, R., "Economic Crisis, Structural adjustment and the State in Africa." In: D. Ghai (ed.) (1991), *op. cit.*

Stokke, O. (ed.), *Aid and Political Conditionality.* London: EADI/Frank Cass, (1995).

Tarp, F., *Stabilization and Structural Adjustment: Macroeconomic Frameworks for Analyzing the Crisis in Sub-Saharan Africa.* London: RKP, (1993).

UNRISD, *States of Disarray: The Social Effects of Globalization.* Geneva: UNRISD, (1995).

Wai, D., "Governance, Economic Development and the Role of External Actors." Oxford, Mimeo, (1991).

Wai, D., *Political Change and Economic Development in Africa.* Washington, D.C.: World Bank, (1994).

Waterbury, J., "The Political Management of Economic Adjustment and Reform." In: J. Nelson (ed.) (1989), *op. cit.*

Widner, J. (ed.), *Economic Change and Political Liberalization in Sub-Saharan Africa.* Baltimore, NJ: Johns Hopkins University Press, (1992).

World Bank, *Accelerated Development in Sub-Saharan Africa: An Agenda for Action.* Washington, D.C.: World Bank, (1981).

World Bank, *Sub-Saharan Africa: From Crisis to Sustainable Growth.* Washington, D.C.: World Bank, (1989).

World Bank, *The Long-Term Perspective Study of Sub-Saharan Africa: Background Papers, Vol. 3, Institutional and Sociopolitical Issues.* Washington, D.C.: World Bank, (1990).

World Bank, *The African Capacity Building Initiative: Towards Improved Policy Analysis and Development Management in Sub-Saharan Africa.* Washington, D.C.: World Bank, (1991).

World Bank, *Governance and Development.* Washington, D.C.: World Bank, (1992).

World Bank, *Governance: The World Bank's Experience.* Washington, D.C.: World Bank, (1994a).

World Bank, *Adjustment in Africa: Reforms, Results and the Road Ahead.* New York: Oxford University Press, (1994b).

Chapter 8

Economic Policy Reforms, External Factors, and Domestic Agricultural Terms of Trade in Selected West African Countries

Tshikala B. Tshibaka

Introduction

The analysis of the effects of price and price distorting policies on the agricultural sector has received a great deal of attention in the economic literature pertaining to Sub-Saharan Africa. This literature suggests that inappropriate price and price-related policies have been the key impediment to agricultural growth and development in most Sub-Saharan countries. This finding appears to have had a significant influence on the design of structural adjustment programs (SAPs) in which getting the structure of relative prices right was considered to be the leading operational objective. Liberalization of input, output and service prices, marketing and trade, privatization of most public enterprises, devaluation of local currencies and disengagement of the state from most support services were primarily meant to achieve this objective.

With respect to privatization of parastatals, policy reforms implicitly assumed that the private sector will take over all the functions performed by these parastatals in a more efficient and cost-effective manner. It is, however, important to observe that the advocates of reforms appear to have paid little attention to some key questions related to the privatization process. Little attempt was made to identify functions that are best performed by government agencies and those that are best handled by the private sector or to assess the private sector base in each country concerned. The failure to examine these and other related key questions has made it difficult for the designers of the structural adjustment reforms to propose appropriate policy measures and actions that could help strengthen and foster the development of the private sector in order to enable it to effectively handle various functions that were previously carried out by parastatals in the economy. In addition, these policy reforms also failed to be specific about the timing and sequencing of the privatization process in order to avoid the disruption of agricultural and other economic activities.

Furthermore, structural adjustment programs appear to have paid limited attention to reforming government public goods, institutional and human capital development policies. These policies are even more central to the long-term development process of the economy than the stabilization and adjustment policies that have been the focus of the structural adjustment programs in Sub-Saharan Africa. Since the main objective of SAP was to get the structure of relative prices right, a question one may wish to address is did these policy reforms achieve this operational objective. What is clear here is that limited attention was paid to the impact of external factors on the structure of relative prices. The whole SAP package seems to have over-estimated the magnitude of leverage national governments in Sub-Saharan Africa have to affect the domestic agricultural terms of trade and hence the growth of agricultural output and income.

In fact, the failure of structural adjustment programs to improve the economic situation in Sub-Saharan Africa is now recognized both explicitly and implicitly by the opponents and designers of this policy reform package. As one of the most seasoned African economists put it "the continent" (Sub-Saharan Africa) has the dubious distinction of being the only developing region of the world that experiences zero average per capita growth over the last thirty years, including negative growth rates over the last two decades (Elbadawi 1995). Embedded in the last two decades of the negative per capita growth is more than ten years of implementation of structural adjustment programs in most of Sub-Saharan Africa. After such a period of time, it is now appropriate to assess the impact these policy reforms have had on the agricultural sector.

This chapter attempts to evaluate the impact of economic policies undertaken since the 1970s including structural adjustment policy reforms on domestic terms of trade of major tradable agricultural commodities and in so doing to establish the extent of leverage governments in Sub-Saharan Africa have to affect the structure of relative prices facing their economies. More specifically, the chapter examines the movements of domestic agricultural terms of trade and its major components in the light of economic policy reforms initiated in Sub-

Saharan Africa; assesses the impact of external factors and domestic policy variables on the real exchange rate, one of the key components of the domestic terms of trade of tradable commodities; and finally estimates the contribution of external factors and domestic policies to change in the domestic terms of trade of agricultural tradables.

This study is based on historical data. Three West African countries, Côte d' Ivoire, Niger, and Senegal, were selected for the study. As can be seen, this sample of countries is made up of one coastal non-sahelian country (Côte d' Ivoire), one coastal sahelian country (Senegal), and one landlocked sahelian country (Niger). These countries are all members of the West African Economic Monetary Union (WAEMU)[1] They have the same currency that is linked to the French franc by a fixed nominal exchange rate. This exchange rate which was pegged at 50 CFA francs for one French franc since the 40' s was changed on January 11, 1994 to 100 CFA francs for one French franc. This policy change has been one of the major macro-economic policy reforms initiated collectively by these and other countries of the Union.

The following crops are selected for the study, cocoa, cotton and rice in Côte d' Ivoire and, groundnut, cotton and rice in Senegal and Niger. It should also be observed that in Niger cotton and groundnut are the major export crops, while rice is the main importable food crop. Likewise in Côte d' Ivoire, cocoa is the major export commodity and cotton is an important raw material for local textile industry, while rice is the major importable food crop. Despite its decline, groundnut still remains a leading export crop in Senegal, while cotton, a relatively new export crop, presents significant potential for growth. With respect to rice in Senegal, it is not only the most important importable commodity, but a leading food crop.

The chapter is organized around six sections including the introduction. The second section reviews the economic policies followed during the last two decades. The third section develops the analytical framework. The fourth section analyses the movement of domestic terms of trade of selected agricultural tradables. The fifth section evaluates the impact of external factors and domestics variables on these terms of trade. Finally, the last section provides some policy implications and concludes the study.

Review of Economic Policies in the Study Countries

Côte D' Ivoire. The agricultural sector has played and continue to play a leading role in Côte d' Ivoire' s economic development and the sector' s rapid growth in the 1960s was the basis of what has been called the "Ivorian economic miracle." Although the Ivorian economy is relatively diversified, it remains dependent on agriculture, which contributes almost one-half of GDP and employs about 54 percent of the economically active population. Two major crops dominate this sector: coffee and cocoa.

Coffee contributes about 50 percent of the country's export revenue and Côte d' Ivoire was the second largest African producer in 1992 after Ethiopia. Coffee farms are 98 percent small in size and the robust type dominates production. Cocoa production doubled in the 1970s and Côte d' Ivoire became the world's largest producer in 1977-1978 when its production over-took Ghana's. A state marketing agency, *the Caisse de Stabilization et de Soutien des Price des Productions Agricoles (CAISSTAB)* traditionally purchased all coffee and cocoa production before its privatization in early 1990s. Furthermore, Côte d' Ivoire has managed to diversify production in rice, cotton and rubber and the country has achieved self-sufficiency in almost all the food crops, except rice.

Côte d' Ivoire has experienced four episodes during the 1965-93 period. The 1965-73 sub-period was marked by high rate of growth of agricultural output and gross domestic product. The government development strategy was based on external borrowing and extraction of agricultural surplus to increase investments in basic infrastructure and other sectors. The CAISSTAB played a major role in this area by paying producer prices which were below the international level.

The 1974-78 sub-period was marked by various policy developments. The most notable development was the proliferation of state-owned companies. The sub-period was also characterized by an unprecedented boom in primary commodity prices and subsequent increasing export earnings. The latter helped the country to enlarge its industrial base in the areas of energy (Kosson dam), agro-based industries (sugar, palm oil, coffee and cocoa processing). In the agricultural domain, the country decided to diversify production into new crops (cotton, rubber, sugar) and extending export and food crops out of the cocoa belt. The 1973-74 oil shock did not force Côte d' Ivoire to reduce investment expenditure, instead the country resorted to heavy external borrowing in order to maintain high investment rate. At the end of this sub-period, the country started to experience difficulties to service its heavy external debt.

This sub-period was characterized economic and financial crisis which prompted the country to apply austerity measures. Incapacity to reimburse external debt was one of the factors which led Côte d' Ivoire to negotiate an economic recovery program in 1980 with the IMF. This helped the country to obtain its first series of debt rescheduling in 1983, 1984 and 1986. But new adverse developments in export revenues led to further deterioration of economic and financial situations. The occurrence of drought in 1983 and 1984 caused production reduction in agriculture with coffee being one of the most affected. Furthermore, a prolonged collapse of primary commodity prices on international markets led to a severe decline in export revenues. Despite this, Côte d'Ivoire decided to increase producer prices in mid-1980s and the replanting program was maintained. Faced by shrinking export revenues, the government reacted to world price collapse by attempting to stockpile cocoa production. This unsuccessful attempt led to further erosion of the country's financial liquidities and subsequent increase in domestic and external debt arrears. Therefore, the country has no choice but to adopt the structural adjustment program.

The 1989-93 sub-period was marked by government's efforts to apply structural adjustment programs and a reverse producers' price policy was adopted. The structural adjustment policies focused on extended financial stabilization measures and structural reforms in order to create a competitive environment for the country. Stabilization measures were reflected in public investment and public employment reduction. Other structural measures included, inter alia, the liberalization of external trade, reduction of corporate tax and reduction of government role in the productive sectors. The latter led to the privatization of some of the largest public enterprises such as water and electricity utilities, opening of private participation in CAISSTAB.

With regard to producer prices, further deterioration in world prices for primary commodities led to reducing coffee and cocoa producer prices by half, while the government encouraged producers to increase food crop production in areas where self-sufficient was yet to be achieved. Despite these measures, improvement in leading economic indicators including the level of the country's competitiveness was rather limited. This led the country to accept along with other member countries in the WAMEU to devalue their commonest currency the CFA in January 1st, 1994.

Niger. During the 1970s, Niger achieved good economic performance largely due to the improvement in primary commodity prices on international markets. Although the leading product with regard to export revenues was uranium, the country seeked to develop an alternative development program based on agriculture. In Niger, agriculture and livestock contribute 37% of GDP (1992) and employ about 85% of the active population. Staple food crops include millet, maize, sorghum, rice and vegetable while cowpeas, cotton and groundnut constitute cash crops. Food production is frequently inadequate to satisfy the needs of the country and drought constitutes a permanent impediment to production. Desertification is a major concern for the country. Niger's economic and agricultural policy reforms delineate three policy sub-periods: 1960-72, 1973-82, and 1983-92.

The 1960-72 sub-period was characterized by optimism with regard to the country's development perspectives, export products were behaving well and the country enjoyed good climatic conditions. The government's development strategy was based on a heavy intervention in the production and processing of agricultural products. An important tool for government policy was the Société nigérienne de commercialisation et de production (COPRO-NIGER), created in 1962, which, in cooperation with other marketing boards, centralized cooperative's activities in the area of marketing of agricultural products. During this period, agricultural development objective aimed to entrance self sufficiency in food production.

During the 1973-82 sub-period, an integrated approach to rural development was adopted. The objective was to reduce food production deficits and the government increased efforts to introduce modern agricultural methods. Significant components of this policy were translated in the development of irrigation infrastructure, distribution of improved seeds and agricultural extension programs. The integrated rural development strategy focused on two sub-sectors: dry grain crops and irrigated crops. The latter includes rice, cotton, and sorghum. Input and equipment as well as credit were subsidized during this sub-period. Niger's overall economic performance in this period has been positively affected by world price increase in the mid-1970s and negatively hit by drought in 1973-74 and early 1980s. After the 1980s, commodity prices collapsed and precipitated the degradation of macroeconomic equilibrium in early 1980s.

The 1983-92 sub-period was characterized by two major developments: (1) several successive debates on how Niger should develop its rural sector and (2) structural adjustment. Structural adjustment reforms were undertaken in an attempt to redress the economic and financial crisis. Three main reforms in agricultural policy were initiated: (1) a review of the incentives to agricultural producer; (2) liberalization of internal trade and prices for crops, elimination of subsidies on inputs and equipment and (3) state retrenchment in the area of agricultural production and trade in order to give room to the private sector. In order to increase production, the country tried to expand area under cultivation and "off-season programmes" were introduced in 1985 in order to compensate for the cereal deficits. Beside, in January 1994, the devaluation of the CFA franc was initiated as the final component of the structural adjustment program.

Senegal. At the time Senegal attained political independence in 1960, groundnut production and trade dominated her agricultural sector. Groundnut occupied 49% of the total cropped area and contributed about 87% of the country' s export earnings. With 70% of the labor force actively involved in agriculture which accounts for more than 20% of total GDP, Senegal continues to be an important producer of groundnut.

Senegal' s economic policy developments and reforms are often perceived and translated through its agricultural policy treatment. These reforms delineate three sub-periods: 1960-78, 1979-84 and 1985-92 sub-periods. During the 1960-78 sub-period, government systematically intervened in the whole economy. Rural cooperatives were established throughout the country, charged with the role of undertaking production and marketing crops with monopoly power. In the 1970s, additional public enterprises and agricultural programs were established to promote the adoption of technological package in the rural sector. Regional Development Agencies were formed along ecological lines to undertake integrated rural development projects.

The 1978-84 sub-period was characterized financial crises. The causes of the financial crisis include: (1) the systematic expansionary policies undertaken in the 1970s (increase in minimum wage, expansion of the public sector, increase in external borrowing); (2) commodity price collapse on international markets (groundnut declined by 37 percent from 1978 to 1980); (3) adverse climatic conditions (drought occurrences in 1973-74, 1978, 1980 and 1983-84) and (4) large deficits of public-owned enterprises (ONCAD, Office National de Cooperation et d' Assistance Pour le Developpement, which centralized cooperative activities, fell bankrupt as it had accumulated an impressive loss of CFA 64 billions in 1981).

The most symbolic sign of Senegal' s financial deadlock was its increasing difficulties to pay salaries to its 68,000 public servants and to reimburse its external debt. As a result, crisis management appears to have dominated the economic policy during this sub-period. In order to obtain its debt rescheduling, Senegal had to sign a stabilization program under IMF and World Bank sponsorship with the objective of stabilizing internal and external deficits. Some of the objectives of the program comprised: reduction of the public deficit to a reasonable level, reduction of inflation, restructuring of a number of public enterprises and compliance to pay fixed annual debt service.

It is important to note that the stabilization was significantly hindered by government concerns to maintain social order. The political class' main concern was the satisfaction of the electorate for the impending first multi-party general elections in 1993. As a consequence, nominal wage continually increased while a number of university graduates were recruited with no significant layoffs in public-owned enterprises. The result was additional increase in public expenditures. In the agricultural sector, IMF/World Bank conditionalities led to the adoption of New Agricultural Policy. Some of the ingredients of this policy were: increase in producer prices; elimination of subsidies on agricultural inputs (except fertilizers); elimination of subsidized urban food prices and restructuring of some agricultural public enterprises.

The 1985-92 sub-period witnessed an increased involvement of foreign creditors and lenders in sponsoring Senegal' s economic policies.
In the agricultural sector, two broad orientations characterized the reforms: a re-organization of relationships between state and other economic agents in the rural and agricultural sector and a new policy with regard to inputs and financing the rural activities. Re-organization of relationships between the concerned agents in the rural sector meant that the government had to pay a minimum role and farmers and other private agents were made responsible for most activities. With regard to input financing, seeds, fertilizers and equipment had to be handled by the private sector, and the financing of agricultural sector had to follow commercial (private) principles.

The policy reforms of 1985-1994 introduced significant change (even if some of them did not receive complete application) in the economy. External trade of agricultural and food products was liberalized, crop prices were free (except for groundnut, cotton, and rice). Many rural development public enterprises were eliminated. Government retrenchment had led to the privatization of SONACOS (Société Nationale de Commercialization de Oléagineux de Sénégal) which was responsible for marketing, processing and exporting of groundnut products in 1995. Observers considered this as one of the most important SAP achievements in Senegal.

Methodology of the Study

Decomposition and movement of relative producer price
For the purpose of this study, the starting point is the recognition that for an average producer, the real producer price of the commodity, also referred to as the commodity domestic terms of trade, is the most relevant domestic price in the production decision making process. Analytically, the real producer price of any product in a given year is the outcome of several influences including domestic and external factors.[2]

It is therefore useful to decompose the movement of the relative price over a given period into factors determined by world market and those related to domestic policies. To this end let the nominal producer price of a tradable product be defined:

(1) $P = P^* E (1 + t)(1 - m)$

where
P = the domestic export commodity price ;
P^* = the border price ;
E = the nominal exchange rate ;
t = the implicit export tax rate on the commodity; and
m = the marketing margin that makes P and P^* comparable.

Considering that the marketing margin reflects normal transport costs and normal (competitive) profits, one can follow the assumption commonly used in the analysis of the movement of commodity prices that the marketing margin (m) remains relatively constant over time (Quiroz and Valdés 1993; Elbadawi 1994; Baustista and Gehlhar 1994). In this case, the marketing margin can be set equal to zero. The expression (1) becomes:

(2) $P = P^* E (1 + t)$

It is established that what matters for a producer is not the nominal price, but the relative price of the product. Expression (2) can then be written as follows :

(3) $P = P^* \; RER \; NPC$

where
$P = P / CPI$ = the relative producer price or domestic terms of trade of the commodity;
$P^* = P / WPI$ = the relative world price or foreign terms of trade of the product;
$RER = (E^* \; WPI) / CPI$ = the real exchange rate;
$NPC = (1 + t) = P / P^* E$ = the nominal protection coefficient of the product;
CPI = the consumer price index; and
WPI = the world price index.

The world price index can be approximated by a weighted average price index derived from the wholesale price indices of the country's major trading partners. The share of each major trading partner in the volume of trade (imports + exports) of the country concerned is used as a weight in computing the world price index facing the country of study.

It is useful to note that the measure of real exchange rate used in this study is an approximation of that used in theoretical literature in which the real exchange rate is defined as the ratio of domestic price of tradables over the price of non-tradables (Edwards 1989). Empirically, what is difficult with this theoretical definition of real exchange rate is how to get the price of non-tradables. Non-tradable goods constitute a large group of heterogeneous consumer and producer goods and services.

Consumer goods such as agricultural non-tradable products often have agricultural tradable products that are close substitutes in consumption. Likewise, agricultural non-tradable inputs such as labor and land have tradable capital and intermediate goods that are close substitutes in production. Services such as transportation, housing, professional services, although non-tradables, often have large components of the costs attributed to tradable capital and intermediate goods.

Using rent for housing or wage rate as some studies have suggested does seem to be also questionable on the basis of the above considerations. Likewise, using the price of a small group of non-agricultural products such as food products as a proxy for the price of tradables in the economy is also difficult to defend, both on theoretical and empirical grounds. As noted earlier, these non-tradable agricultural products often have close agricultural tradable substitutes in consumption. Clearly, the prices of these non-tradables are indirectly affected by sectoral and macroeconomic policies. Furthermore, computing the price of non-tradables on a rather very narrow base would be difficult to defend on statistical grounds.

Given all these difficulties, some studies have found it empirically convenient and useful to use the consumer price index as a proxy for the aggregate price index of non-tradable goods and services in the economy. The drawback is that the consumer price index incorporates both the price indices for tradables and non-tradables alike. Despite this shortcoming, defining the real exchange rate as a product of the nominal exchange rate times the ratio of the world price index over the domestic consumer price index provides a measure of the real price of foreign currency, expressed in local currency, that takes into account both the domestic and world price levels facing the country. As such, this measure of the real exchange rate appears particularly appropriate for macroeconomic policy management.

Expression (3) shows that the relative producer price (P) is made up of three components. The first component is the real border price of the commodity (P^*). This argument—the world terms of trade of the commodity concerned—expresses the purchasing power of the commodity in the world market. The change in this argument is beyond the control of the government (the small country assumption is implied). The second component is the real exchange rate *(RER)* and the third component is the nominal protection coefficient. The change in this second component is partly affected by government policies and by the external shocks operating through the foreign terms of trade and international capital movements (inflows of foreign financial resources). For CFA countries, another external factor that affects the real exchange rate is the nominal exchange rate of the CFA whose change is determined by French monetary authorities. It should be recalled that CFA franc is pegged to French franc. The third component is the nominal protection coefficient *(NPC)*. As a share of border price received by the producer, the nominal protection coefficient is solely determined by government commodity-specific price, marketing and trade policies. These three components are the major sources of change in real producer price.

The above measure of the structure of price incentives—the relative commodity producer price *(P)*—is what Elbadawi (1994) refers to as "the bottom line incentive-wise requirement for achieving a positive supply response, regardless of the external competitiveness of the product." However, it is important to note that because of significant cross-border trade between neighboring countries, it is increasingly difficult to insulate the structure of price incentives in one country from being influenced by those in the neighboring countries. This suggests that a very low *NPC* is likely to reduce both the total and reported commodity supply response.

As can be seen from expression (3), the share of each component in the real producer price can be computed, using a log transformation, in order to identify the leading real producer price component. In addition, the change in the relative commodity producer price resulting from changes in its components can also be computed. The relative contributions of changes in the above components to change in the real producer price can be estimated.

Real world price of tradable crop

The movement of real world price of a tradable commodity is, in the case of a small country, solely determined by world market developments. There is little a small country can do to affect the level of foreign terms of trade of its primary commodities. Actions followed in the process of economic development include: (1) the promotion of processing of primary commodities into semi-finished and/or finished products which have a high value added; and (2) the improvement of resource productivity, among others. These and other actions help the country to stay competitive even in the face of declining foreign terms of trade of its primary products.

The nominal protection coefficient

The NPC is a direct expression of government commodity-specific policies, institutional and structural deficiencies. Measures need to be considered in order to reduce or eliminate these distortions, and hence to improve the structure of relative prices in the economy. However, some of these measures which could be taken with a stroke of a pen will pose a serious dilemma to the government if revenue derived from export tax and other levies represent a substantive share of government income. This becomes even more problematic for a country with low tax effort (tax-GDP ratio) and where taxing through trade and/or marketing arrangements is administratively and politically feasible. This is particularly evident for most Sub-Saharan African countries where trade taxes and/or taxes through marketing channels are not only an important share of the government revenue, but also the most easily collectible taxes.

It is established that raising tax effort is a *sine qua non* for stabilization and growth objectives. Eliminating trade and/or marketing taxes will even further reduce the tax effort in most African countries. Therefore, trade and marketing policy reforms must be accompanied by tax policy reforms. It is worth observing that most trade and marketing studies have paid little attention to budgetary implications of trade and marketing policy reforms. This study does recognize this strong linkage between tax, trade and marketing policy reforms. Tax policy and tax reform—likewise trade/marketing policies and trade/marketing reforms—are essential instruments of economic policy.

Although this study will not examine the impact of structural and institutional deficiencies on the structure of price incentives, suffice it to say that inadequate development of marketing infrastructure and institutions leads to high marketing costs for both tradables and non-tradables. These costs take a significant share of the prices of these commodities and consequently depress the level of price incentives, and hence impede the development of tradable and non-tradable activities. It is also important to note that, in most African countries, the marketing costs include a large share of transport costs. The level of these transport costs are not only a function of development of infrastructure, but also a function of import policy and regulation related to transport equipment as well as fiscal policies (fuel taxes, etc.).

Empirically, it would be an oversimplification to assume, as most studies of agricultural prices, that the marketing margin remains constant over time. This assumption implicitly suggests that the level of development of marketing infrastructure and institutions, including the legal framework, do not change over time in response to changes in government policies. Although empirically elegant, this assumption tends to mask the crucial role of government public goods development policies, including institutional and legal framework, in affecting the prices of agricultural and non-agricultural prices over time.

In short, the level of marketing costs is a function of government trade, public goods development and marketing policies as well as institutional and legal framework. Needless to say that a share of these costs constitutes a normal payment to owners of resources used in performing different marketing and other related functions. It is not this share, although hard to estimate, which constitutes the issue here, but rather the differential between the observed marketing costs and this normal payment to marketing agents. It is this differential that needs to be reduced if not totally eliminated through a careful combination of measures and actions that seek to remove structural, institutional, legal and policy deficiencies.

Furthermore, from an accounting stand-point, the differential between the world and producer price (expressed in local currency) is shared between government and middlemen. It is widely established that this differential is significantly larger than the nominal protection coefficient (the share of world price accrued to domestic producer) where substantive structural, institutional, legal and policy deficiencies exist. It is, therefore, appropriate to consider the share of border price that accrues to government and middlemen as made up of trade tax, other levies, marketing and other transaction costs; while the nominal protection coefficient is that portion of border price paid to the domestic producer.

What is particularly important to stress is that an improvement in the nominal protection coefficient over time suggests that there has been significant improvement in marketing infrastructure, institutions, legal framework, price and commodity-specific trade policies. Conclusively, the importance of the infrastructure, services, and institutions in the development process cannot be over-emphasized. Removing structural, institutional, legal and policy deficiencies that constrain economic agents to perform various marketing and other economic functions, remains one of the key actions government must undertake.

Real Exchange Rate

The real exchange rate plays an intermediary role in transmitting the price effects of domestic and macroeconomic policies and external factors to tradable goods production. This variable is relevant in assessing the relative profitability of the tradable goods relative to non-tradable products. The devaluation of the local currency or the change in nominal exchange rate is a direct attempt by government to affect the level of real exchange rate. It follows from the definition of real exchange rate that movements in this variable are due to movements in the nominal exchange rate and in the general level of both foreign (exogenous to the small country) and domestic prices. Because domestic prices are affected by nominal exchange rate changes (to an extent determined by accompanying monetary and fiscal policies), there is no one-to-one correspondence between the nominal and real exchange rates.

It is, therefore, clear that the change in real exchange rate is associated with changes in macroeconomic policies, as well as with changes in external factors. These two sets of variables, affecting the real exchange rate, could then be introduced into the decomposition of the changes in relative prices of tradables to estimate the net contributions of changes in domestic policies and external shocks to change in relative prices of tradables during the study period.

In the theoretical and empirical literature, the changes in real exchange rate are often explained by changes in broad categories of variables known as long-run and short-run determinants (Edwards 1989). The long-run fundamentals of the real exchange rate are found to include mainly the country's external terms of trade, trade policies, capital flows, government consumption and technological or productivity change. The short-run determinants are found to include primarily the nominal exchange rate and monetary growth. The external terms of trade facing the country and the nominal exchange rate in the case of CFA countries are the two factors that account for the effects of world economic developments on the real exchange rate while other factors account for effects of domestic policies (Tshibaka and Dennis 1996).

Domestic policies, external factors, and real producer price movement

Adverse world market developments which result in unfavorable foreign terms of trade and depressed world market prices of primary commodities, are often blamed by policy-makers in most developing countries for predicaments these countries face in their efforts to improve their economic condition. It is possible to evaluate, on the basic of the above theoretical developments, not only the extent of the impact of these world market developments, but also that of domestic policies on the real producer prices of primary commodities. Therefore, by replacing the value of RER as determined by its relationship with its fundamentals in expression (4), the change in relative producer price over time can then be expressed in terms of changes in external factors and domestic policy variables.

The relative contributions of individual factors to change in relative producer price can be computed and then grouped in relative contributions associated with changes in external factors and domestic policy variables. This analysis provides the magnitude of the effect of external factors and domestic policies on the domestic terms of trade of tradable products.

On the basis of this analysis, it could be inferred that the worsening of the structure of price incentives facing agricultural exports during the last two decades or so in Sub-Saharan Africa, should not be solely attributed to unfavourable world market developments for the region's major exports as most policy-makers claim, but also to inappropriate domestic policies. This analysis provides the magnitude of the effects of external shocks and domestic policies on relative producer prices of export products. The same approach can be followed to examine the effects of domestic policies and external shocks on the real domestic prices of importable commodities and inputs.

Domestic Agricultural Terms of Trade

Movements of Domestic Agricultural Terms of Trade

The review of economic history of Côte d'Ivoire, Niger and Senegal since the 1970s indicates that have gone under several policy episodes characterized by the extent of government interventions in the economy. At the sectoral and commodity levels, different policy treatments were applied. Agriculture as a whole was heavily taxed through distortion of its domestic terms of trade. At the commodity level, various crops faced different policy treatments. These treatments varied from one country to another. In Côte d'Ivoire, some crops were taxed while others were subsidized during the year 1971-73 sub-periods. The real producer prices of rice and cotton were severely depressed through domestic policies to provide both cheap food and raw material to local textile industry. The real producer prices of these two commodities declined at an average annual rate of 16.7% for rice and 6.1% for cotton during the 1971-73 sub-period. The real world prices of the same commodities increased at an average annual rate of 4.1% for rice and 0.7% for cotton during the same sub-period. Cocoa, on the contrary, was protected during the same sub-period. Its domestic terms of trade improved at an average annual rate of 6.2%, while its external terms of trade declined at an average rate 1.9% per year during the sub-period.

As can be seen from Table 8.1 the 1974-78 sub-period was marked by commodity-specific policy changes. The domestic terms of trade of rice improved significantly, so did that of cotton but to a limited extent, while that of cocoa worsened. The third sub-period (1979-88) which coincides with the structural adjustment program, is characterized by improvement in the domestic terms of trade of cocoa and cotton and by a worsening of that of rice. The latter sub-period (1989-90) marked by economic crisis, born out of heavy external debts, among other factors, saw an increase in the level of taxation of export crops. The domestic terms of trade for cocoa and cotton declined at an average annual rate of 49.3% and 15.9% respectively. For the entire period, the real producer prices of agricultural tradables declined faster than their world counterparts.

For Niger, three policy-sub periods were identified. The 1971-72 sub-period was marked by heavy taxation of agricultural tradables. The real producer prices of cotton, groundnut and rice declined at an average annual rate of 7.9%, 16.6% and 34.0% respectively. The second (1973-82) policy sub-period was characterized by an overall improvement in domestic terms of trade of agricultural tradables. The real producer prices of cotton, groundnut and rice increased at an average annual rate of 4.0%, 10.5% and 2.0% respectively during this sub-period. In fact, the domestic terms of these tradable crops increased at higher rates than their external counterparts, suggesting that a clear policy shift to enhance the growth of agricultural sector. The latter policy (1983-92) sub-period saw some changes in government policy. The real producer price of groundnut increased at an average annual rate of 16.5% that of cotton recorded a modest rate of increase of 0.4% per year, while that of rice declined at a rate of 7.9% per year. For the entire (1971-92) period, exportable crops had their real producer prices increasing at a rate of 1.6%, per year for cotton and 10.0% per year for groundnut, whereas rice, an importable crop, had its real producer price declining at an average annual rate of 1.9%.

In Senegal, three policy sub-period were delineated on the basis of policy changes initiated by the government during the 1971-90 period. The 1971-78 sub-period is marked by heavy taxation of agricultural exports. The real producer prices of groundnut and cotton declined at an average annual rate of 10.6% and 15.0% respectively, while that of rice increased at a rate of 4.6% per year during this sub-period. During the second policy (1979-84) sub-period, marked by SAP policy reforms, the real producer prices of both exportable and importable crops increased at an average annual rate of 4.1% for groundnut, 6.1% for cotton and 20.7% for rice, suggesting a reduction in the level of taxation of agricultural commodities. The latter (1985-90) policy sub-period, characterized by a deepening of SAP policy reforms, saw a decline in the rate of increase of the real producer prices of agricultural tradables. Groundnut and rice had their real producer prices increasing at an average annual rate of 2.3% and 2.0% respectively, while the real producer price of cotton declined at an average annual rate of 2.8%. For the entire period, only rice had its real producer price increasing at an average annual rate of 12.7%, while groundnut and cotton experienced a decline in their real producer prices of 0.7% per year and 2.5% per year respectively. It is clear from this analysis that exportable and importable crops were subject to different policy treatments in the study countries even during the SAP sub-periods. In Côte d' Ivoire and Niger, rice faced unfavorable domestic terms of trade relative to export crops, while in Senegal, rice enjoyed a very favorable domestic terms of trade relative to groundnut and cotton.

Movements of Real Producer Price Components

As shown in the methodological part of this study, the real producer price of a tradable commodity can be expressed as a product of the real world price of the commodity, the real exchange rate and the nominal protection coefficient facing the commodity. For all the exports commodities studied, the data suggest that the real world market prices of the cocoa, cotton and groundnut assumed a declining trend over the study period (Table 8.1 -8.3). The real world price of rice also declined for Niger and Côte d' Ivoire, except for Senegal where it displayed a positive rate of change. It appears in the case of Senegal that *la Caisse de Péréquation*, a government monopoly that was in charge of rice imports and distribution in the country, may have locked itself into rice trade arrangements that were not necessarily the most competitive. This seems to explain the differences between the movement of real world price of rice in Senegal and in the two study countries, Côte d' Ivoire and Niger, where rice import and distribution were handled by private economic agents. A general observation that can be drawn is that the overall real world price of export and food crops displayed a declining trend during the 1971-92 period.

Considering the real exchange rate, this macro-economic variable assumed during the 1971-92 period a negative annual rate of change in Côte d' Ivoire and Senegal of 2.5% and 1.8% respectively while in Niger, it increased by an average rate of 1.2% per year. Looking at policy sub-periods, it is worthy to note that the real exchange rate experienced different time profiles. In Côte d' Ivoire, the real exchange rate increased at an average annual rate of 1.2% during the first (1971-73) sub-period and then declined at an average annual rate of 10.2% during the second (1974-78) sub-period. During the third (1979-88) sub-period, it increased at an average annual rate of 1.4 % and declined at an average rate of 5.5% per year during the last 1989-90 sub-period.

In Niger, the real exchange rate declined at an average annual rate of 10.4% and 1.1% during the first (1971-72) and second (1973-82) sub-periods respectively. During the latter (1983-92) sub-period, the real exchange rate increased at a rate of 5.8% per year. This latter sub-period is marked by a deepening of SAP policy reforms. In Senegal, the real exchange rate declined at an average annual rate of 7.4% during the first (1971-78) sub-period, increased at an average annual rate of 6.9% during the second (1979-84) sub-period and then fell at an average rate of 8.4% per year during the latter (1985-90) sub-period. It is important to note that the latter two sub-periods which coincide with the adoption and the deepening of SAP reforms, have not recorded any substantive improvement in real exchange rate. This situation must have prompted the call for devaluation that was finally initiated in January 1994 by all the CFA member states of West and Central Africa.

With respect to the nominal protection coefficient, this commodity-specific policy variable increased at an annual average rate of 2.9% for cocoa and 0.5% for cotton and declined at an average annual rate of 2.3% for rice in Côte d' Ivoire during the 1971-90 period (Table 8.1). These estimates of the nominal protection coefficient suggest that cocoa, cotton and rice faced different commodity-specific policy treatments. These policy treatments were heavily rated in favor of cocoa relative to cotton, while rice was substantially discriminated against. In Niger, the pattern of government direct interventions was similar to that of tradable crops in Côte d' Ivoire. The nominal protection coefficient increased at an average annual rate of 3.6% for groundnut and 1.6% for cotton and declined at an average annual rate of 1.4 % for rice during the 1971-92 period (Table 8.2). In Senegal, all the three crops had their nominal protection coefficients increasing at an average annual rate of 2.1% for groundnut, 1.1% for cotton and 2.6% for rice during the 1976-90 period (Table 8.3).

The positive effects of government direct interventions did not always offset the negative impact of both inappropriate macro-economic policies (indirect interventions) and unfavorable world market developments. In Côte d' Ivoire, the real producer price declined at an average annual rate of 3.1% for cocoa, 4.9% for cotton and 7.9% for rice during the 1971-90 period. In Senegal, the real producer prices of groundnut and cotton declined at an average annual rate of 0.7% and 2.5% during the 1971-90 period respectively, while that of rice increased by 12.7% per year during the same period. In Niger, the real producer prices increased at an average annual rate of 1.6% for cotton and 10.0% for groundnut, while that of rice declined at 1.9% per year during the 1971-92 period.

Effects of External Factors and Domestic Policies on the Domestic Terms of Trade of Tradable Agricultural Commodities

Since among the three components of the real producer price specified above, the real exchange rate is known to be determined by both the external and internal fundamentals, a step was taken to evaluate the effects of these fundamentals on the real exchange rate before assessing the contribution[3] of external and domestic policy factors to change in the domestic terms of trade of agricultural tradable products.

Real exchange rate, domestic policies and external factors

Table 8.4 shows that for the entire period under review, the external factors as a group, have been the leading determinants of the changes observed in the real exchange rate movement in all the study countries.[4] In Côte d' Ivoire, the external factor of significance proved to be the foreign terms of trade, while the internal fundamentals of importance turned out to be the real government expenditure and the trade regime. The foreign terms of trade contributed 81.3% to the observed decline in real exchange rate during the 1971-90 period, while the domestic policy factors contributed only 18.7% to this decline.

In Niger, the key external factors were the foreign terms of trade, the nominal exchange rate and the inflows of financial resources from external sources (capital inflows, aid, grants, and remittances). The domestic policy factors that proved to be significant in affecting the real exchange rate were the growth of money supply and the trade regime. The external factors contributed 63.4% to the change in real exchange rate recorded during the 1971-92 period, while the internal policy factors contributed 37.6%.

In Senegal, the key external factors affecting the real exchange rate were found to be the foreign terms of trade and the nominal exchange rate while the domestic policy factors of importance in this regard were found to be the real government expenditure and the trade regime. The former group of fundamentals contributed about 57.3% to the decline in the real exchange rate recorded during the 1971-90, while the latter accounted for the remaining 42.7% of the fall in the real exchange rate.

This analysis shows that external and internal factors were both critical in shaping the macro-economic environment facing the Sub-Saharan economies, and that, overall, the external factors accounted for 2/3 of the observed fall in the real exchange rate, the key variable describing the macro-economic environment, while the domestic policies accounted for the rest 1/3 in the three CFA countries studied. This finding suggests that government in these countries have limited leverage over the nature of the macro-economic environment facing their respective economies. Efforts on both internal and external fronts have to be initiated if a growth enhancing macro-economic environment is to be created in CFA countries in particular and Sub-Saharan Africa in general.

Real producer price, domestic policies, and external factors

Table 8.5 summarizes the estimates of the effects of external factors and domestic policy variables on the domestic terms of trade of selected tradable crops in the study countries. A general observation from this analysis is that the external factors are the leading determinants of the evolution of domestic terms of trade of tradable agricultural products in the three study CFA countries. This observation suggests that governments in these countries have a limited leverage to affect the movement of real producer prices of tradable products and hence; the structure of reductive prices in their economies.

In Côte d' Ivoire, the relative contribution of changes in external factors to the fall in the domestic terms of trade of agricultural tradables during the 1970 - 91 period was 93.4% for cocoa, 87.5% for cotton and 64.5% for rice. The domestic policy factors, both macro-economic and sectoral/commodity specific policies contributed 6.6% for cocoa, 14.3% for cotton and 35.4% for rice to the decline in the domestic terms of trade of these tradable commodities.

In Niger, the improvement in the domestic terms of trade of groundnut observed during the 1972 - 92 period was 56.0%, an improvement in world market conditions for groundnut and other external factors. The domestic policy factors contribute 44.% to the increase in the real producer price of groundnut recorded during the period under review. Cotton, on the contrary, faced unfavorable world market and external conditions. These external factors negatively contributed 50% to the change in the real producer price of cotton. This negative influence was offset by favorable domestic policy factors and in particular by cotton specific policy. These domestic factors positively contributed 150.0% to the change in the domestic terms of trade of this tradable commodity. For rice in Niger, both external and domestic policy environments were unfavorable. External factors contributed 68.4% to the decline in the real producer price of rice, while the domestic policy factors contributed 31.6% to the decline in the domestic terms of trade of this importable cereal.

In Senegal, export crops; groundnut and cotton faced negative external and domestic policy conditions. External factors contributed to the decline in real producer prices of groundnut and cotton to the tune of 57.1% and 96.0% respectively, while the domestic policy factors contributed the rest, say 42.9% for groundnut and 4% for cotton. Rice, the leading importable crop, enjoyed both favourable external and domestic policy environments. The relative contribution of external factors to the increase in the real producer price of rice was 87.3% while that of domestic factors amounted to 12.7%.

This analysis reveals that traditional export crops—cocoa, groundnut and cotton—faced an unfavorable world economic environment during the study period and that these conditions explain, to a large extent, the observed deterioration of the domestic terms of trade of these commodities. The contribution of inappropriate domestic policies to the worsening of these domestic terms of trade was limited, but not negligible in Côte d' Ivoire and Senegal, while in Niger, these exportable crops enjoyed an enabling domestic economic environment.

The leading importable food crop rice examined in this study faced in Côte d' Ivoire and Niger both unfavorable world and domestic policy conditions. The external factors accounted for two-thirds of the decline in the domestic terms of trade of this commodity in the two countries, while one-third of the decline was associated with domestic policy factors. In Senegal, rice enjoyed during 1971-90 period both external and domestic policy conditions with external factors contributing about 87.3% to the improvement in the real producer price while domestic policy factors accounted for the rest 12.7%. It is clear from this analysis that governments in CFA Countries have limited scope to improve the structure of relative prices in their respective economies. The external factors being the leading determinant of the domestic terms of trade of tradable commodities.

Policy Implications and Conclusions

This study was undertaken to contribute to the views by African scholars on the pertinence of policy reforms initiated to put the Sub-Saharan economies on growth and development path. Among these policy reforms, Structural Adjustment Program (SAP) policy reforms have been the most comprehensive and far-reaching policy reform package implemented in Sub-Saharan Africa.

The dominant economic literature of the 1970s and 1980s pertaining to the Sub-Saharan Africa holds, among other things, that inappropriate price and price-related policies have been the key impediment to agricultural growth and development. On the basis of this literature, getting the structure of relative prices right was retained as the leading operational objective of the structural adjustment program. Liberalization of input, output and service prices, marketing and trade, privatization of most public enterprises, devaluation of local currencies and disengagement of the state from most support services were primarily meant to achieve this objective.

In this vein, this study was set to assess the impact of these policy reforms on the domestic terms of trade of agricultural tradable commodities and then establish the extent of leverage governments in Sub-Saharan Africa have to affect the structure of relative prices facing their economies. The entire SAP package seemed to have overlooked the impact of external factors on the structure of relative prices and in so doing, this policy reform package overestimated the magnitude of the impact of domestic policies on the structure of relative prices and hence the ability of the national governments in the Sub-Saharan Africa region to affect not only the structure of relative price, but also the overall domestic and macro-economic environment facing their economies.

The analysis shows that the domestic terms of trade of a tradable commodity is a product of the world market terms of trade of the commodity, the real exchange rate and the nominal protection coefficient. The first component is solely determined by external factors, the second by both external and internal policy variables and the third solely by the internal policy factors. This shows that the real producer price of a tradable commodity on its domestic terms of trade is the outcome of several influences including external factors and domestic policies. The extent of the leverage national government has to affect the domestic terms is determined by the weight of domestic policy factors relative to external factors.

The analysis shows that the domestic terms of trade of agricultural tradables declined during the entire (1971-90) period in Côte d' Ivoire.

In Niger, the domestic terms of trade of exportable crops assumed an increasing trend during the study (1971-92) period, while that of rice, a major importable crop, declined. In Senegal, the domestic terms of trade of exportable crops assumed a declining trend during the 1971-90 period while that of rice displayed an increasing trend. It is important to observe that no across-the-board improvement of the domestic terms of trade of selected agricultural tradable was observed in the study countries during the subperiods characterized by the deeping of the SAP policy reforms.

The analysis of the impact of external factors and domestic policies on the real exchange rate reveals that the external factors, as a group, are the key determinants of the real exchange rate in all the study countries. In Côte d' Ivoire, the real exchange rate declined at an annual average rate of 5.5% during the 1971-90 period. The relative contribution of the changes in external factor to the decline in real exchange rate was about 81.3%, while the domestic policy factors contributed only 18.7% during the 1971-90 period.

In Niger, the real exchange rate increased at an annual average rate of 1.2% during the 1971-92 period. The relative contribution of external factors to this real exchange rate increase was about 63.4% while the domestic policies contributed 37.6%. In Senegal, the real exchange rate declined at an annual average rate of 1.8% during the 1971-90 period. The external and domestic policy factors contributed 57.3% and 42.7% to this decline respectively. The analysis shows that, the external factors account for two-thirds of changes in the real exchange rate while the domestic policy factors account for the remaining one-third. This suggests that national governments in CFA-zone have a limited scope to alter the level and the evolution of the real exchange rate a key economic variable characterizing the macro-economic environment facing any economy. The world economic conditions including the foreign terms of trade, French government monetary and other macro-economic policies and the flows of capital are the main external factors influencing the real exchange rate which are largely beyond the reach of the national governments of this West African CFA zone.

Other components of the domestic terms of trade of tradable agricultural commodities including the world market terms of trade and the nominal protection coefficients of agricultural tradable are respectively determined by external and domestic policy factors. The world market terms of trade of agricultural tradable produced by Ivorian farmers declined at an annual average rate of 3.4% for cocoa, 3.2% for cotton and 3.1% for rice during the 1971-90 period. In Niger, the world market terms of trade increased at an annual average rate of 5.1% for groundnut, and declined at an annual average rate of 1.2% for cotton and 1.7% for rice. In Senegal, the world terms of trade declined at an annual average rate of 1.05% for groundnut, -1.8% for cotton and increased at an annual average rate of 11.8% for rice.

With respect to the nominal protection coefficient, this commodity specific policy variable increased at an annual average rate of 2.9% for cocoa, 0.5% for cotton and 2.3% for rice in Côte d' Ivoire during the 1971-90 period. In Niger, the nominal protection coefficient increased at an annual average rate of 1.4% for rice.

Distinguishing between the external factors and domestic policy variables affecting the above three components of domestic terms of trade of tradable commodities, the analysis reveals that the external factors are the major determinants of the level and movement of the domestic terms of trade of agricultural tradable in Côte d' Ivoire, Niger and Senegal. In Côte d' Ivoire, the real production price of cocoa declined at an annual average rate of 3.1% during the 1971-90 period. About 93.4% of this decline in domestic terms of trade of cocoa was associated with changes in external factors, while 6.6% of the decline was attributed to domestic policy factors. With respect to cotton, its domestic terms of trade declined at a rate of 4.9% per year during the same period. External factors contributed 85.7% to this decline while domestic policy factors contributed 14.3%. Regarding rice, its real producer price declined at a rate of 7.9% per year. About 64.6% of this decline was contributed by changes in external factors, whereas 35.4% of the decline was ascribed to domestic policy factors.

In Niger, the rural producer of groundnut increased at a rate of 10.0% per year during 1971-92 period. About 56.0% of this increase was attributed to external factors and 44.0% to domestic policy variables. Regarding cotton, its domestic terms of trade increased at a rate of 1.6% per year. External factors negatively contributed 50.0% to the change in the domestic terms of trade while the domestic policy factors positively contributed 150.0% to the change in the domestic terms of trade of cotton. For rice, its real producer price declined at a rate of 1.9% per year during the same period. About 68.4% of this decline is associated with changes in external factors and 31.6% of the decline is due to changes in domestic policy factors. In Senegal, the domestic terms of trade declined at a rate of 0.7% per year for groundnut and 2.5% for cotton during the 1971-90 period. This decline in the domestic terms of trade was due to unfavorable changes in the external factors and domestic policy factors to the extent of 57.1% and 42.9% for groundnut respectively and 96.0% and 4.0% for cotton likewise. Rice, on the contrary, saw its domestic terms of trade improving at a rate of 12.6% per year during the 1971-90 period. External factors contributed 87.3% to this improvement in the domestic terms of trade of this cereal while the domestic policy factors contributed 12.7%.

It is clear from this analysis that both external factors and domestic policies affect the structure of relative prices facing the Sub-Saharan economies and that the external factors play the leading role relative to the domestic policy factors. The study appears to support, to some extent, the claim by African policy-makers that unfavorable external factors are the key impediments constraining the growth of their economies.

Finally, the structural adjustment program failed to recognize the leading role of external factors in affecting both the overall macro-economic environment and the structure of relative prices facing most Sub-Saharan economies. In the specific case of CFA zone, the foreign terms of trade, the French Government's monetary and other macro-economic policies affecting the nominal exchange rate of CFA franc and the flow of capital form the set for these external factors. Considering the domestic policy factors, the analysis shows that trade regime, government expenditure and money supply are the leading domestic policy factors.

Notes

1. Its French acronym is Union Economique et Monétaire Ouest Africaine (UEMOA).
2. The analysis conducted by Tshibaka and Dennis (1996) reveals that the major external factor affecting the real exchange rate in Côte d'Ivoire is the foreign terms of trade. The elasticity of the real exchange rate with respect to this variable is found to be 0.87. Regarding the domestic fundamentals, the analysis indicates that the trade regime and the real government expenditure are the leading domestic variables affecting the real rate. The elasticity of the real exchange rate is about 0.37 with respect to the trade policy variable (the overall trade bias) and about 0.16 with respect to real government expenditure. In Niger, the external factors affecting the real exchange rate are found to be the foreign terms of trade, the nominal exchange rate and the inflows of financial resources from foreign sources. The elasticity estimates of the real exchange rate with respect to the above arguments are about – 0.24 with respect to foreign terms of trade, 0.58 with respect to the rate of change in the nominal exchange rate and – 0.34 with respect to inflows of financial resources from foreign sources. The domestic policies factors of importance are the growth in money supply and the trade regime. The elasticity of the real exchange

rate is about − 0.29 with respect to the rate of growth of money supply and 0.10 with respect to trade policy variable (the overall trade bias). And finally, with respect to Senegal, the key external factors affecting the real exchange rate are the foreign terms of trade and the rate of change in the nominal exchange rate. The elasticity of the real exchange rate is estimated to be about − 0.46 with respect to the external terms of trade and 0.78 with respect to the rate of change in the nominal exchange rate. The domestic policy factors of importance are found to be the real government expenditure and the trade regime. The elasticity of the real exchange with respect to these two policy variables are about 0.20 with respect to real government expenditure and 0.08 with respect to trade policy variable (the overall trade bias).

Table 8.1: Movements of real producer prices of agriculturral tradables and their major components in Cote d' Ivoire

Commodity/ Sub-period	Annual rate of change in real producer price (%)	Annual rate of change in real world price (%)	Annual rate of change in real exchange rate (%)	Annual rate of change in nominal protection Coefficient (%)
Cocoa				
1971-73	6.2	-1.0	1.2	5.9
1974-78	-0.3	6.6	-10.2	3.4
1979-88	6.0	-2.9	1.4	7.5
1989-90	-49.3	-34.8	-5.5	-9.1
1971-90	-3.1	-3.4	-2.5	2.9
Cotton				
1971-73	-6.1	0.6	1.2	-7.9
1974-78	-5.0	-3.5	-10.2	8.7
1979-88	-2.5	-4.5	1.4	0.5
1989-90	15.9	-0.1	-5.5	-10.3
1971-90	-4.9	-3.2	-2.5	0.5
Rice				
1971-73	-16.7	4.1	1.2	-20.0
1974-78	6.6	3.4	-10.2	13.4
1979-88	-14.7	-8.4	1.4	-4.8
1989-90	n.a	n.a	n.a	n.a
1971-90	-7.9	-3.1	-2.5	2.3

Source: Computed by the author.

Table 8.2: Movements of real producer prices of agriculturral tradables and their major components in Niger

Commodity/ Sub-period	Annual rate of change in real producer price (%)	Annual rate of change in real world price (%)	Annual rate of change in real exchange rate (%)	Annual rate of change in nominal protection Coefficient (%)
Groundnut				
1971 - 72	-16.6	0.0	-10.4	-6.2
1973 - 82	10.5	3.7	-1.1	7.8
1983 - 92	16.5	9.3	5.8	1.4
1971 - 92	10.0	5.1	1.2	3.6
Cotton				
1971 - 72	-7.9	-1.8	-10.4	4.3
1973 - 82	4.0	0.9	-1.1	4.1
1983 - 92	0.4	-4.0	5.8	-1.4
1971 - 92	1.6	-1.2	1.2	1.6
Rice				
1971 - 72	-34.0	-18.6	-10.4	-4.9
1973 - 82	2.0	1.4	-1.1	1.7
1983 - 92	-7.9	-1.3	5.8	-12.5
1971 - 92	-1.9	-1.7	1.2	-1.4

Source: Computed by the author.

Table 8.3: Movements of real producer prices of agriculture tradables and its major components in Senegal

Commodity/ Sub-period	Annual rate of change in real producer price (%)	Annual rate of change in real world price (%)	Annual rate of change in real exchange rate (%)	Annual rate of change in nominal protection Coefficient (%)
Groundnut				
1971 - 78	-10.6	-0.7	-7.4	-2.5
1979 - 84	4.1	-1.7	6.9	-1.1
1985 - 90	2.3	-0.5	-8.4	11.2
1971 - 90	-0.7	-1.0	-1.8	2.1
Cotton				
1971 - 78	-15.0	-5.2	-7.4	-2.4
1979 - 84	6.1	1.5	6.9	-2.2
1985 - 90	-2.8	-0.5	-8.4	6.2
1971 - 90	-2.5	-1.8	-1.8	1.1
Rice				
1971 - 78	4.6	11.5	-7.4	0.5
1979 - 84	20.7	-15.2	6.9	-1.4
1985 - 90	2.0	-0.6	-8.4	10.9
1971 - 90	12.7	11.8	-1.8	-2.6

Source: Computed by the author.

Table 8.4: Impact of external factors and domestic policies factors on the real exchange rate in Côte d'Ivoire, Niger, and Senegal

Country/ Sub-period	Rate of change of real exchange rate (%)	Nature and magnitude of the impact of domestic policy factors		Nature and magnitude of the impact of external factors	
		Nature of Contribution	Contribution of changes in domestic policy factors to change in real exchange rate (%)	Nature of Contribution	Contribution of changes in external factors to change in real exchange rate (%)
Côte d'Ivoire					
1971-1973	1.2	positive	35.0	positive	65.0
1974-1978	-10.2	negative	215.8	positive	115.8
1979-1988	1.4	positive	252.9	negative	152.9
1989-1990	-5.5	negative	31.1	negative	68.9
1971-90	**-2.5**	**negative**	**18.7**	**negative**	**81.3**
Niger					
1971-1972	-10.4	negative	8.7	negative	91.3
1973-1982	-1.1	negative	52.0	negative	48.0
1983-1992	5.8	positive	48.6	positive	51.4
1971-92	**1.2**	**positive**	**37.6**	**positive**	**63.4**
Senegal					
1971-1978	-7.4	n.a	0.0	negative	100.0
1979-1984	6.9	positive	6.3	positive	93.7
1985-1990	-8.4	positive	35.2	negative	135.2
1971-90	**-1.8**	**negative**	**42.7**	**negative**	**57.3**

Source: Computed by author
Note: n.a. = not applicable

Table 8.5: Effects of external factors and domestic policies on the real producer price of agricultural tradables in Côte d'Ivoire, Niger, and Senegal

Country/Period/Crop	Rate of change of real producer Price of agricultural tradable (%)	Nature and magnitude of the effects of external factors		Nature and magnitude of the effects of domestic policy factors	
		Nature of Contribution	Contribution of changes in external factors to change in real producer price (%)	Nature of Contirbution	Contribution of changes in domestic policy factors to change in real producer price (%)
Côte d'Ivoire, 1971–1990					
Cocoa	-3.1	negative	93.4	negative	6.6
Cotton	-4.9	negative	85.7	negative	14.3
Rice	-7.9	negative	64.6	negative	35.4
Niger, 1971–1992					
Groundnut	10.0	positive	56.0	positive	44.0
Cotton	1.6	negative	50.0	positive	150.0
Rice	-1.9	negative	68.4	negative	31.6
Senegal, 1971–1990					
Groundnut	-0.7	negative	57.1	negative	42.9
Cotton	-2.5	negative	96.0	negative	4.0
Rice	12.6	positive	87.3	positive	12.7

Source: Computed by the author

Chapter 9

Financial Sector Restructuring Under the SAPs and Economic Development, With Special Reference to Agriculture and Rural Development:
A Case study of Uganda

Germina Ssemogerere

Introduction

Stabilization and Structural Adjustment Policies (SAPs) have been pursued in Sub-Saharan Africa (SSA) since the early 1980s to restore internal and external equilibria by controlling aggregate demand, and by liberalizing markets to reinstate the role of the price mechanism in efficient resource allocation. Within the SAP framework, financial sector reform/restructuring consisted of:

- liberalization of nominal interest rates from a fixed controlled regime to a market-determined regime, to mobilise savings and allocate loanable funds to projects with the highest returns on investment;
- removal of credit allocation quotas and interest rate ceilings to particular sectors, including agriculture, to allow financial services to flow to the most productive uses;
- privatizing parastatal banks and other financial institutions, to remove administrative inefficiencies and bankruptcy due to government bureaucracy, political interference and rent-seeking activities; and
- tightening of banks supervision to ensure profitability by enforcing the recovery of non-performing assets, and solvency by increasing the equity base and tightening prudential regulations.

Implemented as such, financial sector restructuring was expected to promote long-term economic development.

Since the relationship between financial sector restructuring and economic development has been examined by five papers sponsored under the research network, "African Perspectives on Structural Adjustment," it will not be re-examined here (see the chapters by Emenuga; Hussain, Mohammed and Kameir; Inanga and Ekpenyong; Mwega; and Oshikoya and Ogbu. Instead, this chapter concentrates on how financial sector restructuring is affecting long-term agricultural development and off-farm activities in the rural sector, which is the largest single sector in the SSA economies.

An evaluation of the World Bank's early experience over the period 1980-1991 by Jayarajah and Branson (1995) reveals that the approach to financial sector restructuring (chapter 4) is practically *independent* of agricultural sector reform (Chapter 8).

Within agriculture, the reforms focused upon: decontrol of producer prices, removal of the implicit tax on agricultural exports by liberalizing the nominal exchange rate; dismantling state marketing monopolies; and inflation control to improve the rural/urban terms of trade.

For financial sector restructuring, the reforms concentrated on: removing financial repression; rehabilitating the formal banking system; and getting capital markets started.

While it was acknowledged by Jayarajah (1995) that "Financial sector reform in many developing countries is more a process of development than reform" (153), this *development element* was not explicitly incorporated into the restructuring of the financial sector in order to meet the development needs of the single largest real sector of the SSA economies.

In the World Bank Report covering the most recent period, *Adjustment in Africa: Reforms, Results, and the Road Ahead* (1994), chapters 2 and 3 on agriculture and financial sector restructuring do not show any change in the direction of the World Bank's policy thrust. Whereas the need for adequate availability of credit, to sustain the SAPs price-based incentives to promote agricultural sector growth, is acknowledged, in the second volume *Adjustment in Africa: Lessons From Country Case Studies* (1995), chapter 1, the how to provide the needed credit is neither elaborated nor linked to financial sector restructuring.

Meanwhile, Montiel's review (1995) covering seven SSA countries: Ghana, Gambia, Kenya, Nigeria, Malawi, Tanzania and Uganda, is sounding a discordant note. The experience to-date in these countries, according to the review, suggests that the process of financial sector restructuring has not achieved much success: "...financial liberalization, though recently undertaken in various countries of the region, has been tentative, and has thus far not appeared to be very successful. (Montiel 1995:2)

The key researchable issue, therefore, is whether the way in which this process is being conducted can promote long-term economic development, while omitting the largest rural sector, that is, "The issue confronting policy makers in sub-saharan Africa... is not whether further financial liberalization is desirable, but rather when and how it should be brought about." (Montiel1995: 22)

The purpose of this chapter is to examine this issue systematically with respect to agriculture and off-farm activities in the rural sector, taking Uganda as a case study.

Uganda is chosen because it is lately considered the most successful implementer of the SAPs on the SSA continent by the Washington D.C. Bretton Woods Institutions. The macro-economic success indicators over the period 1987-1995 include: a real GDP growth rate of 5 - 6% p.a., a slashed inflation from over 300% to just around 10% p.a., and a parallel market exchange rate premium of over 300% reduced to zero, for example.

The Uganda economy is basically rural. Agriculture provides over 60% of rural household income, with the rest coming from a wide range of off-farm activities: brewing local beer, fishing, handicraft, brick and charcoal making, construction and maintenance of owned-occupied dwellings etc. The Agricultural sector employs around 80% of the entire active labour force: this contributed 45% to real GDP in constant 1991 prices in 1996/97, and 90% to exports revenue. Agriculture is 90% organised by scattered small holders; estates are confined to tea and sugar production.

Amid the successful macroeconomic achievements, Uganda started financial sector restructuring in 1993 to remove financial repression, strengthen the banking system, and open a capital market. Yet the review of this financial sector restructuring program by the Republic of Uganda, Agricultural Policy Committee (1994) states that

> ...this program has no special action to help development of the rural financial market. In fact, the rationalization of the branch network and liberalization of credit allocation can have a negative impact on the role of the banks in rural credit. The liberalization of credit allocation has had the tendency of *crowding out agriculture* from the banks' loan portfolio. (11)

Faced with this brewing crisis by 1996, three years after the start of restructuring the financial sector, a number of proposals and studies are on the table debating how to provide rural financial services. "The challenge facing the banking system is therefore how to expand and broaden the financial intermediation in rural areas since the agricultural sector is crucial for economic development. (11)

This is a serious ongoing debate addressing a major defect in the design of the SAPs. This chapter is set out to contribute to the debate as follows:

First, it describes the importance of financial services for the economic development of the rural sector, which must be appreciated first, in order to put the gravity of the omission of these services in financial sector restructuring into proper perspective.

Second, it outlines the peculiar characteristics of rural SSA, including Uganda, which must be taken into account to deliver financial services effectively; these characteristics put a limit on how much can be learned as relevant lessons from elsewhere in Asia and Latin America with different rural settings.

Third, it reviews the theoretical literature and best practices in rural finance to promote economic development; presents the Uganda case study; and concludes the chapter with a critical evaluation of the likely adequacy and relevance of the current proposals being debated to address the financial needs of Uganda's rural sector.

The Importance of Financial Services for Agriculture and Rural Development

Rural Financial Services and Economic Development

Rural financial services are needed for three purposes: to provide *rural credit* for productive activities and allied services; to provide a *savings facility* for the rural population; and to provide a *payments mechanism* to transfer purchasing power between economic agents within the rural sector itself, and between the rural sector and the rest of the economy.

Figure 9.1: Categories of Rural Credit Needs

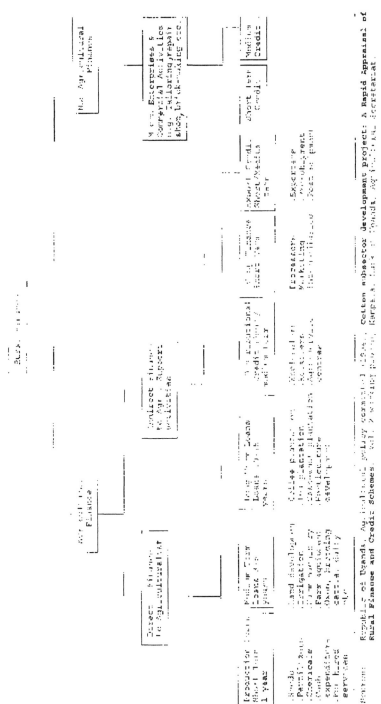

Rural Credit

Following figure 9.1, rural credit needs can be categorized into agricultural and non-agricultural. Within agriculture, credit is needed for *direct production* activities: for *short-term* periods up to one year, to purchase variable inputs and meet working capital requirements; for *medium-term* needs for investments of 3 to 5 years to purchase productive assets and invest in land improvements; and for *long-term needs* 10 to 15 years to establish shambas, irrigation schemes, and other long-term farm infrastructure, etc.

Indirect agricultural credit is needed: *to finance the distribution of inputs* on a timely basis, by wholesalers and retailers; to provide *crop finance* for the purchase, storage, processing and marketing of agricultural produce; and *export credit* to facilitate international shipments.

Non-agricultural credit is needed for both the short and medium term to facilitate micro enterprises that provide *consumer goods and services* to rural areas (e.g., radios, bakeries), and *off-firm employment*, which is particularly important in the utilization of off-season labour, and the diversification of rural income sources.

A Rural Savings Facility

The bulk of the literature that stresses the need for rural credit overlooks that of a savings facility, with the result that the rural sector is forced to save in kind (in animals, birds, etc), or in idle cash in pots, within the informal sector mechanism, which may not bear interest.

To promote economic development, the rural sector needs a savings facility: to earn interest in order to accelerate accumulation; to meet a wider range of contingencies and respond to investment opportunities without the delay caused by looking for a customer to translate savings in kind into cash; and to generate internal loanable funds for credit schemes, rather than rely on donors and governments for continuous injection of funds into these schemes, which is not sustainable.

Payments Mechanism

The rural sector is often inconvenienced by having to carry large sums of cash for payments of school fees, bulky purchases, crop-finance, transfer payments of pension schemes, facilitation of NGO operations, etc. Often this cash is stolen on the way, or poorly counted at the expense of the rural customer.

The rural sector needs a payments mechanism to facilitate all the enumerated transactions.

The Peculiar Problems of the Rural Sector in SSA for the Provision of Financial Services

The provision of rural financial services has three peculiarities in the SSA environment to contend with.

Risk
Agriculture is the single most important sector in SSA economies accounting for between 30% and 60% of real GDP and dominating the rural sector as a source of income and employment, food security, exports, tax revenues, and raw materials for the agro-allied import-substitution industries. Most SSA agriculture is rainfed and yields depend on variation in weather, turning out to fluctuate over wide margins.

Second, there is no effective technology to fight pests and diseases; this opens room for large losses. Institutions for crop and livestock insurance or credit guarantee are particularly hard to manage since they have to cover a wide area of varied climates in order to avoid covariant risk. These institutions hardly exist in SSA, and where they are attempted there is a danger to go bankrupt when all customers fail to repay their loans for no fault of their own.

Third, the best hedge against risk is individual land title of ownership. In many parts of SSA, however, land is communally owned.

Seasonality
A rural financial institution must be able to manage the large cash *inflow* as farmers sell their harvest, the preceding equally large cash *outflows* as traders borrow for crop finance, as well as the demand for *financing inputs* procurement at the time of planting. What makes the management of seasonality even more difficult is the thin or practical absence of financial integration; the rural financial institutions are cut off from their urban counterparts which could have provided the large cash needs for crop finance, and gainfully utilized the large inflows of savings from farmers as they market their harvest.

The Scattered Low-Income Populations
The SSA household settlement patterns are highly scattered, unlike the dense populations of Asia and parts of Latin America. Shifting cultivation or/and transhumance are still practiced. This makes the collection of information on customers to identify bankable projects, supervise and collect loans, and servicing of small deposits, particularly expensive.

Also the required modern physical plants, such as bank offices, computers, etc., to serve customers efficiently become impractical to operate since they have to stay idle or under-utilized in certain migration seasons or have to handle very small businesses where populations are thin, making it impossible to benefit from economies of scale.

Rural infrastructure is particularly thin in SSA, with adverse consequences on the provision of financial services. The high cost of transport due to poor feeder roads lowers profitability and competition, leading to low farm-gate prices, a retreat into subsistence with only a small marketable surplus to meet contingencies, and a retreat by men from Agriculture altogether, leaving food production to women. The reduced volume of business exasperates rural poverty and increases the cost of banking in the rural sector. According to the Food and Agricultural Organization of the United Nations (FAO) Training Manual (1991:220 - 224), farmers in SSA receive only 30% - 60% of the market price for their produce because of the thinner and poorer road network, compared with their colleagues in Asia whose share is 75% - 90% . Over two-thirds (66.1%) of the difference between the consumer and producer prices in SSA is explained by transport costs 39.1% and 27.0% by transaction costs.

The different needs for credit, saving facility and payments mechanism to match the peculiarities of rural SSA are analysed later when we examine the criteria for best-practice in providing financial services.

Financial Sector Restructuring and Economic Development: An Evaluation of the Literature from the Standpoint of Its Relevance to Promoting Long-term Economic Development in Rural SSA

A vast literature sprung up in the mid-1980s and early 1990s regarding the relationship between financial sector restructuring and economic development in rural areas. Extensive reviews include Adams (1988), Desai and Mellor (1993), Simmons (1992), Patten and Rosengard (1991), World Bank (1989), Braverman, Hoff and Stiglitz (1993), Fallas and Owen (1989), for example. This section outlines what the main arguments in this literature imply for the provision of financial services in rural SSA.

The outline is organised around four sub-headings: the pro-Keynesian perspectives; the neoclassical critique; the structural perspective: and the empirical work on best-practices in micro-finance.

The Pro-Kenyesian Perspective

According to the pro-Keynesians, moderately expansionary but regulated financial policies are what promotes higher and more stable economic growth and employment. Under such policies, *institutional finance for Agriculture and rural development should be expanded through the participation of the public sector, along with the private sector*, to increase the *volume* of business in rural areas, *lower transaction costs* and reap benefits from *economies of scale*; lower the *risk of default* by making the environment more conducive to collecting *information*; provide a *variety of forms* of organization and types of service to meet the *rural customers' multiplicity of needs*. Along this reasoning, the following six organizational principles were proposed:

> (1) promoting multiples of RFIs—that is, more than one RFI for a given service area; (2) encouraging a variety of forms of organization of these institutions; (3) ensuring vertical organization of the structure of RFIs from local to regional and national levels; (4) encourage high geographical density of the field-level offices of the RFIs; (5) ensuring that a high proportion of rural clients are reached by them; and (6) promoting diversified and multiple functions that horizontally integrate the agricultural production, input distribution, marketing and processing systems for the benefit of their clients and themselves. (Desai and Mellor 1993:3)

Capital and reserve requirements should be modest, interest rates should have a ceiling and credit should be targeted to socially desirable sectors and projects which cannot attract it on their own from the open market (small holder agriculture, for example). The development rationale underlying the pro-Keynesian hypotheses were:

- that lower interest rates would stimulate investment demand, since the real interest rate is a cost, *and investment is inversely related to the real interest rate*;
- that low interest rates would facilitate the financing of government expenditure for infrastructural and parastatal investments, this being particularly relevant for rural development;
- that credit ceilings, by sector, would assist in the transfer of resources to *sectors with a higher positive social compared to private rate of return*, such as small holder agriculture, which would otherwise be "rationed out" of the credit market because of risk, rurality, and seasonality;
- that overall growth would promote savings from higher incomes.

Evaluation of the Pro-Keynesian Perspective

A major contribution of the Pro-Keynesian School was to identify the organisation principles for the desirable institutional structure. The *multiplicity of institutions* was to ensure competition; the *variety of forms* were to meet the multiplicity of needs of rural customers; the *density of coverage* was to ensure an adequate volume of business to reap economies of scale, and to lower information costs; the *multiplicity of functions* was to promote the emergence of linkages and structural change of the rural economy, with the vertical linkages to facilitate the payments mechanism and to integrate the rural and urban financial sectors into a unified structure to reduce dualism.

Unfortunately, the pro-Keynesians underestimated the negative effects of public involvement, using fixed interest rates and credit quotas, on rural financial institutions. According to Adams (1988), fixing interest rates led to financial institutions bankruptcy because they could not charge rates that would cover the cost of the services; the same fixed interest rates made financial institutions unsustainable because they could not raise voluntary savings and had to rely on donor or budgetary injections for loanable funds.

Political interference led to the award of credit as patronage or rents to supporters with no seriousness to pursue loan repayment: this led to the accumulation of non-performing assets. The *rural poor* in whose favour public involvement was supposed to distribute credit were instead *rationed out of the credit market* since they lacked political connections and clout. The adherence to political criteria destroyed the *incentive to develop professionalism* in the running of rural financial institutions.

What the Keynesians sought was a supply-driven policy of making rural financial services available to stimulate economic development. Unfortunately the methods of implementation, particularly the political involvement into rural sector finance, had so many negative effects that they prevented the emergence of viable and sustainable rural financial institutions; they also excluded the poor rural clients. Because of both of these defects, the pro-Keynesian approach became regarded as a failure in promoting long-term economic development.

However, the development content of *the pro-Keynesian contributions still remain valid* to the SSA context. Financial sector development has to ensure *horizontal and vertical integration* of the entire national financial structure to promote inter-sectoral and inter-seasonal movement of resources to cope with risk and seasonality. *A critical density of financial institutions* is also still required to reduce market failure by increasing information flow, and to reduce unit costs by increasing the volume of business. We shall return to these contributions when we evaluate the Ugandan proposals in the concluding section.

The Neoclassical Perspective

The neo-classical economists, in refuting the pro-Keynesian perspective, focused particularly on the problems created by the repressed interest rates:

- they encouraged *investment* in low-net return projects, thus misallocating scarce capital and reducing potential growth;
- they discouraged the *mobilization of savings*, to finance investment, since savers were not compensated for parting with their liquidity; under inflation and fixed nominal rates, real interest rates were negative;
- they *distributed income* against small economic agents: these were unable to borrow because the cost of collecting information on small loans exceeded the income to be earned from the low fixed interest rates; the small economic agents also earned practically nothing on their savings, which were on-loan to larger economic units.

The overall emerging reaction was for liberalization of the *entire financial sector*, which went beyond the de-regulation of nominal interest rates, to include lifting credit allocation controls to specific sub-sectors, including agriculture, privatization of parastatal rural banks and other financial institutions.

The most provocative neo-classical work that provided the theoretical framework for the financial sector restructuring policy package under the SAPs is edited by Adams, Graham, Pinske (1984); it argues that *rural development should not be undermined with cheap credit*!

The Limitations of the Price Mechanism in Providing Rural Financial Services for Economic Development

The liberalized interest rate, as a price for financial services is essential for viability and sustainability of the financial sector generally since it enables financial institutions to charge a price that is sufficient to cover the direct cost of their services and the opportunity cost of loanable funds.

However, the role of the interest rate under the liberalized SAPs framework while necessary, is not sufficient to ensure the provision of rural financial services. The experience of many SSA countries, reviewed in the work edited by Frimpong-Ansah and Barbara Ingham (1992) indicates the following problems:

Whereas repressed interest rates "crowded out" small economic agents, especially in rural areas, the liberalized interest rates regime, in an overall deregulated environment, is likely to "crowd out" the rural sector even more severely. Commercial banks are no longer required to operate rural: either to collect savings; or provide credit; or administer the payments mechanism.

Instead, commercial banks concentrate on urban and peri-urban sectors, particularly trade and commerce where higher interest rates can be charged, where loan collection can be done at shorter gestation periods than in agriculture subject to seasonality, and where there is no risk from weather, pests and diseases.

Privatization of the parastatal-banks led to closure of rural branches. This raised the risk of rural finance by reducing inter-bank information flow; it also raised the unit cost of doing business and reduced the gains from economies of scale by lowering the overall volume of rural business. In other words, liberalization introduced a private financial institutions' response that run counter to the six organizational principles of rural finance outlined by the pro-Keynesians (Desai and Mellor 1993), with the development content thrown out.

Higher interest rates did not appear to lead to an increase in the volume of financial saving either. Paying interest rates on small deposits is a cost to commercial banks which carefully avoid it by raising minimum deposit requirements. The small savers, with fewer formal financial institutions to turn to, continue to save in kind, perpetuating economic dualism.

Investment did not benefit from higher interest rates either. *The pro-Keynesian hypothesis remains valid, investment is inversely related to the real interest rate.*

Banks that attempted to expand investment in productive sectors hurt the economies in two ways: first, by undertaking riskier projects from which they expected to earn a higher interest rate, without careful evaluation of risk; this has made the financial sector less stable. Second, the new projects paying higher interest rates in an imperfectly competitive setting, by passing on this cost to their customers, contribute to "cost-push inflation."

In the midst of the above criticisms, *the excess demand for credit for productive rural investments, has remained*; the high interest rates under the SAPs have failed to clear the market for loanable funds.

The Structuralist Perspective
According to the structuralists, *the role of the interest rate in clearing rural credit markets is over-stated by the neoclassical school.*

The rural credit market is dualistic: the formal segment provides credit from donors and government sources at below market interest rates; the informal segment relies on credit from private individuals who include, for example, professional money lenders, traders, landlords, friends and relatives (Hoff and Stiglitz 1993).

The informal credit market uses direct rationing to solve the *screening* and *enforcement* problems by lending to only those clients from whom it can collect information and over whom it can exert sanctions to recover the loans. This is to say: the trader lends to those farmers from whom produce can be collected at harvest time to pay the loans; the friend and kin lend within the known circles where information on credit worthiness is available, and within which peer pressure can be exerted to repay the loans. The professional money lender cultivates loyalty from the customers to repay the loans, and collects information on their creditworthiness.

Given the direct mechanisms for screening and enforcement, *informal credit is available to only limited segments of the population: the ability of the lenders to charge high interest rates* does not clear the market.

In the formal credit market, the interest rate is used for screening and enforcement, but only up to a point. Charging higher interest rates increases the profitability of lending: simultaneously, however, *as interest rates go up, the risk of default increases*. It follows that a formal financial institution will only lend up to a point where profitability exceeds the risk of default. Beyond this, the rising demand for credit, as reflected in rising interest rates, will not lead to credit expansion; instead the financial institution will ration out the extra customers by non-price mechanisms such as excessive paper work. *The disequilibrium in the credit market persists despite deregulated interest rates.*

A rise in interest rates in the formal financial market should attract savings from the informal sector, which is not rewarding savers: this movement of resources was expected to increase the volume of loanable funds in the economy to finance investment, according to the prediction from the neo-classical school.

However, the literature summarized by Simmons (1992) indicates that unless the funds from the informal sector are "idle", their flow into the formal sector reduces investment in the informal sector; and, at the margin, there may not be any net expansion in overall investment in the economy.

Owen and Solis-Fallar (1989) have argued that the movement of funds from the informal to the formal financial sector should expand investment, on the grounds that the latter allocates these funds more efficiently in a competitive environment, unlike monopolistic competition in the informal financial market.

However, this critique does not address the problem of risk aversion which prevents rural formal credit markets from clearing.

The Contribution of the Structuralist Critique

The focus of the structuralist perspective on market segmentation, persistent disequilibrium and credit rationing, has contributed the following:

Liberalization of interest rates will not by itself lead to the integration of rural financial markets to equilibrate demand and supply for financial services. The problems of *screening* and enforcement, related to risk aversion due to insufficient information, must be tackled *directly*. The neoclassical view has not been able to resolve this contention.

Public sector involvement will not eliminate the excess demand for rural financial services either, unless it too addresses the information problem directly. Given the possible adverse effects of such involvement, the emerging view from the literature is that public involvement would be more useful if it addressed the structural obstacles to rural finance directly, such as:

i. improve *rural infrastructure* to increase farm-gate prices and farm income by lowering transaction costs: as income increases, the risk of default diminishes;
ii. promote *technological change* through research and extension, which too in turn increases agricultural profitability and farm income, and lowers the risk of default;
iii. improve the *legal framework*, especially the definition and enforcement of property rights, so that collateral can be used more widely to secure loans;
iv. invest directly in the formation of autonomous *cooperatives* and other village *groups* which, once formed, can inculcate the savings culture among its members for savings mobilization, and use peer pressure for screening and loan repayment. This investment should be regarded as *social overhead capital*.

To the structuralist, there is room for public sector involvement in rural finance, but it has to be re-focused on the structural constraints directly.

This conclusion will be re-visited in the section on Uganda, along with the development content of the pro-Keynesian perspective.

Best Practices in Micro Finance and the Problems of Providing Financial Services in Rural SSA

The literature reviewed under this section is based on the following:-

(i) the empirical work of eleven case studies by Christine, Rhyne and Vogel (1994) whose results are summarized in table 9.1;
(ii) the qualitative assessment of four publicly sponsored programs in Asia by Yaron (1994), of which two viz, the unit DESA system and the Grameen Bank, are also included in table 9.1;
iii) another two cases i.e. the Bank for Agriculture and Agricultural Cooperatives (BAAC) in Thailand and the Badan Kredit Kecamat (BKK) in Indonesia;
(iv) the concepts floated by the World Bank Consultative Group to Assist the Poorest (CGAP) to finance the scaling up of successful micro-finance institutions that have proved viable while serving the poor, (1995);
(v) Thillairaja' s World Bank Discussion paper No. 219 (1996) on the development of rural financial markets in SSA;
(vi) the USAID (1995), Principles of Financially Viable Lending to Poor Enterprises;
vii) and the proposal under debate on the role of NGOs in Rural Financial Intermediation in Ghana, by Reed etc (1994).

It is beyond the scope of this chapter to go into the diversity of analytical approaches in these works. The focus here, instead, is on those common themes raised by the works that are relevant to the problems of providing rural financial services in SSA, characterized by risk, seasonality, and rurality.

The common themes covered are: ownership structure; operational self-sufficiency; sustainability; the CGAP proposals for scaling up; financial sector integration; macro and regulatory issues.

Table 9.1: Indicators of Best Practices for Rural Financial Institutions

(1) Name of Institution & Country	(2) Age years	(3) Type Ownership	(4) Urban /Rural	(5) No. of Borrowers	(6) No. of savers	(7) Oper- % self-sufficiency	(8) Return on Average Assets
1. BRK Niger	3	Village	Rural Financial	7,000	n.a	44	- 11.5
2. K.Rep.(Kenya)	4	NGO	Both	5,000	Same	106	- 18.5
3. FINCA (Costa Rica)	10	NGO	Rural	5,000	Same	98	- 6.3
4. Grameen	18	Govt.	Rural	1,587,000	Same	105	- 3.3
5. ADOPEN	12	NGO	Urban	4,000	n.a	94	- 0.8
6. ACEP (Senegal)	8	NGO/ credit Union	Both	2,000	n.a	142	0.1
7. Bancosol (Bolivia)	7	Private Commercial	Urban	46,000	n.a	107	1.0
8. BRI (Unit Desa System)	10	Division Govt. Commercial	Both	1,897,000	11,325	113	1.8
9. BKD(Indonesia)	40+	Village Financial Institution	Rural	907,000	817,000	197	3.2
10. Actuar (Columbia)	6	NGO. Finance Co.	Urban	32,000	n.a		
11. LDP (Indonesia)	10	Village Owned	Both	145,000	379,000	148	7.4

Source: Christen R.P., Rhyne, E., and R.C. Vogel, "Maximizing the Outreach of Microenterprise Finance: The Emerging Lessons of successful Programs," Consulting Assistance for Economic Reform (CAER) Paper, *Havard Institute for international Development*, 1994.

Ownership Structure

With reference to table 9.1 column 3, of the 11 institutions with the "best practices," 5 are NGOs; 2 are village-owned private financial institutions; 3 are either directly government owned or government specially licensed; and 1 is a private commercial bank. It appears that successful provision of rural financial services is possible under a wide variety of ownership in the NGOs, public and private sectors. It follows that *privatization of rural financial institutions is not necessary in every case*; it can be pursued on a selective basis, depending on the circumstances of a country.

The review by Yoran (1994) of the 4 publicly owned success institutions further strengthens the case that *it is possible to meet the criteria of "best practices" in rural finance even in publicly owned institutions*.

This point is very important because it opens room for further discussion of the organizational principles for rural financial services articulated by Desai and Mellor (1993).

The presence of public institutions in the rural areas has specific advantages: firstly it *adds to the density of coverage*, which increases the volume of business and lowers unit costs, as economies of scale are realized; and, secondly, it also adds to the variety of services available to rural customers.

It is not surprising, therefore, that three of the 7 best performance groups in columns (7) and (8) of table 9.1 come from institutions in Indonesia with a variety of ownership.

The public institutions, in particular, can be used as vehicles to *transfer donor funds or budgetary resources* into rural areas where the social rate of return may be higher, provided this can be effected without sacrifice of the organizational criteria that constitute "best practices."

Furthermore, public institutions can be located *"rural-rural"* to *integrate the entire financial sector*, an outcome that is very unlikely to come from private banks in a liberalized environment, but which is crucial to fight the problems of risk, seasonality, rurality and dualism.

Since it is the *practices*, rather than the ownership form that matters, let us examine these practices more critically.

Operational Self-Sufficiency

Operational self-sufficiency is defined to mean that the institution covers its *operational expenses*, irrespective of organizational form, *within 4-7 years*, once it has gone through the teething problems of putting its physical plant in place: offices and equipment, training of staff and establishing logistics of collecting savings, disbursing loans and effecting a payments mechanism. The operating expenses include: salaries of staff, logistics like transport to service clients, the maintenance and repair of office and equipment, and the payment of utilities. Of the 11 institutions listed in columns (2) and (7) of table 9.1, only BRK (Niger), the newest (3 years old), had not yet achieved operational self-sufficiency: it covered 44% of its operating costs.

The rest of the institutions, irrespective of organizational form, had within 4-7 years, been able to cover over 98% of their operating expenses.

Achievement of operational self-sufficiency is regarded as a very important "practice" because one of the key objections stressed in the SAPs literature is that institutions that have to rely on perpetual budgetary subsidies misallocate resources and make the conduct of macro-economic stabilization difficult.

The *interpretation of the determinants of operational self-sufficiency is highly controversial and requires careful sorting, however*, over three issues: *the role of the interest rate; organizational efficiency and the scale of business*; and the *specific economic environments* within which the financial institutions operate.

A. The Role of the Interest Rate and Operational Self-Sufficiency

The neoclassical perspective stresses that interest-rate flexibility is the key determinant of operational self-sufficiency. A financial institution must have the liberty to set interest rates that enable it to conduct business at a profit. The deposit interest rate must be sufficient to attract *voluntary deposits* for on-lending. The *spread* between the deposit and lending interest rates must be sufficient to enable the financial institutions to cover the costs of providing the financial services. If a country is not operating a liberalized interest rate regime, the institution must be allowed to charge commissions, etc., as substitutes to defray the cost of doing business.

The literature on micro-enterprises quotes interest rates on loans as high as 36-56% that the "best practices" institutions have been able to charge to cover the costs of their services.

The fact that very high interest rates *destabilize the financial system* by encouraging loans to riskier businesses is not answered in the microfinance literature, however.

It must be noted that the "best practice" institutions do *not* collect the high interest rates quoted on loans in every case. In the Grameen Bank, for example, the "*effective interest rates are kept low*" by programmed *weekly payments* on the *declining principle*. The problem is that the projects must yield *continuous cash flows*, and this "crowds-out" Agriculture in favour of the informal sector, particularly petty trade, which has a shorter gestation period and is not subject to *seasonality*. This is a key issue to which SSA must pay special attention. In fact, the Indonesian "success" did not start until micro-finance got out of Agriculture!

B. Organizational Efficiency Practices

An alternative way to minimize the interest rate charges is to streamline the organization structures to achieve *efficiency*. At the micro level, the literature documents the practices to keep operational expenses to the minimum:
- simplified workplaces with minimum furniture and equipment to keep office costs low;
- a highly motivated and well-paid staff, equipped with low cost transport e.g. bicycles and motorcycles, and resident in the communities served, to collect deposits and supervise loans, to ensure high recovery rates;
- inculcation of a voluntary savings and credit repayment culture so that the financial institution is regarded as a business, not a charity or a source of political patronage.

These practices are non-controversial and very worthwhile to adopt in rural financial institutions. However, they do not tell the entire story.

C. The Scale at Which Business is Done

A third route to lower interest charges is to *scale-up* the volume of business. There are *two* interpretations in the literature over this question. The neoclassical interpretation is that the rural financial institutions with "best practices" in table 9.1 should be run as *mass* institutions: for example, the minimum clientele serviced is 2,000 loan accounts by ACEP in Senegal and 379,000 savings accounts in LDP in Indonesia. The literature surveyed by Christen etc. (1994), and that in the CGAP (1995) proposals put the critical figure at 3,000 clients to realize economies of scale and minimize unit costs.

Minimizing unit cost is part of the quality of service. In order for the services to reach so many people, the following particulars are necessary:
- the services must be kept simple with streamlined passbook accounting procedures that can be followed by illiterates to attract more clients;
- the services must be flexible where loans can be rescheduled to meet

unforeseen problems not caused by willful delinquency, this keeps down the default rate and sustains a high volume of business;
- the services must be accessible i.e., as near as possible to the workplace of the clients, to minimize transport costs; and
- the services must be almost automatic so that a transaction can be completed in one visit, to minimize transaction costs to the client.

In order to have a high enough volume of loanable funds for the *mass* institution, voluntary savings are attracted, and kept liquid on simple savings accounts to induce customers to translate their savings from in-kind to cash. This requirement needs careful liquidity management to maintain solvency and prevent a run on the financial institution. In addition, *access to credit on completion of a loan repayment is rated the most important motive for keeping the delinquency rate low and for staying with a financial institution.*

The pro-Keynesian interpretation, while not refuting the need to streamline and standardize operational procedures to offer services on a mass scale, argues that there are rural-rural businesses with *development components* to which financial services cannot be provided at full cost, initially, until a level of development is reached.

The Ghana paper identifies and costs the development components. These consist of varieties of *technical assistance needed by entrepreneurs with financially bankable projects but without prior bank experience*: to process loan applications; and to time the loan maturities to coincide with the highest flow of revenue from the funded projects etc.

The paper gives two examples: a community-owned oil mill; and storage of a harvest to sell at off-season higher prices, both in *rural* Ghana.

The *frequency of visits* to *rural-rural* customers, who must be *served on site*, makes it impossible for the provider of financial services cum-technical assistance to recover full-costs.

Yet when the enterprises mature, they increase the *volume* of rural business and lower unit costs for the entire rural financial sector; they increase profitability and lower the risk of default; they encourage the ploughing back of rural savings into enterprises, which contributes to sustainability; they improve the linkages between the rural and urban economy, reducing dualism; and they reduce rural/urban migration by improving rural productive employment prospects.

The mass institutions, as viewed by the *neo-classical literature,* take bankable projects as given and only provide loans and collect savings. They minimize transaction costs by concentrating on densely populated areas where it is easy to form credit and loan groups, village associations, etc, and use peer pressure for screening loan applicants and enforcing repayments. The businesses are foot-loose, and the financial institution needs not go on site. The turn over is quick, and the cash flow continuous.

If the SAP framework does not restructure the financial sector to develop bankable projects to increase the volume of business, then it must admit that it is ignoring its own acknowledgement that "Financial sector reform in many developing countries is more a process of development rather than reform" Thillairajah (1995: 153).

Sustainability, the CGAP Proposals and Financial Sector Integration

Sustainability is defined as the ability of a financial institution to earn a positive return on capital comparable to the going rate of interest in the financial market. A financial institution with this capability no longer needs injection of donor or public grants; it generates a positive income on its loans portfolio *to cover not only its operating expenses, but in addition, to pay for the cost of loanable funds*; either by paying a positive real interest rate on voluntarily mobilized savings, or by raising equity on the open market in competition with other financial institutions; or by charging itself the market going interest rate for donor or public funds used.

Table 9.1 column 8, shows that not all the institutions with the "best practices" have been able to achieve financial sustainability over the periods covered.

Some of the obstacles to achieving sustainability are specific to the institutions, such as the neglect of mobilizing and managing a large enough volume of voluntary savings for on-lending. However, there are two types of other obstacles that are commonly shared: rurality and integration.

A. Rurality
The section on Operational Self-sufficiency presented the arguments that those institutions developing bankable projects located in rural areas which require frequent visits on site do *not* achieve operational self-sufficiency; they indeed cannot achieve sustainability, despite their critical role in rural development.

B. Integration
The literature reports that the institutions with the "best practices" : are *autonomous*, having grown up from the grassroots to meet the needs of rural clients.

The links to urban institutions are only used for the purpose of keeping deposits in safe custody. This observation agrees with the experience documented by the studies in SSA, by Aryeetey (1992), Hyuha (1993), Chipeta and Mkandawire (1992) and Aredo (1993) sponsored by the African Economic Research Consortium (AERC), which addressed the problem of financial sector integration by investigating whether there are links between the formal, semi-formal, and informal financial sectors.

According to the proposal in a World Bank sponsored study by Thillairajah (1995), there should be three types of institutions in a properly functioning financial market, which would ideally play the roles of, wholesaler, intermediary, *and* retailer, of financial services to the ultimate customers at the grassroots, with the following advantages in the linkage chain:

i) The *retailers* would cope with the rurality of the small scattered customers and minimize the cost of collecting information on creditworthiness, as well as loan recovery and savings mobilization.

ii) The *intermediary role* of the semi-formal sector would enable the informal and semi-formal customers to earn interest by depositing their savings in the formal sector. This would assist the mobilization of savings and help the formal sector to create more money. The intermediaries would also save the formal sector the extra cost of dealing with small deposit and loan accounts. The same intermediaries would pass on loanable funds to the retailers, relieving the excess demand for credit at the grassroots.

iii) The *wholesaler* formal sector intermediaries would be able to meet cash shortages and utilize idle funds. These operations would be crucial in meeting the problems of seasonality, and mitigating covariant risk.

Unfortunately, the AERC sponsored studies concluded that the linkages are either extremely weak or indirect on both the credit and deposit side, or practically non-existent.

The neoclassical logical argument would run that once the returns on assets in rural institutions are positive, the entire financial sector would integrate itself for funds to flow to the most profitable enterprises, urban and rural. This is yet to be seen, given the perceived prejudices of urban-based banks against the more humble organizational set-ups required to keep costs to the minimum in micro-enterprise banking, and the unorthodox clientele served without collateral.

c) The CGAP Proposals

The CGAP proposals intend to raise donor funds for on-lending to those micro-enterprise institutions that have achieved operational self-sufficiency, *to scale up* the process to convert into mass institutions that become self-sustaining.

The possibilities for CGAP funds open up three more points for the debate on rural finance:

First, the donors are admitting that micro-finance institutions, even when properly run, are not easily integrated into the national financial sector, otherwise they would raise enough of their own capital for scaling up.

Second, it is hoped that the distribution of CGAP funds will not discriminate against those rural-rural NGO institutions that are trying to increase the volume of rural business by stimulating the development components, as documented in the Ghana case. To do so would be to systematically "crowd-out" Agriculture. In SSA this would amount to a *negation* of rural development.

Third, some CGAP money should be deliberately lent to streamlined programmes to integrate the entire financial sector, since liberalization by itself is unlikely to bring about the desired linkages to accomplish this.

The Macro-Economic and Regulatory Framework

The focus of the macro-economic discussions is on maintaining positive real interest rates, by controlling inflation. There is no dispute on the desirability of this enabling environment for rural finance, provided it is understood that *positive real interest rates while necessary, are not sufficient to stimulate rural business.*

For the regulatory framework, the literature extensively highlighted the necessity for setting up nation-wide procedures that each institution should follow: to rate the quality of assets, whether performing, doubtful or non-performing; to document the amounts of saving and loan accounts and the average amount per account; and to rate customer credit-worthiness. These regulations would assist the entire financial sector to monitor its health and avoid bankruptcy; use the performance rating to facilitate the institutions identified as the most credit-worthy to borrow from the open financial market; assist customers to have their credit-rating easily checked to enable them to use a multiple of institutions to meet their credit needs; assist the monetary authorities to integrate and regulate the entire financial sector by being able to generate performance indicators across the entire spectrum of financial institutions.

Rural Financial Services and Economic Development Under the SAPs: A Case Study of Uganda's Experience

Introduction

This section evaluates Uganda's experience in providing rural financial services under the SAPs, in the light of the literature reviewed in the preceding section. Ideally the different theoretical perspectives and propositions should be tested empirically. Unfortunately, the time of frame of this research project does not allow the collection of the detailed microeconomic data required for rigorous testing. Instead, the evaluation is qualitative, but it is based on a wide enough variety of secondary data, to enable us to reach some tentative conclusions.

The following section gives a brief history of Uganda's rural financial sector inherited at independence, which was severely damaged during the civil strife and economic mismanagement; the removal of this damage provided most of the justification for the policies pursued since 1987. The subsequent section presents the persistent demand for rural financial services, despite financial sector restructuring; the limited effects of public interventions, on the one hand, and the liberalized interest rate, on the other, to reduce the excess demand; and the persistence of this excess demand, which has sparked off the ongoing debate looking into the future.

Rural Financial Services in Uganda's Historical Context

Whatever rural financial institutions Uganda inherited at the time of independence in 1962 suffered serious decline in the 1970s and most of the 1980s due to: civil strife which demonetized the economy and led to escape into semi-subsistence agriculture, and the poor macroeconomic policies which denuded the value of financial assets and reduced rural incomes.

The indigenous saving-credit clubs were not strong, compared to the case in other countries in SSA like Cameroon or Ghana; they consisted of "welfare" and "mutual-help" arrangements among associates, villages, or clans, to tidy over emergencies and bulky expenditures at funerals, wedding, etc.

The emerging modern rural financial sector consisted of: traders, the cooperative movement, and the Post Office Savings Bank.

A. The Traders' Provision of Financial Services

The Asian community provided crop-finance at harvest time, and consumer credit to farmers to purchase iron sheets, general merchandise etc. On the savings side, farmers repaid in instalments, or made prior deposits with the traders for specific purchases. The expulsion of the

Asians in 1972 completely disintegrated this rural savings-credit network; the personal trust on which it was based will be very difficulty to reconstruct.

B. The Cooperative Movement

The bulk of cooperatives were marketing institutions which gathered cash crops from farmers and sold them to the cooperative unions for processing, from where they were forwarded to the marketing boards for export. In the 1985 Survey, there were 2,922 marketing cooperatives for cotton and coffee. These cooperatives provided purchased inputs in kind to the farmers on an annual basis, but they did not mobilize savings.

A different group of cooperative savings and credit societies which numbered 373 in 1985 received both compulsory and voluntary savings from members and extended small loans for both consumption and investment. The bulk of the membership came from lower-cadre civil servants wage earners: teachers, nurses, midwives and clerical workers. Some of the societies would have thrived well, except for the poor policy of paying salaries late by the public sector that placed these societies under strain.

All cooperatives suffered from the Cooperative Act of 1970 which took away the autonomy of the cooperative movement, and run cooperatives like state-parastatals, with political interference in the election of leaders, and accounting for finances. The monopolies of the Coffee Marketing Board, and the Lint Marketing Board for cotton, made the situation worse by paying farmers late, cash whose value was denuded by inflation. Although the Cooperative Act was reformed in 1992 to restore some autonomy, and cash crop marketing has been liberalized, most marketing cooperatives continue to suffer from lack of trust by farmers and are financially insolvent. (An account of the cooperative movement in the savings and credit markets is given by Ssemogerere, 1992).

C. The Post Office Savings Bank

This provided a nation-wide network for collecting savings in simple-passbooks on which interest was paid. High inflation made the real interest rates negative and discouraged savings. The Post Office service itself was badly rundown by looting, theft of packages, and degenerated along with other infrastructure during the civil strife. The currency reform of 1987 struck off two zeros and took off 30% conversion tax on all financial balances. This practically wiped out the small post office savings accounts.

Attempts to revive the Post Office Savings Bank network are underway but they will have to overcome the traumatic memories of the previous lost savings.

The Excess Demand for Rural Financial Services
This section considers the excess demand for rural credit, savings facilities, and a payments mechanism in Uganda.

The Excess Demand for Rural Credit
Table 9.2 presents an estimate of the credit requirements for the Agricultural sector alone, without the requirements of the other off-farm activities. The estimates, therefore, should be regarded as *minimum requirements.*

According to the Agricultural Secretariat, total commercial bank credit estimated to be available for the entire economy in 1994/95 was USh 194 billion, which only represented 87% of the agricultural sector requirements of USh 222 billion. The actual total flows into the Agricultural sector was only USh 4.9 billion: this figure included existing obligations, which meant that net flows into the sector were much less.

Therefore, not only was the agricultural sector practically dry of credit, but the rural sector as a whole, including non-farm activities, and the entire Ugandan economy was starved of formal credit.

In Table 9.3, Mayanja (1995) attempted to estimate the size of the informal financial sector as a while and put the figure, in the last column, at around 35% on the average.

On the credit side, if we took the USh 194 billion and augmented it by an average of 35%, the total figure of loanable funds in the entire economy would be USh 252 billion; this is an augmentation by only USh 68 billion. The government, therefore, should not take solace in the capacity of the informal sector to solve Uganda's rural credit requirements: the contribution of the informal sector is too small, relative to the overall demand.

Table 9.4 shows that financial sector liberalization is not improving matters for agriculture: the flows of commercial bank lending into agricultural production and crop finance, in columns 2 and 3, have been dwindling since 1991, while commerce, the largest gainer, and to a small extent industry, are increasing their percentage shares.

Table 9.2: Estimated Credit Requirements for the Agricultural Sector

Year	Crop Finance	Short Term Investment	Livestock	Fisheries	Total
1994/95	163.19	34.20	20.00	5.00	222.39
1995/96*	420.63	37.80	25.00	6.00	309.43
2000*	474.01	53.10	28.50	7.00	562.61

Source: Bank of Uganda, 1995.
* estimates

Informed circles contend that crop finance to agriculture is actually larger than that implied by column 2 of table 9.4. Part of the crop-finance is borrowed privately from abroad at 7% nominal interest rate because Uganda commercial bank lending interest rates are proving sticky downwards around 20% to 28% despite interest rate liberalization and the decline in the rate of inflation.

However, the trickle down of funds from abroad goes first to large-scale coffee exporters; how much trickles through to small-scale rural customers, who constitute the majority, is yet to be established, particularly the extent to which it reduces the excess demand for rural credit.

Table 9.3: Total Nominal Deposits to Money Supply in Uganda (In billion Shs.) and the IFS (1973-1994)

Year	Total Deposits	M2	Deposits/M2	IFS as a % of the Financial Sector
1973	0.020	0.028	0.71	0.29
1974	0.028	0.039	0.72	0.28
1975	0.033	0.042	0.79	0.21
1976	0.040	0.062	0.65	0.35
1977	0.045	0.074	0.62	0.38
1978	0.058	0.093	0.62	0.38
1979	0.079	1.137	0.58	0.42
1980	0.112	0.184	0.60	0.40
1981	0.239	0.353	0.67	0.33
1982	0.259	0.385	0.67	0.33
1983	0.355	0.544	0.65	0.35
1984	0.676	1.115	0.61	0.39
1985	1.569	2.620	0.60	0.40
1986	3.608	7.184	0.50	0.50
1987	9.479	18.540	0.51	0.49
1988	19.169	39.189	0.49	0.51
1989	42.389	79.104	0.54	0.46
1990	66.595	117.203	0.57	0.43
1991	109.860	190.937	0.58	0.42
1992	164.524	262.354	0.63	0.37
1993	221.300	354.000	0.63	0.37
1994	267.200	402.600	0.66	0.36

Sources: V. Mayanja. *The Role of the Informal Financial Sector In Providing Financial Services to Small Scale Producers: A Case Study of Mawokota-County in Mpigi District.* Dissertation submitted in partial fulfillment of the requirements of the MA in Economic Policy and Planning of Makerere University, 1995.

Table 9.4: Commercial Bank Outstanding Loans and Advances to the Private Sector (% of total lending)

Year (as on Dec)	Agriculture	Crop Finance	Industry	Trade & Commerce	Others
1985	6.3	44.9	18.7	21.7	8.4
1987	11.2	46.5	22.1	13.2	7.0
1988	14.9	29.5	12.7	32.4	10.5
1989	11.5	17.8	13.8	35.9	21.0
1990	13.9	20.3	11.9	38.5	21.0
1991	8.7	25.7	13.3	37.6	14.7
1992	4.5	20.8	14.4	35.8	14.5
1993	3.9	20.7	18.1	45.2	12.1
1994	2.5	21.5	21.0	43.5	11.5
1995	2.0	20.5	22.5	45.5	9.5
1996	1.5	18.0	27.0	45.5	8.0

Sources: Bank of Uganda, 1995 and *Statistical Abstract. 1997.*

Some local commercial banks are also said to be lending rural, but the figures are not made available by the claimants, among whom is the local office of the World Bank.

The Excess Demand for a Rural Savings Facility

The case for savings is not documented as that for credit. The available data from the "Background to the Budget 1995 - 1996" is for aggregate savings as a % of GDP.

The trend in the public sector is from a negative - 5.3% in 1991/92 and - 0.8% in 1993/94, to a positive +0.7 in 1994/95. For the private sector, the trend is 8.0%, 8.8% and 11.6% percentages for the corresponding years.

The figures are estimates from national accounts and do not show the savings facility used by the private sector; whether these are internal retained earnings, or commercial bank deposits. It is more likely that these figures are from the formal private sector in the cities using commercial banks and provide no answer to the excess demand for the rural savings facilities. This aspect of the problem remains *blank* and undocumented.

The Excess Demand for a Rural Payments Mechanism

The Uganda Commercial Bank (UCB), a parastatal, has been the only institution with a branch network in every district in the country. The

UCB has been offering services for the payment of teacher's salaries and other upcountry civil servants, school fees, and more recently, income tax. The UCB together with the rest of the banking system has been assisted by the new regulation to put in prison any author of a bouncing personal cheque. The UCB bank cheque can be cashed anywhere in the system, just as good as cash.

Since financial restructuring, however, the UCB Branch network has been substantially closed down in an attempt to reduce operating expenses and to recover non-performing assets. The effect has been to reduce the number of commercial bank branches which has shrunk from 250 in 1992 to just 65 in October 1997.

Table 9.5 shows how financial sector restructuring has reduced the branch network and increased the percentage concentration of commercial bank branches in the central region, where the capital, Kampala is located.

More radically, the UCB is being privatized. It is most unlikely that a private commercial owner, to replace the parastatal, will maintain the rural payments network. Rather, the likely outcome is more closure of the rural branches and increased concentration in cities. The actual outcome awaits to be seen. As of now, the demand for a payments mechanism in rural areas remains equally undocumented, just like the demand for saving facilities.

Table 9. 5: Geographical Distribution of Uganda Commercial Bank Branch Network 1992-1997

Region	Central	Western	Eastern	Northern	Total
No. of Branches					
1992	99	67	55	29	250
1996	74	36	22	18	150
1997	28	15	13	9	65
Percentage share of the regions in the number of branches					
1992	40	27	22	12	100
1996	49	24	15	12	100
1997	44	24	20	12	100

Sources: The 1992 Figures are from J. Ddumba-Ssentamu,. The Role of Commercial Banks in Deposit Mobilization in Uganda. PhD Thesis submitted to Makerere University, Department of Economics, 1993. The 1996 and 1997 figures are from the Bank of Uganda.

Public Intervention in the Provision of Rural Financial Services

Public interventions have been in the form of administered credit schemes of funds from donors and, most recently the budget (in the form of the Entandikwa scheme), the credits are listed in Table 9.6.

There are two outstanding features in Table 9.6 which show poor performance of public interventions in relieving the excess demand for rural financial services.

First, the emphasis has been on the provision of credit. A savings facility is practically non-existent in the programs. The result is that once the public injections by donors and government dry out, the programs are not sustainable since they cannot generate own funds for on lending.

A second feature is the extremely low rate of utilization recorded in the last column of Table 9.6 for most projects. Given the excess demand for credit in Tables 9.2 and 9.4, this low rate of utilization indicates gross inefficiency in the administration of the systems, and it further implies inability to achieve operating self-sufficiency and financial viability, because of the poor outreach of the programs.

The limited information available in Table 9.6 suggests, therefore, that the public programs offer *poor quality service* and are unable, as designed, to ameliorate the excess demand for financial services.

It must be recalled from the literature, however, that it is not the public ownership of the funds or the facilities through which they are administered, which create the inefficiency: rather, it is the way the programs are administered that is at fault.

Financial Liberalization and the Role of the Interest Rate:

Table 9.7 presents the trend in interest rates since liberalization. The Treasury Bills auction rate has been used as the indicative nominal interest rate in column 3.

The first feature of Table 7 are the highly negative real interest rates on savings deposits in column (5), despite a decline in inflation, especially over the recent 3 years. This cannot encourage voluntary savings especially by small rural customers who, in addition, have to travel distances to make and withdraw deposits.

The second feature is the wide gap between the lending rates in column 1 and the deposit rates in column 2. A number of explanations have been advanced: first, that the banking sector is oligopolistic and segmented with no competition for deposits to offer better deposit rates; and second, that the presence of a large number of non-performing assets forces banks to survive on a few good loan customers from whom they extract high lending rates to cover inefficiency.

Table 9.6: Administered Agricultural Credit Programs

St. No.	Projects	Source of Funding	Implementing Agency	Currency	Amount committed	Amount Shs Equiv (millions)	Amount Disbursed (mil. Sh)	Utilization rate (% p.a.)
1	ADP	IFAD	MAAIF	US$	1.50	1425.0	587.0	44
2	NURP	IFAD	MAAIF	US$	0.71	674.5	422.4	68
3	SCRP	IFAD	MAAIF	US$	2.50	1983.2	258.8	13
4	SWRARP	IFAD	MAAIF	US$	0.50	475.0	108.6	23
5	Livestock Service Project	IDA	MAAIF	US	2.00	1900.0	"	"
6	R.C. Tractor Scheme	MAAIF		USHS			"	"
7	CSDP	IFAD	BOU/DFD	US$	10.00	9500.0	"	"
8	Hoima Int. Dev. Proj.	IFAD	MLG	US$	2.0	1900.0	"	"
9	RFS	ADB	UCB	FUA	15.0	18525.0	14573.0	79
		EEC		ECU	1.0	1187.5	1187.5	100
		DANIDA		DRK	8.0	4560.0	4560.0	100
		USAID		SHS	413.0	413.0	75.0	18
		DANIDA		SHS	260.0	260.0	203.0	78
		WFP/DDC		SHS	1500.0	1500.0	179.0	12
		Government		SHS	1445.0	1445.0	1200.0	83
		UCB		SHS	1445.0	1445.0	1445.0	100
10	CCS	USAID	CB	SHS	500.0	5000.0	2300.0	46
11	BOU Refinance Schemes	DFF	BOU/DFD	SHS	3940.0	3940.0	2800.0	71
12	Entandika	Government	Govt.	SHS	6000.0	6000.0		"
13.	NGO's		NGO's	SHS	600.0	302.98		50
	TOTAL					73058.20	34839.63	48

Source: Uganda, Agricultural Policy Committee (1994), *Cotton Sub-sector Development Project: A Rapid Appraisal of Rural Finance and Credit Schemes*: Vol. 2, 1994, Working Papers, p.204.

Note:

1. ADP and RPE project period expired and amount available will be amount disbursed.
2. Since most of the advances were for short-term production loans, the amounts were expected to be revolved for lending.
3. Utilization rate is the amount disbursed as % of amount committed.

ADB	=	African Development bank
ADP	=	Agricultural Development Project
CCS	=	Cooperative Credit Scheme
CB	=	Cooperative Bank
CSDP	=	Cotton Sub-Sector Development Project
DFF	=	Development Finance Fund
IFAD	=	International Fund for Agricultural Development
IDA	=	International Development Agency
NGO	=	Non-Government Organization
RC	=	Resistance Council (Local government)
RFS	=	Rural Finance Scheme
RPE	=	Rehabilitation of Productive Enterprises (USAID)
NURP	=	Northern Uganda Rehabilitation Program
SCRP	=	Small-holder Cotton Rehabilitation Project
SWRARP	=	Southern Region Agricultural Rehabilitation Project
UCB	=	Uganda Commercial Bank
WFP/DDC	=	World Food Program
MAAIF	=	Ministry of Agriculture, Animal Industry and Fisheries
MLG	=	Ministry of Local Government
BOU	=	Bank of Uganda
USAID	=	United States Agency for International Development

Table 9.7: Bank Interest Rates, Yield on Treasury Bills and Inflation

Period	Lending Rate Average	Deposit Rate Average	Yield on Treasury Bills	Inflation Annual Rate (1989=100)	Real Inter-Deposit Average	Spread Between Deposit & Lending Rates
	(1)	(2)	(3)	(4)	(5)	(1)-(2)
1987	30.2	15.7	31.5	..	..	14.5
1988	31.8	17.0	33.0	..	..	14.8
1989	40.9	26.5	43.0	..	..	14.4
1990	37.3	25.3	41.0	34.0	..	12.0
1991	37.2	21.8	34.0	27.7	-6.4	15.4
1992	30.1	19.1	37.8	54.5	-35.4	11.0
1993	24.7	12.3	24.4	5.1	+7.2	12.4
1994	22.1	5.9	15.2	10.0	-4.9	16.2
1995	20.2	2.8	12.8	6.6	-3.8	17.4
1996	20.3	2.3	13.5	7.0	-4.7	18.0

Source: Republic of Uganda, Satistical Abstract 1987

It is beyond the scope of this paper to give a conclusive explanation of the interest rate anomalies in Table 9.7. What can be said is that liberalization is not sufficient to make the interest rate play its part in the allocation of financial resources, and in attracting savings. Other market imperfections such as non-performing assets, and the oligopolistic commercial bank market structure, require additional direct policy actions.

Rural Financial Services and Economic Development: A Summary of Uganda's Experience Under the SAPs

The demand for rural financial services *has persisted* despite financial sector restructuring under the SAPs.

The excess demand for credit appears widening: Agriculture is increasingly "crowded out" with its share of production and crop-finance credit falling from 11.2 and 46.5, respectively, in 1987 at the beginning of the SAPs, to 1.5 and 18.0 percent in 1996.

Although private sector savings in the formal banking system have risen from 8.8% of aggregate savings in 1991/92 to 11.6% in 1994/95, the rise appears to be coming from the cities in the formal sector, rather than in rural areas.

The geographical distribution of commercial bank branches is increasingly eskewed from 40% of the branches in the Central Region, where the capital city Kampala is located, in 1992 to 44% in 1997; private banks have no incentive to locate rural.

The spread between deposit and lending interest rates has ranged between 11% and 18%; this is attributed to inefficiency of an oligopolistic banking structure which, despite the liberalization of interest rates is not competing for the modolization of deposits. Average real deposit rates are largely negative.

Most of the publicly administered Agricultural Credit Schemes, on which data is available, have a very low rate of utilization, reflecting inefficiency.

The results, though limited, do suggest that the multiplicity of market failures in rural finance persists, despite financial sector restructuring. It is this persistence which the new proposals in the following section are trying to address.

A Proposed System for the Provision of Rural Financial Services in Uganda

The proposed system, under active discussion, is sketched in figure 9.2. This section evaluates the key features of the system in terms of their suitability, in the light of the literature, to promote rural development. The evaluation is intended to further the debate for more penetrating policy reforms.

The proposed system is divided into 6 layers, each with comparative advantage to provide financial services at a particular level. Each layer is first evaluated separately. The overall policy recommendations are then given.

The Retailor Providers of Financial Services at the Grass Roots, Level 5 to Level 6

The comparative advantage of the retailer institutions proposed is to minimize transaction costs to serve small scattered farmers and off-farm enterprises. Four types of institutions are proposed at this level: the Cooperative Societies; the Private Sector Operators; the NGOs; and the National Farmers Association.

A. The Primary Cooperative Societies

The primary cooperative societies have been singled out as the most promising candidates to retail rural financial services to the grassroots. They have a legal organization structure and can be sued for default. They have documentation procedures which can be used for performance rating. They need, however, to improve the savings culture, administrative efficiency, and strive for operational self-sufficiency and financial sustainability if they are to become mass institutions with large scale outreach. The historical political interference must stop in order to restore public trust in cooperatives. Furthermore, government should pay civil servants on time so that those who belong to cooperatives can regularly save and meet their loan obligations.

Figure 9.2: A Rural Financial Services System: Operational Diagram

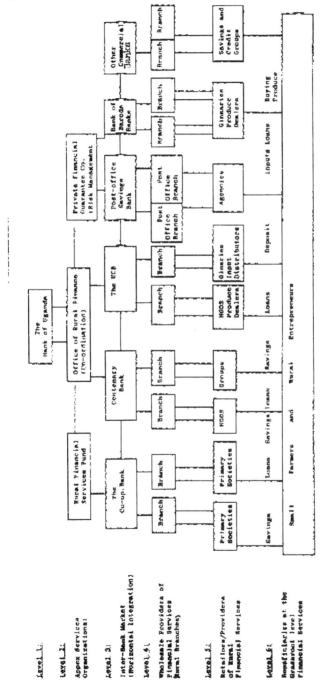

B. The Private Sector Dealers, Processors and Marketers

In the private sector, produce dealers, processors and input distribution traders, have been earmarked as retailers of financial services. Whereas these are suited to provide variable inputs, credit and crop finance to contract farmers, their role in savings mobilization needs to be clarified. Their ability to document and measure performance is also a problem. Besides, the trust that existed with the Asian traders and the rest of Ugandans needs to be re-established. Whether these providers are low-cost, cannot can be taken for granted since they are in the business to make a profit. What is at stake are the terms of the contract. There is a real possibility for monopsony to develop where the farmers are exploited and have no alternative outlet to get credit, sell their crops, get inputs or appeal for redress.

C. The NGOs/Local and Foreign

Some NGO groups provide credit, along with other services to the grassroots, particularly to women clients, and are earmarked to expand these services. There are many problems with NGOs, however. First, many provide a multiplicity of services which drive up unit costs and require donor subsidy. Second, the training functions of the staff should carefully be separated from the provision of financial services, and made uniform to national standards. Third, NGOs need to inculcate a voluntary savings culture, and to use the interest rate as a price at which they should sell services to customers for sustainability. Fourth, NGOs have to be persuaded to adopt record-keeping procedures that can be used to measure performance using nation-wide indicators; most NGOs only keep records for their own donors requirements.

Fifth, NGOs do reach out to poor communities: for this function, they should be assisted to make social capital investment a public good. Social capital is required to animate the communities to a level where they become capable of forming financial groups for saving and receiving credit. This assistance is critical for local NGOs who cannot afford to meet the subsidy element to cover the social capital investment.

D. The Indigenous Financial Institutions

Indigenous financial institutions include: hire-purchase schemes from better off to poor farmers; money lenders; clan and kin arrangements; "Munno-Mu-Kabi" emergency assistance schemes and trader-to-trader assistance. The heterogeneity of organizational form and objective, together with lack of a culture for keeping records, makes this retail channel difficult to develop. It is not ruled out, but it is likely to be thin on the ground for the foreseeable future, and it needs a large injection of social capital to become viable.

E. The Uganda National Farmers Association

The Uganda National Farmers Association wants to retail credit to its members through 16 appointed Business Agents (BAs) under a credit guarantee fund provided by them up to a sum of USh 300m. against which they would borrow USh 400m. The BAs are to onlend (retail) to the members (farmers), and also provide tangible securities to cover 30% of the borrowing. This proposal is worth considering on the following key issues:

It saves the commercial banks the cost of retailing services to scattered farmers at the grassroots such as screening credit-worthy customers, supervising and recovering loans.

It shoulders the risk of default through the guarantees at both levels, the UNFA and the BAs.

It is a registered body that can be sued.

What is not assured is the required record keeping for regulating this part of the financial sector by the BOU. This has to be provided for.

It is not clear, whether this association will assist members to have a saving facility.

The coverage nation-wide as to how many farmers to be reached is not clear either. The NFA can turn out to be an elite institution of a few rich farmers.

Common Problems at the Retail Level

Whereas the diversity of institutions is advocated in the Pro-Keynesian literature, as a way to meet the multiplicity of rural customers needs, the implementation of the Ugandan proposal should pay attention to three critical problems:

Adequacy of Coverage to reach rural customers needs to be encouraged and monitored: Since all proposed retailing institutions are *not* public, they have no a priori statutory obligation to canvas the rural area.

A second problem is *record keeping*: records need elements of uniformity to enable the Bank of Uganda to monitor and supervise the activities of all retailers. The NGOs, the NFA, and the Private Sector providers need a push to comply with this requirement.

A third problem is protection of rural customers against exploitation, particularly in the case of private sector providers where the relationship with grassroot customers can easily degenerate into monopsony.

The Wholesaler/Providers (Level 4 in Figure 9.2)

This level should consist of intermediaries that collect savings from Level 5 and deposit them with commercial banks branches at Level 4, and that pass on loans back to level 5. The intermediary role would minimize the costs of dealing with small scattered customers in a rural setting.

Unfortunately it is becoming clear that private commercial banks under liberalization are *unwilling* to extend their rural branch networks to perform this intermediary role; in other words, *the emerging financial sector is disconnected between Level 4 and Level 5.* The likely exception is the Centenary Bank.

A. The Centenary Rural Development Bank

The Centenary Rural Development Bank is a religious founded institution by the Catholic Church, with a network of Church Parishes as shareholders, which are also used to screen credit-worthy customers. By organizational form this bank is suitable to intermediate between Levels 4 and 5.

B. The UCB Rural Branch Network

The UCB is the largest Commercial Bank in Uganda with the bulk of Rural Branches.

In the view of many critics, the government of Uganda made a mistake, which is extremely hotly debated, to privatize UCB, even if the moves to recover non-performing assets, restore financial viability and profitability are welcome.

The experience of the public institutions in Table 9.1 by BRI and LDP in Indonesia demonstrate that operating self-sufficiency and financial viability are possible to achieve in a variety of organizational forms, including public institutions. In Uganda itself the experience of the Uganda Grain Millers which was run as a profitable parastatal for over 10 years attests to this.

The special advantages of a public institution's presence in the rural area are: to increase the diversity of coverage and the volume of business which makes it possible to realize economies of scale, increase information flow and lower risk; to move donor and public resources into rural areas *where the social rate of return is higher than the private rate of return*; and in the case of UCB, in addition, to provide a rural payments facility. Government is proposing to subsidize those loss making branches which offer the only banking services in certain districts. Whether this is a better option than retaining the UCB as a parastatal awaits to be seen.

C. The Rural Farmers Scheme (RFS) as Wholesaler

The RFS was part of UCB that proposed to hive itself off when UCB is sold and to become an autonomous rural bank.

The history of RFS of poor loan recovery is blamed on how it was merged with the rest of the UCB and deprived of its resources, especially motor-cycle transport to monitor its customers. This complaint and its implications on the cost structure of the RFS and loan recovery rate were not examined in detail prior to privatization.

However, many customers were retailers at Level 5, i.e., the cotton and coffee cooperatives. If the RFS wants to be autonomous, it would have to operate as a wholesaler.

But the RFS was also proposing to retail USh 1m -2m per customer. This micro-credit is better left to level 5 retailers: the RFS would find it too expensive to administer.

Common Problems at Level 5

The most serious challenge government faces at the wholesaller end is the refusal of Commercial banks to open enough rural branches for Retailers-Institutions to bank with. Under privatization, unless commercial banks are given special incentives or unless they are of a particular institutional design like Centenary, they have no incentive to bank rural.

With the disconnection between Levels 4 and 5, the Level 5 institutions will have to be developed separately as wholesalers until such a time as they have achieved operational self-sufficiency and financial sustainability to operate on the open financial market. In other words, financial sector integration would have to come *from below, with the disadvantage that it will be slow—projected to take 4 to 7 years, as the minimum.* The projects such as CGAP being floated by Donors, are intended to speed up this process by making more funds available to those Level 5 institutions which have achieved operational self-sufficiency to replicate their experiences rapidly to achieve economies of scale and financial viability so that they can integrate themselves into the financial sector at Level 4.

The Inter-Bank Market

Under financial restructuring, horizontal integration is being attempted, to shift funds from surplus to deficit banks, to increase the overall efficiency of the financial system, at Level 3.

However, as noted previously, the vertical links between Levels 4 and 5 are necessary to cope with seasonality and covariance of risk arising from agriculture by making more resources available, and by increasing the utilization of these resources by the entire financial sector.

The co-existence of a liberalized interest rate regime and the excess demand for financial services in the rural sector strongly suggest that for the Inter-bank Market to operate efficiently, specific incentives are required.

The Apex Organizations at Level 2

The office of Rural Finance Co-ordination is slated to co-ordinate Level 3 institutions down to the rest of the financial sector. The workings of this institution are yet to be articulated, however.

The Private Finance Guarantee Company addresses directly the problem of "risk" and is proposed to play the role of a Nation-wide Credit Guarantee Scheme. Unfortunately, again, Uganda's commercial banks have not yet shown enthusiasm to joint the Credit Guarantee Scheme, Government has to entice the banks. Furthermore government needs to develop an efficient legal framework for handling risk claims.

The Rural Financial Services Fund is proposed for capacity building given the importance of following sound accounting and record-keeping procedures in order to achieve "best-practice" in rural finance. *Capacity building is essential throughout the entire system, particularly at Level V and VI.*

The Regulatory Framework at Level 1

The Bank of Uganda has issued draft guidelines which are intended to establish minimum uniform performance prudential standards for financial institutions. They cover: capital adequacy, assets quality, limitation of credit to a single borrower, limitation of credit to insiders, compliance with the Bank of Uganda licensing procedures, reporting procedures for inspection purposes, and minimum liquidity requirements as percentage of total deposits.

This is one of the most healthy developments in the financial system: to guard against bankruptcy, develop an early warning system against non-performing assets; and develop reporting procedures that allow the entire set of national financial institutions to be measured by uniform performance criteria.

Although the Level 5 institutions do not yet fall under the guidelines, the capacity building project should assist them to adopt as much of the accounting procedures as are possible for performance rating. The government is engaged in a dialogue between these institutions and the Bank of Uganda supervision department. It is hoped that the regulatory procedures will emerge from this participatory process for Level 5 institutions. Some tentative recommendations from the dialogue are published by the Private Sector Foundation (1996).

The "Entandikwa" for Social Overhead Capital

One of the newest public interventions is the "Entandikwa" scheme, financed by the Treasury annual vote, and administered by NGOs and local government councils in every district, for poor customers who cannot afford collateral.

The first controversy with Entandikwa is the control of the lending interest rate at 10-12%, which the critics have observed distort the interest rate structure by imposing a ceiling without sufficient demonstration of its relationship to the cost of providing the service, and in direct contradiction to the interest rate liberalization policy in place.

Other criticisms include public interference in credit administration by local council authorities that participate in the screening of customers.

One of the proposals being floated is to use the "Entandikwa" to form groups to cultivate a credit/savings culture, particularly given the fact that poor communities need social capital to qualify for financial services.

The role of groups in reducing transaction costs, particularly information gathering, savings mobilization and enforcement of credit repayment, is recognized in the literature. The agents which form groups: individuals or NGOs, incur private costs for providing "social capital" from which they cannot collect returns to compensate themselves (Section IV above) (1993). This is a public good that "Entandikwa" is suited to finance.

Works Cited

Adams, D., "The Conundrum of Successful Credit Projects in Floundering Rural Financial Markets,". *Economic Development and Cultural Change*, 1988.

Aryeetey, E., *The Relationship Between The Formal and Informal Sectors of the Financial Market in Ghana*. AERC Research Paper No. 10. Nairobi: Kenya, 1992.

Aryeetey, E. and Gockel, F., *Mobilizing Domestic Resources for Capital Formation in Ghana: The Role of Informal Financial Sectors*. AERC Research Paper No.3, Nairobi: Kenya, 1991.

Bank of Uganda, *Summary of the Bank of Uganda Guidelines to Financial Institutions*. Presented to the Core Working Group of the Micro Credit Development Trust. (Draft), 1996.

Bell, C., "Interactions Between Institutional and Informal Credit Agencies in Rural India: A Case Study," ed. Hoff, K., etc., *The Economics of Rural Organization: Theory, Practice and Policy*. New York: Oxford University Press, for the World Bank, 1993.

Braverman A., and J.L. Guasch, "Administrative Failures in Government Credit Program s" ed. Hoff, K. etc., *The Economics of Rural Organization: Theory, Practice and Policy*, 1993.

Chipeta, C. and Mkandawire, M.L.C., *Links Between the Formal/Semi-formal Financial Sectors in Malawi*. African Economic Research Consortium, Research paper 14. Nairobi: Kenya, 1992.

Christen, R.P., Rhyne, E. and R.C. Vogel, *Maximizing The Outreach of Microenterprise Finance: The Emerging Lessons of Successful Programs*. Consulting Assistance for Economic Reform (CAER) Paper. Harvard Institute for International Development, 1994.

Consultative Group to Assist the Poorest (CGAP), "The Consultative Group to Assist the Poorest: A Micro-Finance Program," *Focus*. Note No.1, October 1995.

Dejene, Aredo, *The Informal and Semi-Formal Financial Sectors in Ethiopia: A study of the Iqqubi, Iddir, and Savings and Credit Co-operatives*. African Economic Research Consortium, Research Paper 21. Nairobi: Kenya, 1993.

Desai, B. M. and J.W. Mellor, *Institutional Finance for Agricultural Development: An Analytical Survey of Critical Issues*. Washington, D.C.: International Food Policy Research Institute. Food Policy Review No.1, 1993.

Dimery, L., Sen, Binayak, and Tara Vishwanath, *Poverty, Inequality and Growth*. Education and Social Policy Discussion Paper Series, No. 70. The World Bank: Human Resources Development and Cooperation Policy, 1995.

Ellis, F. (2nd ed.), Peasant Economics: Farm Household Models and Agrarian Development. Cambridge University Press, 1993.

Emenuga C., *The Outcome of the Financial Sector Reforms in West Africa*. Evaluation Seminar, Organized by CODESRIA, 1996.

Financial Sector Adjustment Credit Implementation Secretariat, *Study of the Implications of the Potential Closure of UCB Rural Branch Network on the Provision of Financial Services*. Workshop Presentations of Early Findings, by the Bank of Uganda: Agricultural Policy Secretariat, 1996.

Food and Agriculture Organization of the United Nations (FAO), *Agricultural Price and Marketing Policy: Government and the Market in Africa: An FAO Training Manual*, 1991.

Gibbon P., Haunevik. K.J. and K. Hermele, *A Blighted Harvest: The World Bank and African Agriculture in the 1980s*. London: James Currey, 1993.

Government of Uganda, Ministry of Planning and Economic Development, Agricultural Policy Committee, *Operationalization of the Medium-Term Plan for Modernization of Agriculture 1997/98 -2001/2002. Final Draft*, 1997.

Gudger, M., *The Development of Self-sustaining Credit Guarantee Facility for Rural*

Sector Finance. A Consultancy Report. Kampala: Bank of Uganda, 1995.

Hussain M.N., Mohamed N.A, and kameir El-Wathig, *Resource Mobilization, Financial Liberalization and Investment: The Case of Some African Countries*. Evaluation Seminar, organized by CODESRIA, 1996.

Hussain I., and R. Faruqe ed., *Adjustment in Africa: Lessons from Country Case Studies*. A Regional and Sectoral Study. Washington D.C.: The World Bank, 1995.

Hussi, P. etc., *The Development of Cooperatives and Other Rural Organizations: The Role of the World Bank*. World Bank Technical Paper No. 199, African Department Series, 1993.

Hyuha, M., Ndashall, M.O. and J.P. Kipokola, *Scope, Structure and Policy Implications of Informal Financial Markets in Tanzania*. African Economic Research Consortium, Research Paper 18. Nairobi: Kenya, 1993.

Inanga E.L. and D.B. Ekpenyong, *Financial Liberalization in Africa: Legal and Institutional Framework and Lessons From Other Less Developed Countries*. Evaluation Seminar Organized by CODESRIA, 1996.

Kasterton, A., *Sustainability of Financial Service Provision for the Low Income Sector in Kenya*. Oxford Centre for Mission Studies, 1992.

Matovu, J. and Okumu L., "Credit Accessibility to the Rural Poor in Uganda," *Economic Policy Research Centre, Bulletin*. Vol.2, No.1, 1996.

Mayanja, V., *The Role of the Informal Financial Sector in Providing Financial Services to Small-Scale Producers: A Case Study of Mawokota County in Mpigi District*. A dissertation submitted in partial fulfillment of the requirements for the award of the MA Degree in Economic Policy and Planning of Makerere University, 1995.

Migot-Adholla, G.S., Hazal, P., Blare, B., and F. Place, "Indigenous Land Rights Systems in Sub-Saharan Africa: A Constraint on Productivity?" *World Bank Economic Review*, Vol. 5, No.1, 1991.

Montgomery, R., *Disciplining or Protecting the Poor? Avoiding the Social Costs of Peer Pressure in Solidarity Group Micro-Credit Schemes*. University of Wales, Centre for Development Studies, Papers in International Development No. 12, 1995.

Montiel, Peter, J., *Financial Policies and Economic Growth: Theory, Evidence and country-specific Experience from Sub-Saharan Africa*. African Economic Research Consortium, Special paper 18. Nairobi: Kenya, 1995.

Mwega F.M., *Financial Sector Reforms in Eastern and Southern Africa*. Evaluation Seminar Organized by CODESRIA, 1996.

Oshikoya T.W., *Financial Liberalization: Emerging Stock Markets and Economic Development in Africa*. Evaluation Seminar Organized by CODESRIA, 1996.

Patten, R.H, etc., *Progress with Profits: The Development of Rural Banking in Indonesia*. San Francisco, California: International Centre for Economic Growth Study No. 4, 1991.

Private Sector Foundation, *Uganda's Micro-Enterprise Finance: Urgent Need for Reform* (Issues and Recommendations on the Reform Agenda), 1996.

Reed, L., etc., *The Role of NGOs in Rural Financial Intermediation in Ghana*. A study prepared for USAID/Ghana, 1994.

Republic of Uganda, *Program for the Alleviation of Poverty and the Social Costs of*

Adjustment (PAPSCA): Progress Report, April 1993-June 1994.

Republic of Uganda, *Background to the Budget 1996/97-1998/99*, Kampala: Ministry of Finance and Economic Planning.

Republic of Uganda, Prime Minister's Office, *Revised Criteria for Selection of an Intermediary Entity for the Poverty Alleviation Project (PAP)*.

Republic of Uganda, Agricultural Policy Committee, *Cotton Sub-Sector Development Project: A Program to Build Rural Financial Capacity: Draft Final Report*. Agricultural Cooperative Development International, 1994.

Republic of Uganda, Agricultural Policy Committee, *Cotton Sub-Sector Development Project: A Rapid Appraisal of Rural Finance and Credit Schemes*. Volume 1. Main Report. Volume II, Working Papers. Kampala, Bank of Uganda, Agricultural Secretariat, 1994.

Simmons, R., "The Mobilization of Domestic Resources for Development: Some Current Theoretical Issues." ed. Frimpong - Ansah J.H. and B. Ingham, *Saving For Economic Recovery in Africa: Theory, Policy, Case Studies*. London: James Currey and Heinemann, 1992.

Solis-Fallas, O. and Owen P.D., "Unorganized Money Markets and `unproductive' Assets in the New Structuralist Critique of Financial Liberalization," *Journal of Development Economics*. 31, 1989.

Ssemogerere, G., "Mobilization of Domestic Financial Resources in Uganda: Commercial Banks Versus the Uganda Cooperative Savings and Credit Union," ed. Frimpong-Ansah J.H. and B. Ingham, *Saving for Recovery in Africa: Theory Policy, Case Studies*. London: James Currey and Heinemann, 1992.

Stiglitz E.J., "Peer Monitoring and Credit Markets," ed. Braverman, A.K. etc., *The Economics of Rural Organization: Theory, Practice and Policy*, 1993.

Stiglitz J.E. etc., "Imperfect Information and Rural Credit Markets: Puzzles and Policy Perspectives," ed. Braverman A. K., Hoff K. and J.E. Stiglitz, *The Economics of Rural Organization: Theory, Practice and Policy*, 1993.

Thillairajah, S., *Development of Rural Financial Markets in SSA*. Discussion Paper No. 219, World Bank, African Technical Department Series, 1996.

United States Agency for International Development, "Principles of Financially Viable Lending to Poor Enterprises," *Microenterprise Development Brief No.3*, 1995.

Van Holst Pellekaan etc., "The Role of Infrastructure and Credit in Rural Development and Poverty Reduction: Lessons from Experience and Analysis," Paper prepared for a Seminar on *The Challenge of Growth and Poverty Reduction in Uganda*, organized by the Government of Uganda and the World Bank, Kampala, 1996.

Von Braun, J., *Rural Credit in Sub-Saharan Africa: Enabling Smallholder Production Growth and Food Security*. Paper prepared for the Workshop "Agricultural Development Policy Options for Sub-Saharan Africa," organized by the International Food Policy Research Institute: Washington, D.C., 1992.

World Bank Operations Evaluation Study, *Structural and Sectoral Adjustment: World Bank Experience 1980-1992* by Carl Jayarajah and William Branson. Washington D.C.: World Bank, 1995.

World Bank, *Adjustment in Africa: Reform, Results and the Road Ahead.* A World Bank Report. Oxford University Press, 1995.

World Bank, *Uganda Financial Sector Strategy Note: Rural Financial Markets,* 1995.

World Bank, "Financial Systems and Development," *World Development Report,* 1989.

World Bank, *A Review of Bank Projects in Africa Involving Microenterprises.* AFTPS: Draft 12/05/94.

Yaron, J., "What Makes Rural Financial Institutions Successful," *World Bank Research Observer,* Vol. 9. No.1, 1994.

Chapter 10

Financial Sector Reforms in Eastern and Southern Africa

Francis M. Mwega

Introduction

There has been a resurgence of interest on factors that influence economic growth and for policies that would reduce poverty and promote economic development in Sub-Saharan Africa. The region's growth performance in the last three decades has been described as a tragedy, with the Africa dummy in the standard core growth regressions negative and highly significant. This suggests that the region grows more slowly that elsewhere even after controlling for macroeconomic factors (financial deepening, exchange rate and fiscal stance), human capital formation, political instability and conditional convergence (Easterly and Levine 1994). In the Easterly and Levine study, the Africa dummy becomes non-significant only after introducing neighborhood (contagion) growth effects. Poverty also seems to have worsened (Ali 1995).

Financial liberalization is one of the key pillars of SAP in most of Africa. A lot of work has been done on the relationship between financial deepening and economic performance. Many studies find a close link between financial deepening, productivity and economic growth and conclude that policies affecting the financial sector have substantial effects on the space and pattern of economic development (Goldsmith 1969; King and Levine 1993). It is for example estimated that policies that would raise the M2/GDP ratio by 10% would increase the long-term per capita growth rate by 0.2-0.4% points (World Bank 1994). The financial sector is involved in the mobilization of resources among savers and their allocation to borrowers as well as transformation and distribution of risks and maturities over time. It facilitates saving and the efficient allocation of these savings to investment. In the process, it plays an important role in reducing risks and in the transformation of maturities in the saving-investment nexus (Nissanke et al. 1995). Financial institutions lower the cost of investment when they evaluate, monitor and provide financial services to entrepreneurs. They promote productivity and growth through improved efficiency of intermediation, a rise in the marginal product of capital and or an increase in the savings rate (Montiel 1994).[3]

According to Callier (1991), the performance of the financial sector in Sub-Saharan Africa has an important bearing on the overall economic performance because: (i) the region continues to be in economic crisis and the financial system is relatively underdeveloped compared to any other developing region; (ii) structural adjustment programs require more reliance on the private sector and hence its financing; (iii) the debt crisis and reduction in external savings translates to the need to increase the mobilization of domestic savings for investment; (iv) reform is needed if the financial system is to overcome and avoid the problems of financial distress and restore confidence; and (v) the need for international competitiveness requires that the financial system be as adaptable and flexible as possible.

Financial sector reforms in the region and elsewhere have mainly been motivated by the financial repression paradigm promulgated by the McKinnon (1973) and Shaw (1973) who emphasized the role of government failures in the sector. Accordingly, the objective of financial reforms is to reduce or reverse this 'repression' .

According to the McKinnon-Shaw hypothesis, financial repression arises mainly when a country imposes ceilings on nominal deposit and lending interest rates at a low level relative to inflation. The resulting low or negative real interest rates discourage savings mobilization and the channeling of the mobilized savings through the financial system. While the low and negative interest rates facilitates government borrowing, they discourage saving and financial intermediation, leading to credit rationing by the banking system with negative impacts on the quantity and quality of investment and hence on economic growth (Mwega *et al.* 1990).

Advocates of this hypothesis postulate that many financial systems in Africa have been subjected to financial repression characterized by low or negative real interest rates, high reserve requirements (sometimes of 20%-25% compared to 5%-6% in developed countries) which lead to high spreads thereby imposing an implicit tax on financial intermediation; mandatory credit ceilings; directed credit allocation to priority sectors which undermine allocative efficiency; and heavy government ownership and management of financial institutions. The latter suggests that much credit is given on political rather than commercial considerations, giving rise to a huge pile of non-performing loans in the banks' portfolios. There is also limited competition, with government and parastatals major borrowers due to the large deficits that they experience. According to Camen *et al.* (1996), financial repression in Africa has hindered the development of the capacity of financial institutions in carrying out their informational and resource mobilization role.

As a consequence, it is postulated that the financial sectors in many African countries (i) are segmented, fragmented and dualistic; (ii) are mainly bank-based, with few NBFIs; (iii) serve the short end of the market; (iv) are heavily regulated, with much of their services geared towards servicing the public sector deficits, leading to a crowding-out of the private sector; and (v) they face limited competition or innovations, with many of them dominated by oligopolies (Soyibo 1994).

An important component of financial sector reforms is interest rates liberalization. It is argued that this would raise real interest rates, increase saving and the supply of investable resources in the economy. The productivity of investment also rises as these resources are channeled to projects that have higher rates of return than hitherto. Liberalization of interest rates would also discourage capital flight and help to stabilize the economy.

Financial sector reforms have also included (a) reducing direct and indirect taxation of financial institutions through reserve requirements, mandatory credit ceilings and credit allocation guidelines; (b) reducing barriers to competition in the financial sector by scaling down government ownership through privatization; and facilitating entry into the sector by domestic and foreign firms; and (c) restructuring and liquidation of solvent banks (Inanga 1995).

The objective of this chapter is to analyze the nature, extent and outcomes of financial sector reforms in Eastern and Southern Africa based on case studies of Kenya, Tanzania, Malawi and Zimbabwe. These countries have been selected because they bring out variety in the structure of financial sectors in the region, with the financial sectors in Tanzania and Malawi relatively underdeveloped compared to those of Kenya and Zimbabwe based on the number and variety of financial institutions. Kenya, Tanzania and Malawi launched their reforms in the 1980s while Zimbabwe did not initiate reforms until 1991. These countries are also members of the Common Market for Eastern and Southern Africa (COMESA).

The paper examines whether the financial sector reforms have achieved their objectives of increasing the mobilization of resources and their efficient allocation for investment. Specifically, the study critically reviews the World Bank (1994) interpretation of the African experience with financial sector reforms based on these four cases.

Table 1 shows some indicators of financial repression at the onset of reforms in these countries. These countries experienced low or negative real interest rates, low M2/GDP ratio and high fiscal deficits/GDP ratio compared to the more advanced countries. The degree of financial deepening for example compares very unfavourably with the M2/GDP ratio for the USA (67.0%), Japan (183.1%), Singapore (126.1%), Portugal (74.1%); Greece (79.3%) and Spain (68.8%) (ADB 1994).

Table 10.1. Some indicators of financial repression in Kenya, Malawi, Tanzania, and Zimbabwe

	Year of Reform (%)	M2/GDP Ration Rate (%)	Real Deposit Rate (%)	Real Lending GDP	Fiscal Deficit Ratio (%)
Kenya	1986	28.2	0.3	3.3	-3.8
Malawi	1988	19.5	-18.0	-9.2	-6.0
Tanzania	1986	32.6	-28.3	-21.3	-4.5
Zimbabwe	1991	42.4	-2.5	1.6	-11.0

Source: Seck and El Nil (1993) and author's calculations.
Note: Figures apply to the year of reform, except Kenya (1985), Tanzania (1985), and Zimbabwe (1989).

Financial Institutions and Financial Sector Reforms in Kenya, Tanzania, Malawi, and Zimbabwe

Kenya[2]

The financial system is dominated by the central bank (preceded by the East African Currency Board and established in 1966), commercial banks and deposit-taking NBFIs. The country also has a wide range of other more specialized financial institutions.

Kenya has a long history of commercial banking, with the predecessors of the three major commercial banks set up before the 1920s. By independence in 1963, Kenya had 10 commercial banks with the "big three" — National and Grindlays Bank, Barclays Bank, and Standard

Bank — holding about 80 per cent of the total commercial bank deposits. All these banks (except one) were branches of foreign banks, and mainly financed foreign trade.

After independence, there was a felt need to develop a more locally, indigenously-oriented banking system. Three banks were established by 1970: the Cooperative Bank of Kenya, National Bank of Kenya, and Kenya Commercial Bank. The latter tookover National and Grindlays to become the biggest bank in the country. By the early 1970s, the structure of commercial banking in Kenya had been transformed with the two public banks—Kenya Commercial Bank and the National Bank of Kenya—accounting for 35 percent of paid and assigned capital; and Barclays Bank and the Standard Bank accounting for about 22 percent each.

The Kenya commercial banking system continued to grow in the 1970s and 80s so that by the onset of financial reforms in the mid-1980s, the number of licensed commercial banks had doubled to 24, about 15 foreign-owned, 3 state banks and 6 locally-owned private banks.

While the Kenya commercial banking system in dominated by an oligarchy of four commercial banks, they face competition from smaller banks and deposit-taking NBFIs which reduce their market power. Fears have however been expressed that they collude to determine the structure of interest rates following the liberalization of the financial system.

"Specified" NBFIs are institutions that are established under the Banking Act of 1968, even though their history goes back to the period well back before independence. By 1969, the number of these NBFIs had risen to 10 and by the end of 1979 to 19. By mid-1980s, the number of these NBFIs had risen to 53, with a total of 100 branches mainly located in urban areas.

The country therefore experienced a very rapid growth in these institutions in the 1980s. Between 1973 and 1986, the outstanding credit of these NBFIs to the public and private sectors increased more than 19-fold while the proportion of their credit to that of commercial banks credit increased from 20.3% in 1973 to peak at 52.1% in 1986 following a shift in some of their deposits to the more established commercial banks because of solvency problems in some NBFIs.

The rapid growth in NBFIs in the 1980s may be attributed to several factors. First, the coffee-tea boom of the 1976-77 which was not sterilized with all the proceeds passed on to the farmers. Second, these institutions were cheaper to establish, with the minimum capital and gearing ratios required to start an NBFI smaller than that for commercial banks. Lastly, favorable policy also enabled NBFIs to offer higher rates of interest making them more competitive than commercial banks while they were not subject to credit ceilings. Hence, NBFIs operated in a more liberal legislative framework than commercial banks. However, in the 1985 amendments to the Banking Act, further consolidated in 1989, controls on NBFIs were tightened. They were barred for example from acquiring or holding share capital or to have direct interest in commercial, industrial or other undertakings where their financial contribution would exceed 25 percent of their paid capital or unimpaired reserves as well as holding for commercial purposes immovable properties such as land and owning equity in commercial banks.

Besides the central bank, commercial banks, and "specified" NBFIs, Kenya has numerous other more specialized institutions. By late 1980s, these included:

- 207 hire purchase companies (from 3 at independence) with outstanding balances amounting to about 12 per cent of the outstanding loans and advances of commercial banks. A large proportion of these funds are used to finance the purchase of vehicles as well as household consumer durable goods.
- About 10 DFIs that provide medium and long-term finance. The major ones service industry and commerce (Industrial Development Bank, Development Finance Company of Kenya, Kenya Industrial Estates, and Industrial and Commercial Development Corporation); agriculture (Agricultural Finance Corporation); building and construction (National Housing Corporation); and the tourism sector (Kenya Tourist Development Corporation).
- 22 building societies providing housing finance.
- 39 locally incorporated insurance companies whose total assets amounted to about two-fifths those of commercial banks.
- A Post Office Savings Bank whose deposits as a proportion of the banking systems' deposits was about 3 per cent (from about 5% in the late 1960s), later converted to a commercial bank.
- Numerous pension schemes, including the National Social Security Fund (NSSF).
- About 4000 savings and credit cooperative societies in both rural and urban areas; and
- The Nairobi Stock Exchange, handling 55 publicly quoted companies.

The country formally adopted financial sector forms in 1989, supported by a $170 million World Bank adjustment credit. Financial reform proposals were first incorporated in the 1986-90 structural adjustment program. The main features of the program was full interest rate liberalization which was achieved in July 1991 after a gradual increase in nominal rates in the 1980s. In the first half of the 1980s for example, nominal deposit rates were increased by about 100 percent and lending rates by about 50 percent, from relatively low levels. Before this period, the government followed a low-interest-rate policy whose main objective was to promote investment. From the time the Central Bank was established in 1966 until 1980 interest rates were only adjusted upwards once in 1974 by 1 to 2 percentage points.

Other reforms included (i) liberalization of the treasury bills market in November 1990; (ii) setting up a Capital Markets Authority in 1989 to oversee the development of the equities market with a view to enhancing availability of long-term resources for investment; (iii) abolition of credit guidelines in December 1993 (which were in existence since 1975 in favor of agriculture); and (iv) improving and rationalizing the operations and finances of the DFIs, though against the wishes of some donors who urged for their dissolution or privatization. In 1988 and 1994, the two parastatal banks were partially privatized, selling 30% of their shares to the public.

A component of reforms has been the restructuring of financial institutions. The country experienced a bank crisis in 1986 when a number of 'specified' NBFIs and a small commercial bank collapsed. To avoid a repeat, eight financial institutions were taken over and merged into a state bank in 1989 - Consolidated Bank of Kenya Ltd. The central bank has also strengthened the supervision and the inspection of financial institutions and introduced a Deposit Protection Fund which guarantees deposits upto Ksh 100,000. The initial capital for setting up financial institutions has been increased both for commercial banks and "specified" NBFIs.

Malawi[3]

In mid-1990s, the financial system is dominated by the Reserve Bank of Malawi, two large commercial banks (the National Bank of Malawi and the Commercial Bank of Malawi) and a small new one, a merchant bank (the Investment and Development Bank - INDEBANK), four DFIs, two finance houses, the New Building Society, the Post Office Savings Bank, a number of insurance companies and the Malawi Union of Savings and Credit Societies (MUSCCO).

The two large commercial banks account for about 80% of total deposit liabilities, and more than 50% of the total credit (excluding the central bank credit). The commercial bank loans are concentrated in a few conglomerates. The dominant holding company in the country, the

Press Corp, is the major shareholder of the two dominant banks and a number of non-financial firms.

The funds of the four DFIs come largely from government and foreign sources. The Malawi Development Corporation, the largest DFI, is state owned and functions as a holding company with investment in the major Malawian firms. The second largest DFI, Agricultural and Marketing Corporation (ADMARC) is privately owned, although a large government parastatal owns about a quarter of its equity. It mainly invests in agriculture and agro-industry. The third DFI, Investment and Development Fund (INDEFUND), an offshoot of INDEBANK, makes loans in foreign exchange to finance small and medium enterprises. The fourth DFI, Small Enterprise Development Organization of Malawi (SODEM) is also directed at financing small-scale enterprises.

The financial system has been repressed through administered interest rates, credit ceilings and reserve requirements. Before 1985, the Reserve Bank of Malawi prescribed interest rates in bank deposits, loans and government securities. There were preferential rates for agriculture and other favored activities.

Solvency of the financial system is often periodically threatened, because of financial difficulties of its major borrowers particularly the estate tobacco sector. To contain the problem, the government restructured the two main bank customers; a large parastatal (ADMAR) and a privately owned national conglomerate (the Press Corp.). The government also aided recapitalization of banks through retained earnings by widening interest rates spreads.

The Malawian financial reforms were launched in 1987, with an increase of interest rates (by about 3% points) and the abandonment of credit ceilings. The practice of setting preferential rates for banks lending to agriculture was discontinued in August 1989 and controls on interest rates were fully lifted in May 1990. The government since then has encouraged the development of NBFIs to promote competition. Consequently, two NBFIs, the INDEBANK and the Lease and Finance Companies have branched into new activities. Besides, INDEBANK and the Malawi Finance Corporation has been granted license to accept deposits, issue letters of credit and bankers acceptances and handle other merchant banking activities.

The legal framework of the financial system (the Reserve Bank and the Banking Acts) was revised in 1989, together with improvement in prudential regulations and strengthening of the reserve bank's supervisory role. The reforms also created the statutory basis permitting monetary policy to be conducted through the discount window variation in reserve requirements and open market operations. The Capital Markets

Act of 1990 provides the legislative framework for revamping the securities and equities markets.

Tanzania[4]

Between 1967 when banks and financial institutions were nationalized under the Arusha Declaration until 1991, the financial system was entirely owned and controlled by the state. The system was extremely narrow comprising of the central bank, three commercial banks, five DFIs, two insurance companies, two contractual savings institutions and one hire purchase company. The state owned many of these institutions including the three commercial banks, the two insurance companies, the single social security institution and three of the five DFIs. The National Bank of Commerce was the only commercial bank of any significance, with 90% of all deposit liabilities of deposit-taking institutions. These institutions were subject to neither competition nor adequate supervision.

The system was therefore subject to massive financial repression and was geared mainly towards the provision of cheap credit to the central government, parastatals and cooperatives, with the Bank of Tanzania acting as the lender of first resort. Within the credit ceilings set annually, there would be targets for these institutions, with the private sector taking the residual (usually about 5% of the total credit supply). Interest rates were fixed for much of the period, with real rates negative up to 1988, reaching a low in 1984.

Many branches of the National Bank of Commerce and some of the DFIs made large losses mainly due to bad project choice and poor management. They were subject to inadequate supervision, auditing, and legal protection for both debtors and creditors.

Financial reforms began slowly in the 1984/85 budget but were intensified in 1986 with Economic Recovery Program. The objectives of ERP (among others) were to reduce the monetization of the deficit, reduce credit expansion and to direct more credit to the private sector.

A Presidential Commission of Enquiry into the Monetary and Banking System in Tanzania was established in 1988 and a Banking and Financial Institutions Act was passed in 1991 to effect financial sector reform through the restructuring of the existing financial institutions, to promote private banking, deregulate the capital market and rationalize and strengthen the legislative and supervisory powers of the central bank. Private banks and financial institutions (domestic and foreign) were now free to enter the market. Since 1992, banks are also free to determine both the deposit and lending interest rates. Open market operations have been introduced with weekly auctions of short and long-dated treasury bills to absorb liquidity, to finance government expenditures and to determine government rediscount rate.

Provision has been made for restructuring the existing financial institutions and for tighter regulatory control of the financial system. Non-performing loans were estimated at about 60% of total assets or 50% of government expenditure in 1993/94. The restructuring of balance sheets started in 1991 under the provisions of the Loans and Advances Realization Trust Act (LART) whose function is to clean the banks' balance sheets and to collect the bad debt on their behalf. The non-performing portfolios of the banks would be transferred or sold to the LART, whose liabilities would be to the Bank of Tanzania and would only be fully guaranteed by the government. This was completed for commercial banks in October 1993, before moving on to DFIs. The cost of the partial structuring is estimated at roughly 40% of GDP (World Bank 1994).

To prevent future mismanagement and financial distress, the Bank of Tanzania issued new guidelines in 1991 to govern the licensing of banks, as well as prudential guidelines for the management of assets, provision for losses and the accrual of interest. The minimum capital requirements to obtain a license was increased and the applicants were required to demonstrate ability to operate efficiently, profitably and prudentially. Financial institutions are also required to diversify in order to spread risk and are not allowed to lend more that a certain percentage of their core capital to individual borrowers. To oversee the system, the Bank of Tanzania has strengthened its Bank Supervision Directorate.

According to ADB (1994), several beneficial effects have resulted from financial liberalization in Tanzania. These include: (i) restructuring of National Bank of Commerce, thereby restoring some public confidence in it, leading to some financialization of savings; (ii) establishment of new financial institutions (two banks by 1994); and (iii) repatriation of capital flight due to the liberalization of the financial and foreign exchange markets.

Zimbabwe
In mid-1990s, the country had six major commercial banks, six merchant banks, four discount houses, five registered financial institutions, four building societies, an export credit insurance corporation, a number of insurance companies and the state controlled Zimbabwe Reinsurance Corporation. DFIs included the Zimbabwe Development Bank which mainly concentrated its activity in the manufacturing sector as well as small- and medium-sized businesses; and the Agricultural Finance Corporation (EIU 1996). The Zimbabwe Stock Exchange deals

with shares of more that 60 companies.

Financial liberalization was implemented in Zimbabwe in 1991. As in the other three countries, the flagship of liberalization was the deregulation of interest rates and occurred at a time of severe negative shocks in the real economy, compounded by a tight monetary policy. The drought of 1992 reduced agricultural output by 24%, with a decline in output in 1993 partly due to the credit squeeze (Ncube *et al.* 1995). Financial liberalization was therefore followed by a sharp increase in nominal interest rates as these data show:

	1989	1991	1993
Bank rate	9	20	29
Interbank rate (3 months)	9	17	34
Treasury bill rate	8	13	33
Deposits rate-time:3-months	9	14	29
Commercial lending rate	13	16	36

Source: IMF, *International Financial Statistics*.

Before 1991, the credit market operated under administrative allocation procedure with controls on interest rates, mainly driven the availability of foreign exchange.

Other reforms include efforts to develop the money market following the introduction of indirect monetary policy policies, especially open market operations; as well as the introduction of regulatory focus with the supervisory power of the Reserve Bank strengthened.

Outcome of Reforms

The World Bank (1994) Report

According to the Report, structural adjustment programs have been more successful in achieving macroeconomic stabilization and trade and agricultural reforms than with privatization of public enterprises and financial sector reforms. The actual experience of most African countries with financial reforms has been that of limited success, mainly as a result of failure of real deposit rates to remain consistently positive. This is because of the relatively high fiscal deficits which have characterized these countries. Interest rate liberalization and a decline in inflation have however helped eliminate extremely negative interest rates although they continued to be negative in a few countries, e.g. Zimbabwe in the study period (1987-91).

The financial sectors are therefore still heavily burdened by public sector demands for credit, with the central government alone (excluding public enterprises) absorbing more that 30% of domestic credit. This is compounded by government interference in the management of the financial sectors. A large share of bank lending still goes to the public enterprise sector, making it difficult for the private sector to borrow. Also as a result of poor macroeconomic management and the slow pace of reform in the public enterprise sector, the financial position of the banking sector is weakened, reducing its capacity to monetize the fiscal deficits, especially if the reforms are not accompanied by increased mobilization of resources but by reduced interest rate spreads. The increase in real interest rates following financial liberalization also exacerbates fiscal deficits as the cost of government debt increases.

The Report for example judges only Tanzania to have "good" fiscal stance in the study period (1987-91) with a budget deficit of less that 1.5%. Malawi had a "fair" fiscal stance with a budget deficit of 1.5%-3.5%, Kenya a "poor" fiscal stance of with a deficit of 3.6%-7.0%, and Zimbabwe a "very poor" fiscal stance with a deficit of 7.1% or more. Monetary policy stance was not any better. Kenya, Malawi and Zimbabwe were judged to have a "fair" monetary policy, based on the degree of seignorage, inflation and real interest rates while that of Tanzania was "poor." Overall Tanzania and Zimbabwe experienced a "large" improvement and Malawi and Kenya a "small" improvement in the macroeconomic policy stance between 1980-86 and 1987-93.

The Report urges that the little progress in reforming the financial sector perhaps also reflects political reality, there being much less consensus on how to proceed compared to other reforms. Financial sector reforms are postulated to be particularly difficult because of the powerful vested interests that have been created through high government intervention. A strong social consensus on the need to improve governance is hence recommended as a prerequisite for progress.

The Report posits that adjustment programs have been overly hasty in cleaning balance sheets and recapitalizing banks in an environment where institutional capacity is weak and the main borrowers (the government and public enterprises) are financially distressed. Many programs were therefore based on the assumption that banks could improve their performance simply by removing the bad loans from their balance sheets, replacing managers, and injecting new capital to bring assets up to international standards. This usually was insufficient for several reasons: reforms were not accompanied by needed macroeconomic and structural changes, bank managers continued to be exposed to political interference and regulatory and supervisory capacities were inadequate.

In summary, the Report postulates that the major constraints to successful financial reforms in Sub-Sahara Africa are: (i) inappropriate sequencing, (ii) continued state involvement and interference in the management of financial institutions that has over time created powerful vested interests that undermine reforms, and (iii) inadequate regulation and supervision of these institutions.

Besides better sequencing, the Report therefore recommends a strategy to improve bank solvency that involves downsizing publicly owned banks, privatizing them where possible, and encouraging new entrants. And because most African countries lack the capacity to regulate and supervise, the challenge is to devise a financial system that offers extra cushion against risk — by setting higher that normal capital adequacy ratios, relying more on foreign banks, and limiting entry to reputable banks with a solid capital base. In this case, countries must strike a balance between the need to increase competition and the need to ensure the solvency of financial institutions.

An Evaluation

As emphasized by the Report, financial sector reforms in African countries have been undertaken in the context of pervasive macroeconomic instability, contrary to the advice that "postponing the removal of interest rate regulation may be appropriate until...the situation has been stabilized and banking supervision strengthened" (Villanueva and Mirakhor 1990). Successful sequencing for financial liberalization requires achievement of macroeconomic stability (control of inflation) and fiscal discipline (reduction of fiscal deficit) in addition to improved legal, accounting and regulatory systems for the financial sector; a tax system that does not discriminate against the sector; and care of sequencing for example such that capital inflows from liberalization does not offset macroeconomic stability (World Bank 1989).

While Kenya and Malawi for example are judged by the Report to have a "fair" and Tanzania and Zimbabwe a "poor" macroeconomic policy stance in 1990-91, the situation deteriorated in Kenya (to "very poor"), Malawi (to "poor") and Zimbabwe (to 'very poor') in 1991-92, with only Tanzania improving its macroeconomic policy stance (to "fair"). These countries macroeconomic stance are therefore "fragile, slow and often reversal-prone" (Bouton et al. 1994).

This has made it quite difficult to raise real interest rates before (1980-86) and during (1987-93) financial reforms as Table 2 shows. Real deposit interest rates were generally negative and declined between the two periods in Kenya, Malawi and Zimbabwe. It was only in Tanzania where the real deposit rates increased but from very low levels, with the same pattern for the real lending interest rates. The lending-deposit interest rate margin also seems to have increased in three of the four countries, declining only in Zimbabwe.

Table 10.2: Interest rates in Kenya, Malawi, Tanzania, and Zimbabwe

	1980-86	1987-93
Real deposit rates		
Kenya: 3-6 months	-1.6	-7.5
Malawi: 3-month fixed rate	-2.5	-5.0
Tanzania: Fixed 3-6 month	-25.8	-7.3
Zimbabwe: time 3 months	-3.8	-5.8
Real lending rates		
Kenya	1.4	0.5
Malawi	4.9	2.7
Tanzania	-17.4	5.4
Zimbabwe	6.0	-2.1
Nominal lending-deposit interest rate margins		
Kenya	3.1	8.0
Malawi	4.9	2.7
Tanzania	-17.4	5.4
Zimbabwe	6.0	-2.1

Source: IMF, *International Financial Statistics* (interest rates) and World Bank, *World Tables* (1995) (CPI).

The World Bank (1994) assessment of financial reforms in Africa was conducted to a large extent within the McKinnon-Shaw framework, with the overall approach to financial development — removing financial repression, dismantling directed credit programs, introducing better accounting, legal, and supervisory frameworks, continuing with institution building, deepening and developing capital and money markets — is postulated to be on target and in the right direction. As shown in Section II, the reforms carried out by the four countries under study have varied, but all have entailed interest rate liberalization as well as bank restructuring and liquidation. The Report does not address itself adequately to the basic tenets of this paradigm, with the presumption that if the problems described above were resolved, then reforms would achieve the objectives postulated by Mckinnon-Shaw.

However, the argument that private saving has a significant positive real interest elasticity is not supported by either economic theory or empirical evidence. In economic theory, high real interest rates have two effects on private saving that work in opposite directions. They have a substitution effect, in which saving increases as consumption is postponed to the future, and a wealth effect, in which savers increase current consumption at the expense of saving, so that the impact of real interest rates on private saving is ambiguous. The McKinnon and Shaw doctrine therefore postulates that under conditions of financial repression the substitution effect dominates the wealth effect and that there is a portfolio effect in which an increase in real interest rates induces a shift in the composition of the wealth portfolio from non-financial to financial assets, thereby enhancing financial intermediation (Mwega *et al.* 1990).

There is almost a consensus that real interest rates have a small or non-significant impact on aggregate or private saving rates, and therefore an increase in real interest rates does not support a higher level of investment (ADB 1994, Mwega 1995). According to Nissanke et al (1995), "it has been increasingly recognized that an adoption of a financial liberalization policy has not been sufficient to generate a strong response in terms of increased savings mobilization and intermediation through the financial system, wider access to financial services and increased investment by the private sector." [5] Mwega *et al.* (1990) and Oshikoya (1992) for example found non-significant coefficients for Kenya after controlling for a range of factors.[6]

The Report mentions that savings have traditionally been low throughout the region and adjustment has had little impact on them so far. It then proceeds to postulate that savings largely follow economic growth, so that the key is not forcing up the saving rate, but establishing an environment that follows rapid accumulation, efficient resource use and rapid productivity growth. Empirical evidence however suggests that domestic savings are highly correlated with investment (Bosworth 1993). Studies on the other hand usually find the investment rate one of the most robust determinants of economic growth across countries (Levine and Renelt 1992).

Table 3 shows the saving, investment and growth performance in the four countries. Only Zimbabwe experienced an increase in the average gross savings rate between the two periods and only Tanzania and Zimbabwe experienced an increase in average rate of gross investment (the large increase in Tanzania supported by foreign capital inflows).

The average saving rates in Malawi and Tanzania declined between the two periods (mainly because of a large increase in external transfers to these countries) and are still very low, below the average in Sub-Saharan Africa (13%). The decline in Kenya was caused by a reduction in public savings (Mwega 1995). However, there is a discernible improvement in per capita income in the liberalization period which the Report attributes to both external factors and an improvement in policies. The per capita income growth is nevertheless from very low or negative levels so that the foundation for long-term growth is still very fragile. It is only Zimbabwe that did not experience an increased growth in real per capita income.

Table 10.3. Savings, investment and growth in the four countries

	1980-86	1987-93
Gross domestic savings (% of GDP)		
Kenya	20.3	19.1
Malawi	13.0	6.9
Tanzania	10.8	8.3
Zimbabwe	17.6	20.7
Gross domestic investment (% of GDP)		
Kenya	24.0	21.9
Malawi	18.6	18.1
Tanzania	20.1	42.6
Zimbabwe	19.5	22.0
Growth in real GDP per capita		
Kenya	-0.08	0.28
Malawi	-1.89	-1.49
Tanzania	-1.42	1.52
Zimbabwe	-1.42	-0.52

Source: World Bank, World Tables (1995).

Studies on the impact of financial liberalization usually show some positive impact on financial savings (Nissanke et al. 1995). However, while Lipumba *et al.* (1990) found a positive and significant relationship between real deposit interest rates and financial savings, Mwega et al. (1990) and Chipeta and Mkandawire (1992) did not find such a relationship for Kenya and Malawi, respectively, so that safety rather that returns has been the major reason for keeping savings with financial institutions.

Table 4 shows the data pre (1980-86) and during (1987-93) financial liberalization for M2 as well as deposits in "other banking institutions" as a proportion of GDP, which can be taken as measures of financial deepening.

Table 10.4. Degree of financial deepening in the four countries.

	1980-86	1987-93
M2/GDP, %		
Kenya	29.1	32.0
Malawi	21.2	21.7
Tanzania	37.9	33.6
Zimbabwe	30.5	28.0
M2 plus Deposits in other banking institutions/ GDP, %		
Kenya	41.2	46.4
Malawi	24.0	24.9
Tanzania	38.4	33.7
Zimbabwe	51.4	50.9

Source: IMF, *International Financial Statistics* and World Bank, *World Tables* (1995).

The M2/GDP ratio declined in Tanzania and Zimbabwe and slightly increased in Kenya and Malawi. Hence there is no definite pattern on the impact of financial liberalization policies on financial deepening in the four countries. This does not change when deposits in 'other banking institutions' are included except for Zimbabwe where the combined ratio increases slightly.

The share of time and savings deposits to total deposits in deposit money banks increased in Kenya (from 59.3% to 65.4%) and Tanzania (from 42.4% to 47.1%) and declined in Malawi (from 66.9% to 63.1%) and Zimbabwe (from 60.8% to 56.9%) between 1980-86 and 1987-93. In Malawi, Chipeta and Mkandawire (1994) report that commercial banks adopted a policy of lower interest rates on longer term deposits relative to shorter term deposits to discourage the former.

The following data also show the growth in real domestic credit where real credit is measured by nominal monetary survey credit divided by the GDP deflator:

	Kenya	Malawi	Tanzania	Zimbabwe
1981-86	6.5	4.2	-1.6	12.0
1987-93	1.7	-2.3	13.6	7.4

Only in the case of Tanzania did the supply of credit from the country's banking system increase during the reform period, when the loans-deposits ratio jumped from 84% in 1986 to 130% in 1989-91 (Nissanke *et al.* 1995). Table 5 also shows that the stabilization policies pursued in this period increased the credit allocated to the private sector, except in the case of Kenya where the general elections in 1992 led to a massive increase in government expenditure, the budget deficit and money supply.

Table 10.5: Percent Allocation of Domestic Credit by the Monetary System in the Four Countries

	1980-86	1987-93
Kenya		
Central government (net)	33.7	34.7
Local Governments	0.1	0.1
Non-financial public enterprises	3.6	5.1
Private sector	60.9	55.7
Other financial institutions		
Malawi		
Central government (net)	45.5	40.1
Official entities	14.1	14.4
Private sector	40.5	45.4
Tanzania		
Central government (net)	63.1	43.5
Official entities	31.4	25.0
Private sector	5.4	34.2
Other financial institutions	0.2	0.0
Zimbabwe		
Central government (net)	37.3	24.0
Non-financial public enterprises	37.9	18.4
Private sector	24.7	57.7

Source: IMF, International Financial Statistics

The World Bank Report (1994) also largely ignores or does not adequately incorporate other analytical frameworks on the behavior of credit markets which provide important insights on the design and consequences of financial liberalization.

First, borrowing and liquidity constraints are pervasive in the kind of economies being analyzed here. The extent to which individuals can actually dissave depends on their ability to borrow. If the borrowing constraints are binding, for example due to banks unwillingness to lend because of uncertainty of future incomes or risk of moral hazard by the borrowers, those who would like to borrow to increase present consumption cannot do so and are constrained to their current incomes and assets. If the borrowing constraints are less stringent, present consumption will increase and saving will decrease. Financial reforms, by easing the borrowing constraints and lowering uncertainty of future income stream, may therefore reduce private saving, investment and ultimately the rate of economic growth. If the reforms however are also successful in raising incomes by improving the quality of the intermediation process, the adverse effects on private saving may be mitigated (Mwega 1995).

It is not clear to what extent this hypothesis applies to these four countries. The share of private consumption to GDP increased in all the countries except Zimbabwe (Table 6). Real private consumption also increased at a faster rate during the reform period. It is however difficult to tell the extent to which this was due to improved economic growth or a relaxation of borrowing and liquidity constraints. Using a structural model, Schmidt-Hebbel et al. (1992) used beginning of period money balances as an indicator of the stringency of the borrowing constraints to explain household saving behavior in a sample of 10 developing countries and found strong negative effects, suggesting the importance of these constraints. Jappelli and Pagano (1993) also found some support for the hypothesis in OECD countries, where the saving rate increased with the size of the minimum mortgage down-payment for housing.

Table 10.6. Consumption in the Four Countries

	Private consumption/GDP (%)		Growth in real private consumption	
	1980-86	1987-93	1980-86	1987-93
Kenya	61.3	63.8	2.8	3.9
Malawi	69.3	76.2	1.5	4.5
Tanzania	75.2	81.1	n.a.	n.a.
Zimbabwe	62.4	56.1	1.3	5.6

Source: World Bank, *World Tables* (1995).

Second, the neo-structuralists argue that a high interest rate policy may be stagflationary by increasing the cost of working capital and by reducing real wages, aggregate demand and investment. They may also induce capital inflows and an overvaluation of the real exchange rate, with the consequent Dutch effects. High interest rates may also worsen the budget deficit by increasing the cost of debt service. They may also increase bank losses and the distress of the financial system as clients become unable to service loans, leading to curtailment of bank credit (ADB 1994). Hence, they may increase the probability of default by borrowers, and in the extreme could lead to a situation where borrowing is done to pay interest on previous loans or to stave off bankruptcy, rather than to invest or to finance working capital. It may therefore not be the case that financial liberalization may improve the quality of financial intermediation and portfolio choices and hence investment and productivity of capital. The Report only briefly acknowledges this, railing against highly positive real interest rates which could put the financial soundness of the banks at risk.

An increase in financial savings may also be at the expense of the informal financial sector, and adversely affect the supply of credit as the informal sector is more efficient and experiences lower reserve requirements. Wijnbergen (1983) and Buffie (1984) for example argue that financial savings need not be interest elastic, and even if they are, may not be translated to increased credit to the private sector if they are used to raise the cash and foreign-asset reserves held by these institutions, or to finance fiscal deficits.

Third, the imperfect information paradigm propagated by Stiglitz (1989, 1993) and Stiglitz and Weiss (1981) argues that liberalizing interest rates would not necessarily eliminate credit rationing — denying loans to some borrowers at prevailing interest rates or restricting loan sizes to below desired levels. Because of information asymmetry between lenders and borrowers (information possessed by the borrowers but not availed to the lenders), higher rates would tend to attract more risky projects, worsen the portfolio of financial institutions and increase the cost of monitoring the loans. Lenders may therefore use interest rates as a mechanism for affecting the quality of loan portfolio rather than clearing the loan market by inducing investment in less risky projects.

This paradigm hence focuses on market failures that are specific to the credit market which may give outcomes that are contrary to McKinnon-Shaw. High real interest rates may aggravate the problems of adverse selection and moral hazard. They may worsen the financial soundness of banks if they are induced to lend at these high rates to remain profitable, thereby assuming more risk than prudent financial practices would warrant.

Adverse selection occurs when borrowers with investment projects that are viable at reasonable risk are discouraged from seeking credit and give way to borrowers with projects with higher returns but involving more risk. The increased risk of default to financial institutions may however offset the gains from the higher interest rates. Moral hazard arises when borrowers increase loan repayment default rates as interest rates increase. Financial institutions may therefore prefer not to increase interest rates when faced with the excess demand for credit partly to influence the behavior of the borrowers and the uses made of the loans as this may increase the average return of their portfolio.

Given asymmetric information between lenders and borrowers, credit rationing may persist even with financial liberalization, perhaps accounting for the pervasive incidence of financial institutions holding excess reserves and liquidity even after liberalization (Nissanke *et al.* 1995). This is particularly the case in LDCs where information flows between debtors and lenders are quite limited. With the ability to raise collateral likely to be positively associated with riskier borrowers, its use to a large extent does not overcome these problems but restricts the supply of credit to the better-off socio-economic groups in the country. The presence of deposit insurance— to reduce the risk of systematic failure and hence stabilize the financial system— may also exacerbate these problems due to moral hazard on the part of financial institutions which select more riskier projects with higher returns as deposits are now guaranteed (Villanueva and Mirakhor 1990).

Another important issue is the ownership and credibility of these programs. Financial reforms are mainly implemented as part of conditionalities of international financial institutions and may therefore be perceived to be time inconsistent. Time inconsistency occurs when there is uncertainty on whether the government is a true liberalizer or just favors temporary reform. The government might for example adopt financial liberalization policies in order to get program aid, after which it will revert to controls. Lack of credibility has a negative impact on the effectiveness of financial sector reforms by reducing the responsiveness of both savers and investors until this uncertainty is resolved. This is particularly the case if these reforms are not widely accepted ("owned") by the populace or understood by the people who are supposed to implement them (Lipumba 1994).

Experience with intervention and liberalization policies suggest that while financial repression have retarded the development of financial sectors in Africa and Latin America, financial repression and directed credit schemes advocated by the imperfect information paradigm have been skillfully utilized to promote economic growth and transformation in South East Asia. The outcomes therefore depend on the nature of interventions and implementation, with positive and welfare-enhancing effects or negative and deleterious effects (Nissanke *et al.* 1995).

It is therefore important to view the role of the state in the financial sector in a pragmatic manner taking into account the institutional and other aspects in designing and sequencing reforms including the credibility of policy environment and motivations for financial reform. Not all state interventions in the financial market are unwelcome when there market failures, including imperfect information, information asymmetries, externalities and economies of scale and scope.

Financial reforms motivated by the McKinnon-Shaw framework for example (i) ignore rural-urban dichotomy and the need to extend financial institutions to rural areas; (ii) do not take into account the problem of credit access by households and small- and medium-sized enterprises usually neglected by the formal sector; (iii) focus on credit market to the neglect of capital market; and (iv) ignore the informal financial sector which may be larger that the formal sector (ADB 1994; Inanga 1995). These "failures" may justify an active role for the state in the financial market.

According to ADB (1994) for example, a well developed financial system has four main requisites, each of which can contribute to the process of financial deepening, and to raising the level of saving and investment, the productivity of capital and growth of output. These are (i) the monetization of the economy; (ii) the development of a commercial banking system with adequate central bank supervision; (iii) the development of other financial intermediaries including the capital market; and (iv) the integration of the formal and informal financial markets.

A range of financial intermediaries are therefore needed to intermediate between savers and investors with different requirements and time horizons. Special development banks for example are needed to stimulate leading sectors of the economy where the risks are higher than commercial banks are willing to lend. These are unlikely to develop out of *laissez faire*. Similarly, active government interventions are needed to facilitate the growth of the capital market at least through implementation of a tax policy that does not discriminate against equity financing, design of an appropriate legal framework for share operations and regulations on minimum informational standards.

Despite the growth of the formal financial sectors and the implementation of financial liberalization programs in the latter half of the 1980s, the informal sectors continues to thrive (ADB 1994, Nissanke *et al.* 1995). This is attributed to (i) inadequate access of rural areas to formal financial institutions; (ii) institutional and other barriers to use of these institutions (where they exist) by peasant farmers, households and small-scale enterprises in the form of rules of procedure for obtaining financial assistance, long distances, inconvenient banking hours etc; (iii) the informal financial sector complements the formal sector in providing credit; and (iv) structural adjustment programs have reduced the supply of credit from the formal sector, leading to a substitution to the informal sector (Aryeetey and Hyuha 1991). In Tanzania, for example, Bagachwa (1994) finds that the informal financial sector has expanded with liberalization as the rise in interest rates have induced increased participation in ROSCAs and saving and credit societies.

Policies are therefore needed to provide a better link between the formal and the informal financial sectors to exploit their comparative advantages and specialization. The high interest rates charged in the some segments of the informal sector for example add to costs and to debt, and these could be reduced if the informal sector was exposed to more competition from the formal sector. Some informalization of the formal sector may also be achieved through utilization of the informal sector as a conduit for funds to take advantage of the relatively low transaction costs, local knowledge and greater flexibility and to provide mechanisms to guarantee loans from the informal sectors and provide insurance against risk (ADB 1994). Designing appropriate schemes to provide a better linkage between formal and informal financial sectors however require an in-depth knowledge of their markets, their operational capacity and the constraints that they encounter both on the supply and demand sides (Nissanke *et al.* 1995). The World Bank (1994) Report gives scant attention to informal financial markets, only urging for a strategy that should encourage the formal financial sector to cooperate with the informal sector, especially in areas where high transaction and information costs put the former at a disadvantage.

Conclusions

In an attempt to improve economic performance, many African countries have adopted structural adjustment programs since the early 1980s. The objective of this study was to analyze the nature, extent and outcomes of financial sector reforms in Eastern and Southern Africa based on case studies of Kenya, Tanzania, Malawi, and Zimbabwe. The paper examines whether the financial sector reforms have achieved their objectives of increasing the mobilization of resources and their efficient allocation for investment. Specifically, the study critically reviews the World Bank (1994) interpretation of the African experience with financial sector reforms.

After a brief review of financial institutions and the financial sector reforms implemented in Kenya, Tanzania, Malawi, and Zimbabwe, the paper discusses the outcome of these reforms. According to the World Bank (1994) Report the actual experience of most African countries with financial reforms has been that of limited success, mainly as a result of failure of real deposit rates to remain consistently positive. This is because of the relatively high fiscal deficits which have characterized these countries. It posits that the little progress in reforming the financial sector perhaps also reflects the political situation as there is less consensus on how to proceed compared to other reforms. Powerful vested interests have also been created through high government intervention that undermine reforms. Efforts to clean the banks' balance sheets have not been very successful.

The Report postulates that the major constraints to successful financial reforms in Sub-Saharan Africa are inappropriate sequencing, continued state involvement and interference in the management of financial institutions, and inadequate regulation and supervision.

In the evaluation of this report, several issues are raised while taking into account the fact that an assessment on the basis of a few years data is problematic and it is difficult to separate the effects of financial reforms from those of other structural adjustment policies (Inanga 1995).

The first issue raised is the extent to which it is appropriate to undertake these reforms in the context of pervasive macroeconomic instability, contrary to the conventional advice that macroeconomic stability and bank supervision should be strengthened before interest rate regulations are completely removed. Inappropriate sequencing has made it difficult to raise real interest rates and to reduce lending-deposit interest rate margins in the four countries.

Second, the World Bank (1994) assessment of financial reforms in Africa was conducted to a large extent within the McKinnon-Shaw framework. The Report does not address itself adequately to the basic tenets of this paradigm, with the presumption that if the problems described above were resolved, then reforms would achieve the objectives postulated by Mckinnon-Shaw. However, the argument that private and financial savings have a significant positive real interest elasticity is not supported by either economic theory or empirical evidence. No clear pattern emerges with respect to the likely impact of financial liberalization on aggregate or financial savings, growth of credit and investment in the four countries. The results however show that the stabilization policies pursued in this period increased the credit allocated to the private sector (except in the case of Kenya).

Third, the World Bank Report (1994) also largely ignores or does not adequately incorporate other analytical frameworks on the behavior of economic agents in the credit market. These frameworks provide important insights on the design and consequences of financial liberalization.

Financial reforms, by easing the borrowing constraints and lowering uncertainty of future income stream, may reduce private saving, investment and ultimately the rate of economic growth. The share of private consumption to GDP increased in all the countries except Zimbabwe, but it is difficult to tell the extent to which this was due a relaxation of borrowing and liquidity constraints and or due to improved per capita economic growth.

The neo-structuralists also argue that a high interest rate policy may be stagflationary by increasing the cost of working capital and by reducing real wages, aggregate demand and investment; may induce capital inflows and an overvaluation of the real exchange rate, with the consequent Dutch Disease effects; may worsen the budget deficit by increasing the cost of debt service; and may increase bank losses and the distress of the financial system as clients become unable to service loans, leading to curtailment of bank credit. An increase in financial savings may also be at the expense of the informal financial sector, adversely affecting the supply of credit (Wijnbergen 1983; Buffie 1984).

The imperfect information paradigm on the other hand argues that liberalizing interest rates may worsen the portfolio of financial institutions and increase the cost of monitoring the loans. Financial institutions may continue to ration credit even with liberalization to influence the behavior of the borrowers and the uses made of the loans as this may increase these institutions average rate of return from their portfolio. Use of collateral or introduction of deposit insurance may exacerbate the adverse selection and moral hazard problems.

The fourth issue raised is that of ownership and credibility of these programs. Financial reforms are mainly implemented as part of conditionalities of international financial institutions and may be perceived to be time inconsistent. The resulting lack of credibility may reduce the responsiveness of both savers and investors until the uncertainty is resolved.

All these issues suggest that the role of the state should be viewed in a pragmatic manner taking into account the institutional and other aspects in designing and sequencing reforms as not all interventions are likely to be adverse. This is particularly the case when there are pervasive market failures, including imperfect information, information asymmetries, externalities and economies of scale and scope that characterize these countries.

Financial sector reforms should for example take into account the rural-urban dichotomy; the neglect of households and small- and medium-sized enterprises in the supply of credit by the formal financial sector; and the role of informal financial sector which may be larger that the formal sector and continues to thrive even with financial liberalization. These type of "failures" may justify an active role for the state to facilitate the development of a range of financial institutions to intermediate between savers and investors with different requirements and time horizons.

Notes

1. Financial deepening may however be driven by economic growth, be vulnerable to international shocks as well as downturn in economic performance and therefore not be a sufficient condition for improved productivity and growth.
2. This account is mainly based on Mwega 1992.
3. This account draws on Soyibo 1994, Chipeta and Mkandawire 1992, 1994, and Nissanke et al. 1995.
4. This account draws on ADB 1994, Bagachwa 1994, and Nissanke et al. 1995.
5. The responsiveness was tested for 1970-1993 by relating the aggregate saving rate (S/Y) to per capital real income (y); growth in per capita income (y/y); a representative real deposit rate of interest (d-ι) as given by the difference between the nominal deposit rate and the rate of inflation; the net foreign savings inflow as given by the ratio of current account deficit to GDP (FS/Y) and the lagged dependent variable (Mwega et al. 1990).

    In no case was the real deposit rate significant, with the t-values much below one. Controlling the data for non-stationarity and using an error-correction model did not fundamentally change these result. The standard tests (SBDW, DF and ADF) showed that the gross saving rate in Kenya and Tanzania were basically I (0) but with ambiguous results on whether per capita income was I (1) or I (2) in the two countries. The other variables were all I (1).
6. Azam 1995 finds a significant coefficient after controlling for terms of trade growth and financial repression, omitting other important conditioning variables.

References

African Development Bank (ADB), *African Development Report,* 1994.
Ali, A.G.A., "The Challenge of Poverty Alleviation in Sub-Sahara Africa." Paper Presented at the XI World Congress of the International Economic Association in Tunis on December 18-22, 1995.

Aryeetey E. and M. Hyuha, "The Informal Financial Sector and Markets in Africa: An Empirical Study." In A.J. Chhibber and S. Fischer, (eds.) *Economic Reform in Sub-Saharan Africa*, Washington D.C., The World Bank, 1991.

Azam J., "Saving and Interest Rates: The Case of Kenya." in *Savings and Development*, Vol. 20, No. 1, pp 33-44, 1996.

Bagachwa S.D.M., "Financial Integration and Development in Sub-Saharan Africa: A Study of Formal financial Institutions in Tanzania." Mimeo, June, 1994.

Bagachwa S.D.M., "Financial Integration and Development in Sub-Sahara Africa: A Study of Informal finance in Tanzania." ODI Working Paper, 1995.

Bosworth B., *Saving and Investment in an Open Economy*, Washington D.C., The Brookings Institution, 1993.

Bouton L. C. Jones, and M. Kiguel, "Macroeconomic Reforms and Growth in Africa: Adjustment in Africa Revisited." World Bank Policy Research Working Paper No. 1394, 1994.

Buffie, E.F., "Financial Repression: The New Structuralists and Stabilization Policy in Semi-Industrial Countries." *Journal of Development Economics*, Vol. 14, no. 3, 1984. Pp. 305 – 302

Callier P., "Financial Systems and Development in Africa." Economic Development Institute of the World Bank, Washington D. C., World Bank 1991.

Camen U., M. Ncube and L. Senbet, "The Role of Financial Markets in the Opåration of Monetary Policy. March." Mimeo, 1996.

Chipeta C. and M. Mkandawire, "Domestic Savings for African Development and Diversification." International Development Centre, University of Oxford, Mimeo, 1992.

Chipeta C. and M. Mkandawire, "Financial Integration and Development in Sub-Sahara Africa: A Study of Formal Finance in Malawi." Report of a Study Prepared for the Overseas Development Institute, London, 1994.

Easterly W. and R. Levine, "Africa's Growth Tragedy." Paper Presented at an AERC Plenary in Nairobi in May, 1994.

The Economist Intelligence Unit (EUI), *Zimbabwe 1994-95*, London, 1996.

Fry M., *Money, Interest and Banking in Economic Development*. Johns Hopkins University Press, Baltimore, 1988.

Gelb A., "Financial Policies, Growth and Efficiency." Policy Research Working Paper 202, The World Bank, Washington D.C., 1989.

Goldsmith R.W., *Financial Structure and Development*. Yale University Press, New Haven, 1969.

Inanga E.L., "Financial Sector Reforms in Sub-Sahara Africa." Paper Presented at the XI World Congress of the International Economic Association iri Tunis on December 18-22, 1995.

King R. and R. Levine, *Finance, Entrepreneurship, and Growth*. Paper presented at World Bank Conference on "How Do National Policies Affect Long-term Growth", Washington D.C., February, 1993.

Jappelli T. and M. Pagano, "Saving, Growth and Liquidity Constraints." *Quarterly Journal of Economics*, Vol. 109, no. 1, February,. Pp.83109, 1990.

Levine R. and D. Renelt, "A Sensitivity Analysis of Cross-Country Growth Regressions." *American Economic Review*, Vol. 82, no. 4, pp. 942-63, 1992.

Lipumba N.H.I., "Africa Beyond Adjustment, Overseas Development." Council Policy Essay No. 15, 1994.

Lipumba N.H.I., N. Osoro, and B.M. Nyagetera, "The Determinants of Aggregate and Financial Saving in Tanzania." Final Report Presented at an AERC Workshop, May, 1990.

McKinnon R., *Money and Capital in Economic Development*. Washington D.C., Brookings Institution, 1973.

Montiel P.J., "Financial Policies and Economic Growth: Theory, Evidence and Country-Specific Experience from Sub-Saharan Africa." Paper Presented at an AERC Plenary, May, 1994.

Mwega F., "Mobilization of Domestic Savings for African Development and Industrialization: A Case Study of Kenya." Oxford University International Development Centre Discussion Paper, May, 1992.

Mwega F.M. (1995), "Private Saving Behaviour in Less Developed Countries and Beyond: Is Sub-Saharan Africa Different?" Paper Presented at the XI World Congress of the International Economic Association in Tunis on 18-22 December, 1995.

Mwega F. M., N. Mwangi, and S.M. Ngola, "Real Interest Rates and the Mobilization of Private Savings in Africa: A Case Study of Kenya." AERC Research Paper 2, 1990.

Ncube M, P. Collier, J. Gunning and K. Mlambo, "Trade Liberalization and Regional Integration in Zimbabwe." Paper Prepared for the AERC Collaborative Project on Trade Liberalization and Regional Integration in Sub-Saharan Africa. March, 1995.

Nissanke, M., E. Aryetey, H. Hettige, and W.F. Steel, *Financial Integration and Development in Sub-Saharan Africa*. Washington D.C., World Bank. Mimeo, 1995.

Oshikoya T. W., "Interest Rate Liberalization, Savings, Investment and Growth: The Case of Kenya." *Savings and Development*, Vol. 26, no. 3, pp. 30520, 1992.

Seck D. and Y.M. Nil, "Financial Liberalization in Africa." *World Development*, vol. 21, no. 11, November, pp. 186781, 1993.

Shaw E., *Financial Deepening in Economic Development*, New York, Oxford University Press, 1993.

Schmidt-Hebbel K., S. Webb and G. Corsetti, "Household Saving in Developing Countries: First Cross Country Evidence." *The World Bank Economic Review*, Vol. 6, no. 3, pp. 52947, 1992.

Stiglitz J.E., "Financial Markets and Development." *Oxford Review of Economic Policy*, Vol. 5, No. 4, pp. 5568, 1989.

Stiglitz J.E., "The Role of the State in Financial Markets." Proceedings of the World Bank Annual Conference on Development Economics, 1993.

Stiglitz J.E. and A. Weiss, "Credit Rationing in Markets with Imperfect Information." *American Economic Review*, Vol. 71, no. 3 pp. 393410, 1981.

Soyibo A., "Financial Liberalization and Bank Restructuring in Sub-Saharan Africa: Some Lessons for Sequencing and Policy Design." Paper Presented at an

AERC Plenary, December, 1994.
Villanueva D. and Mirakhor A, "Strategies for Financial Reforms." *IMF Staff Papers*, vol. 37, no. 3, pp. 50936, 1990.
Wijnbergen van S., "Interest Rate Management in LDCS." *Journal of Monetary Economics*, vol. 12 no. 3 pp. 43352, 1983.
World Bank, *World Development Report*. Oxford, Oxford University Press, 1989.
World Bank, *Adjustment in Africa: Reforms, Results and the Road Ahead*. World Bank Policy Research Report, Oxford University Press, New York, 1994.

Chapter 11

Resource Mobilization, Financial Liberalization, and Investment:
The Case of Some African Countries

Mohammed Nureldin Hussain, Nadir Mohammed and Elwathig M. Kameir

Introduction

The role of interest rate in the determination of investment and, hence economic growth, has been a matter of controversy over a long period of time. Yet, what constitutes an appropriate interest rate policy still remains to be a puzzling question. Until the early 1970s, the main line of argument was that because the interest rate represents the cost of capital, low interest rates will encourage the acquisition of physical capital (investment) and promotes economic growth. Thus, during that era, the policy of low real interest rate was adopted by many countries including the developing countries of Africa. This position was, however, challenged by what is now known as the orthodox financial liberalization theory. The orthodox approach to financial liberalization (McKinnon-Kapur and the broader McKinnon-Shaw hypothesis) suggests that high positive real interest rates will encourage saving. This will lead, in turn, to more investment and economic growth, on the classical assumption that prior saving is necessary for investment. The orthodox approach brought into focus not only the relationship between investment and real interest rate, but also the relationship between the real interest rate and saving. It is argued that financial repression which is often associated with negative real deposit rates leads to the withdrawal of funds from the banking sector. The reduction in credit availability, it is argued, would reduce actual investment and hinders growth.

Because of this complementarity between saving and investment, the basic teaching of the orthodox approach is to free deposit rates. Positive real interest rates will encourage saving; and the increased liabilities of the banking system will oblige financial institutions to lend more resources for productive investment in a more efficient way. Higher loan rates, which follow higher deposits rates, will also discourage investment in low-yielding projects and raise the productivity of investment. This orthodox view became highly influential in the design of IMF-World Bank financial liberalization programmes which were implemented by many African countries under the umbrella of structural adjustment programs.

The purpose of this chapter is to provide a theoretical and empirical examination of the question of resource mobilization in the context of African countries as envisaged by the theory of financial liberalization. The chapter begins by developing the conceptual framework for the whole study. This involves the examination of the theory of financial liberalization, and the development of an analytical framework which exposes the theory and its critique. The chapter concentrates on examining the empirical relationship between the real interest rate, saving and investment. It draws a distinction between total saving and financial saving and estimates separate functions with special emphasis on the role of the real interest rate in the determination of each category of saving. For the relationship between the real interest rate and investment, this section employs a 3-equation investment model which tests for the effect of below equilibrium and above equilibrium interest rates on investment. The model also allows the calculation of the net effect of the real interest rate on investment after taking into account the effect of the real interest rate on the provision of credit and the cost of investment.

Resource Mobilization and Financial Liberalization

Resource Mobilization and Financial liberalization: A Conceptual Framework

The accumulation of capital stock through sustained investment is indispensable for the process of economic growth. In a closed economy, investment itself can only be financed from domestic saving. Because the acts of saving and investing are usually conducted by different people, the financial sector is entrusted with the functions of channeling resources from savers to investors. The relationships between domestic saving and economic growth can be examined through the Harrod-Domar Result:

$$g = \pi(S/Y) = \pi(I/Y)$$

where g is the rate of growth of real output, π is the productivity of capital and (S/Y) is the ratio of total domestic saving to income which, in equilibrium, is equal to the ratio of investment to income (I/Y). Accordingly, given the productivity of capital, the growth rate should increase the higher the ratio of saving (investment) to income. Conversely, if the ratio of saving (investment) to income is given, the growth rate can be increased by improving the efficiency of investment which will raise the productivity of capital (π). To do this, it is necessary to promote investment that support efficient production in sectors where rapid growth in effective demand can be expected (Okuda 1990).

The orthodox approach to financial liberalization suggests that, financial liberalization will both increase saving and improve the efficiency of investment (Shaw 1973). By eliminating controls on interest rates, credit ceilings and direct credit allocation, financial liberalization is said to lead to the establishment of positive interest rates on deposit loans. This, in turn, is said to make both savers and investors appreciate the true scarcity price of capital, leading to a reduced dispersion in profits rates among different economic sectors, improved allocative efficiency and higher output growth (Villanueva & Mirakhor 1990).

Figure (1) provides a diagrammatic illustration of the theory backing financial liberalization programs. The figure exhibits the behavior of savings (S) and investment (I) in relation to the real rate of interest (r). The savings schedule slopes upwards from left to right on the (classical) assumption that the rate of interest is the reward for foregoing present consumption. The investment schedule slopes downwards from left to right because it is assumed that the returns to investment decreases as the quantity of investment increases, which means that a lower real rate of interest is therefore necessary to induce more investment as the marginal return to investment falls. If the interest rate is allowed to move freely (i.e., no interest rate controls), the equilibrium rate of interest would be r^* and the level of saving and investment would be at I^*. If the monetary authorities impose a ceiling on the nominal saving deposit rate, this will give a real interest rate of, say, r_1. If this rate is also applicable for loans,[1] *saving will fall to S_1 and investment will be constrained by the availability of saving to I_1.* At r_1 the unsatisfied demand for investment is equal to AB. According to the financial liberalization theory, this will have negative effects on both the quantity and the quality of investment. That is, credit will have to be rationed, consequently many profitable projects will not be financed. There will also be a tendency for the banks to finance less risky projects, with a lower rate of return, than projects with a higher rate of return but with more risk attached.

If the ceiling on interest rate is relaxed, so that the real interest rate increases to r_3, saving will increase from I_1, to I_3, and the efficiency of investment also increases because banks are now financing projects with higher expected returns. Unsatisfied investment demand has fallen to A_1B_1 and credit rationing is reduced. It is argued that savings will be «optimal» and credit rationing will disappear, when the market is fully liberalized and the real rate of interest is at r^*.

Although it appears convincing, the financial liberalization theory suffers from major shortcomings. As it has been argued by Warman & Thirlwall (1994), the financial liberalization theory makes no clear distinction between financial saving and total saving. To be sure, the saving symbol which appears in equation (1) stands for total saving and not financial saving. The relationships suggested by the Harrod-Domer result, between saving, investment and growth, are complicated by the fact that a significant portion of domestic saving may be held in the form of real assets (e.g., real estate, gold and livestock), exported abroad in the form of capital flight, or claimed by informal markets such as the informal credit market, the underground economy and the black market for foreign exchange. The fact that financial saving is only one form of saving, raises many important issues regarding the theory of financial liberalization. In what follows, a simple conceptual framework is developed to restructure the debate on financial liberalization and to articulate the arguments against the financial liberalization theory. It puts into focus some of the worries, criticisms and limitations of the financial liberalization theory which are important to bear in mind when evaluating the implementation of policies in the context of African countries.

Total Saving, Financial Saving, and the Leakage

The flow of total national saving can be decomposed into public saving and private (household and enterprise) saving:

$$S_T = S_G + S_P \quad (2)$$

Where S_T, S_G, and S_P are total, public, and private savings respectively. The flow of private saving can be divided into two major components: private financial saving which comprise the portion of private saving that is kept in the form of financial assets in the formal financial sector (F_P) and private saving residue which comprises the portion of private saving which is kept in non-financial forms or put into other uses (L). That is:

$$S_P = F_P + L \quad (3)$$

Substituting equation (3) into (2), we get:

$$S_T = S_G + F_P + L \quad (4)$$

The flow of total financial saving (F_T) comprise public financial saving (F_G) and private financial saving (F_P). That is:

$$F_T = F_P + F_G \quad (5)$$

On the assumption that all government saving is kept in the form of financial assets (so that $F_G = S_G$) and substituting equation (5) in (4), and rearranging we have:

$$L = S_T - (F_G + F_P) \quad (6)$$

and,

$$F_T = S_T - L \quad (7)$$

Dividing equation (6) by S_T, we obtain:

$$F_T/S_T = 1 - \sigma \quad (8)$$

Where, $\sigma = L/S_T$, which measures the proportion of total saving that is leaked out of, or not captured by the formal financial sector. If equations (6), (7) and (8) are expressed in stock rather than flow terms, they can be interpreted as giving the condition for the case of what can be called *full financial deepening* where the whole stock of total saving is kept in financial forms and the leakage, L, is zero. The degree of financial deepening at any point in time, can be measured by equation (8), where the smaller, σ, the higher will be the degree of financial deepening. The equations can also be used to clarify the confusion in the literature between total saving and financial saving. Total saving and financial saving are identical only in the case of a zero leakage (i.e., L=O).

In their flow forms the equations can be interpreted as giving the 'dynamics' of the process of financial deepening. The case of a zero leakage with $L = \sigma = 0$, corresponds to what can be called *full financial augmentation* where all the additions to total saving are kept in the form of financial vessels (so that $S_T = F_G + F_P$ in equation (6) and $F_T/S_T=1$ in equation (8)). A reduction in the leakage (i.e., $\delta L < 0$) implies an improvement in the process of financial deepening while an increase in the leakage (i.e., $\delta L > 0$) implies an increase in the process of financial shallowing. The equations also illustrate the important result that even though total saving might be stagnant (i.e., $\delta S_T=0$) financial saving can increase if $\delta L<0$. To elucidate this result, the leakages of saving outside the formal financial sector may be divided into the following main components:[2] the portion which is kept in the form of real assets including livestock and gold (R); the portion which is claimed by the informal financial sector (N), the proportion which is claimed by the underground economy including the black market for foreign exchange (U), the portion which goes into capital flight (C) (this is usually kept abroad in the form of foreign currency deposit accounts, financial assets or physical assets) and; the portion which is hoarded by households in the form of domestic or foreign currency holdings (H). Using these definitions in equation (8), we obtain:

$$F_T = S_T - (R + N + U + C + H) \quad (9)$$

Equation (9) is a restatement of the fact that financial saving is only one type of saving. The main other types of saving are represented by the components of the leakage on the right hand side of the equation. According to the equation, if saving is stagnant (i.e., $\delta S_T = 0$), financial saving can still increase - keeping other things constant - by reducing the stock of saving which is kept in real assets (i.e., $\delta R < 0$); by reversing the process of capital flight (i.e., $\delta C < 0$); by attracting the resources of the informal sector into the formal sector (i.e., $\delta N < 0$); by reducing the amount of saving claimed by the underground economy and; by encouraging dishoarding of foreign and domestic currency by households (i.e.,$\delta H < 0$). Conversely, large additions to total saving might not increase financial saving if they are offset by an equal increase in any of the components of the leakage, say, capital flight. Also, substitution among the components of the leakage might occur without affecting financial saving. An increased dishoarding of domestic and foreign currency, for instance, might be offset by an equal increase in capital flight (i.e., $-\delta H = \delta C$) leaving financial saving unchanged, and so on.

The Orthodox Versus the New Structuralist

Equation (9) allows us to reinterpret the controversy between the proponents of the orthodox financial liberalization theory and their opponents from the New Structuralist School. In the context of our model, this controversy concentrates mainly on the interactions between financial saving and the components of the leakages shown by the equation. According to McKinnon (1973), in most developing countries, a significant proportion of working capital for the financing of inventories, goods in process, trade credit and advances to workers is obtained through bank financing. The supply of bank credit determines the level of available working capital, and thus of net output. An increase in deposit interest rates, is thus expected to increase holdings of financial assets (broad money) and hence working capital and output. This increase in financial saving will occur through the reduction in the leakage.

For the new structuralist, an increase in deposit rates is likely to result in an increased holdings of financial assets, but the outcome will not necessarily be a positive increases in working capital and output. This depends on which component of the leakage is reduced and on whether the reduced component will cause an offsetting reduction in output. Two basic reasons (assumptions) are usually given to explain the position of the new structuralist: first, the intermediation of the informal credit market is said to be complete while that of the formal market is not and; second, the informal credit market is assumed to support equally productive and efficient activities, while the other components of the leakage do not. As for the first reason, the funds in the informal market are said to be transmitted, in full, to production entities with no holding of reserves, while in the formal sector the transmission is less than full. The required reserve ratio and the holding of excess reserves constitute another leakage in the flow of funds between savers and investors. This can be illustrated by assuming that financial savings are equal to bank time deposits. The relationship between financial saving and the supply of bank loans may be written as:

$$B_C = (1 - \Theta) F_T \quad (10)$$

where B_c, is the supply of bank loans, Θ is the required and excess reserves ratio held by banks and F_T is financial saving. Thus, while reductions in any of the components of the leakage in equation (9), keeping all other parameters constant, will bring about an equal increase in financial saving, the supply of bank loans will not increase by the same amount because of the leakage caused by banks' holdings of reserves.

As for the second reason, according to Van Wijnbergen's (1983) new structuralist model, if the increase in financial saving occurs through the reduction in hoarded cash balances [-H in equation (9)] and other intrinsic unproductive assets, then it will have a positive effect on output. If, however, it is at the cost of informal credit market (-N) it will lead to a fall in total private sector credit, working capital and output. The loss of informal sector credit without an equal compensating increase in formal sector lending, is said to bid up the informal sector lending rate and reduce net working capital causing a decline in output.

The new structuralist argument rests, therefore, on the contentions that the intermediation of the formal sector is not full and that informal sector resources, and not the other components of the leakage in equation (9), are likely to be the closest substitute for time deposits. However, it has been argued by Serieux (1993), that less than full intermediation can only occur if we assume away the money creation capacity of banks. That is, most informal sector loans are essentially cash loans. It follows that a shift from informal sector resources to bank resources would imply a surrender of cash to the formal banking sector. This, will increase banks' reserves and hence their credit creation capacity. Accordingly, bank intermediation might be complete or even multiplicative. Also, an increase in bank deposit rates, may lead to shifts among the components of the leakage such that the available informal credit remains intact.

The Real Interest Rate and the Determinants of Saving and Investment in Africa

As outlined in the conceptual framework, the financial liberalization theory, is based crucially on three postulates concerning the relationship between the real interest rate, saving and investment:

(i) that saving is positively related to the real rate of interest;
(ii) that investment is determined by prior saving; and
(iii) that the effect of the real rate of interest on investment will depend on whether the real interest rate is below or above the equilibrium rate.

Although the financial liberalization theory places more emphasis on the desirable effects of raising the real interest rate towards equilibrium, it also postulates that the impact of the change in the real interest rate on investment depends on whether the actual interest rate is below or above equilibrium. Below the equilibrium interest rate, investment is constrained by saving. An increase in the real interest rate towards equilibrium, will increase saving and investment. Hence, as long as the equilibrium interest rate is not reached, investment is positively related to the real interest rate (see Figure 1). Beyond this equilibrium, an increase in real interest rate will have a negative effect on investment as the economy moves along the negatively-sloped investment demand curve. The relationship between the real interest rate and investment as postulated by the financial liberalization theory is depicted by Figure (2).

Against this theory which is based on classical notions, we have the Keynesian framework which postulates that saving is positively related to income and investment is negatively related to the price of credit for which the interest rate stands as a proxy. However, the proponents of this Keynesian view concede that the real interest rate might also have a positive effect on investment through the provision of credit. That is, as financial saving and the rate of interest are positively related, interest rate may also have a positive effect on investment through the process of financial deepening and the provision of credit to the private sector (Warman & Thirlwall 1994). Thus the net effect of the real interest rate on investment will depend on the relative strength of its negative effect through the cost of investment and its positive effect through the provision of credit. In what follows, we discuss the empirical models that we are going to use to examine the validity of the postulates of the financial liberalization theory.

Total Saving and Financial Saving: Empirical Models

The financial liberalization theory postulates that saving is positively related to the real interest rate. The theory, however, does not make clear distinction between total savings and financial saving. Total domestic saving consists of private and public savings of which financial savings is a part. While financial saving is the portion of total saving that is channeled through financial assets which comprise short and long-term banking instruments, non-bank financial instruments such as treasury bills and other government bonds and commercial paper. It is prudent, therefore, to examine the role of the real interest rate in the determination of both total saving and financial saving. To this end, the following two equations for total saving and financial saving are specified:

$$T_S = \sigma_0 + \sigma_1 r + \sigma_2 Y \quad (12)$$

$$F_S = \Theta_0 + \Theta_1 r + \Theta_2 Y + \Theta_3 (C/M_1) \quad (13)$$

It is hypothesized that, total real domestic saving (T_S) is a function of real income (Y) and the real interest rate (r). It is expected that total saving is positively related to real income (Y). Whether the total saving is positively or negatively related to the real interest rate, will depend on the relative strength of the income and substitution effects of changes in the rate of interest [see Warman & Thirlwall (1994)]. The substitution effect of a higher interest rate is to encourage agents to sacrifice current consumption for future consumption, but the income effect is to discourage current saving by giving agents more income in the present, and the two effects may cancel each other out.

As saving is a flow concept, financial saving is measured by the change in the stock of such financial assets. It is hypothesized that real financial saving (F_S), is positively related to real income (Y), and also positively related to the real rate of interest (r) as postulated by the financial liberalization theory. Also, because of the important role played by the informal credit market in many African countries, an attempt is made to capture its effect on formal financial saving. This is done by including as an independent variable, the ratio of currency outside banks to narrow money M_1. The proportion of narrow money (M_1) held as currency is commonly accepted as a measure of the size of the informal credit market (see Shaw 1973 and Serieux 1993), on the grounds that the higher this ratio the larger will be the size of the informal credit market.

An Investment Model

As it has been noted before, the financial liberalization theory suggests a differing effect of the real interest rate on investment, depending on whether the real interest rate is below or above the equilibrium rate. It is also noted that, on Keynesian grounds, the real interest rate might affect investment negatively through the cost of investment and positively through the provision of credit. The foregoing arguments suggest the use of a switching regime model which differentiate between the effect of above and below equilibrium real interest rates. They also suggest that attempts should be made to separate the positive and negative impact of the interest rate on investment. To this end, we use a switching investment model:

$$I = \alpha_0 + \alpha_1 B_C + \alpha_2 r + \alpha_3 [(r-r^*)D] + \alpha_4 \delta Y \quad (14)$$

$$B_C = \Omega_0 + \Omega_1 F_S \quad (15)$$

$$F_S = \Theta_0 + \Theta_1 r + \Theta_2 Y + \Theta_3 (C/M_1) \quad (16)$$

where I is real gross fixed investment; r is the real interest rate; r^* is the real equilibrium interest rate; B_C is the supply of bank credit; D is the switching-point dummy variable which takes the value of zero when the interest rate is below equilibrium (i.e., $r < r^*$) and the value of unity when the interest rate is above equilibrium; and δY is the change in real income as a measure of the income accelerator effect on investment. Substituting equation (16) into equation (15) and the result in (14), and by totally differentiating equation (14) with respect to the real rate of interest and rearranging, the investment/interest rate (dI/dr) multiplier will be given as follows:

$$dI/dr = \alpha_2 + \alpha_1 \Theta_1 \Omega_1 + \alpha_3 D \quad (17)$$

In the case where the interest rate is above the equilibrium with D=1, the negative effect of interest on investment is assumed to be at work and the effect of the real interest rate on investment will be given $\alpha_2 + \alpha_1 \Theta_1 \Omega_1 + \alpha_3$. While below the equilibrium rate of interest, with D=0, the impact of α_3 will disappear and there will remain only the impact measured by $\hat{a}_2 + \alpha_1 \Theta_1 \Omega_1$.

The product $(\alpha_1 . \Theta_1 . \Omega_1)$ represents the chain effect that goes from the real interest rate to investment, through the supply of credit. An increase in interest rate is expected to stimulate saving in financial forms (Ω_1), which is expected to increase the supply of credit (Θ_1), which is expected, in turn, to increase investment (α_1). The question of whether the final effect of interest rate on investment is negative or positive, depends on the relative magnitude of the parameters $\alpha_2, \alpha_3, \Theta_1, \Omega_1$ and α_1.

Empirical Results: Saving Equations

The saving equations (12) and (13) are estimated for 25 African countries over the period 1972-1992. A single equation OLS procedure is employed and in the case of the presence of autocorrelation in OLS estimates, a General Autoregressive and Moving Average Error Correction Process (GAMAECP), with a first-order autoregressive and/or first-order moving average is used. Different experiments with time lags on the real interest rate variable were conducted. Tables 11.1 and 11.2 give a summary of the results for the total saving and financial saving respectively.

Table 11.1: Summary of the results of total domestic savings equation in 25 African countries, 1970-1992

Percentage of the total

Independent Variable	Positive and Significant	Positive and Insignificant	Negative and Insignificant	Negative and Significant
r	20	20	32	28
GDY	72	24	4	0

Table 11.2. Summary of the results of the financial savings equation (F_s) in 25 African countries, 1970-1992

Percentage of the total

Independent Variable	Positive and Significant	Positive and Insignificant	Negative and Insignificant	Negative and Significant
r	4	40	16	40
GDY	16	44	8	32
RATIO	4	24	32	40

It can be observed from Table (11.1) that of the 25 countries in the sample, the real interest rate has a positive impact on total saving in the case of 15 countries (60 percent of the total). The coefficient on the real interest rate is, however, positive and statistically significant, at the 10 percent level of confidence, in only 8 cases (32 percent of the total), these are Burkina Faso, Gabon,

Mauritius, Nigeria, Swaziland, Zaire, Zambia, and Zimbabwe. This positive and significant relationship between the real interest rate and total saving, indicates that in the case of these countries the positive substitution effect of real interest rates outbalances the negative income effect. This also implies that, in the case of these countries, there is a strong substitution effect between present and future consumption. The real interest rate has a negative and significant effect on total saving in five cases (28 percent of the total). These are Benin, Côte d' Ivoire, South Africa, Togo, and Tunisia. Thus, in these five countries the implication with regard to the substitution and income effects is exactly the opposite to that of the eight cases above.

In conformity with Keynesian theory, real income proved to be the most important determinant of saving. The coefficient on income has the expected positive sign in all cases with the exception of Tanzania, and it is positive and statistically significant in 18 cases. For some African countries (e.g., Ghana, Senegal, and Swaziland) the estimated propensity to save is very small not exceeding 0.01 for one unit of income. The largest propensities to save are recorded by Gabon (0.92) and Mali (0.61). In the majority of all the other countries, the estimated propensity to save ranges between 0.10 and 0.35.

Contrary to expectations, the interest rate plays no significant role in the determination of financial saving (Table 11.2). The real interest rate and financial saving are positively related in 11 cases. However, this positive relationship is statistically significant in one case, Mauritius. The relationship between the real interest rate and financial saving is negative and significant in four cases, namely Tanzania, Tunisia, Zaire and Zimbabwe. Similarly, the impact of gross domestic income on financial savings is statistically insignificant in three quarters of the countries in the sample. It is positive and statistically significant in four cases (Kenya, Mali, Mauritius, and Togo) and negative and statistically significant in Gambia and Madagascar. The activities of the informal credit market as proxied by the ratio of currency outside banks to narrow money, proved to be the most important determinant of financial saving. The relationship between financial saving and (C/M_1) ratio is negative in 18 cases. This indicates that as this ratio increases (indicating an increase in the activities of the informal credit market) financial saving decreases. This negative relationship between financial saving and the activities of the informal credit market is statistically significant in eight cases. These are Burkina Faso, Côte d' Ivoire, Ghana, Kenya, Mali, Mauritius, Morocco, and Nigeria.

Thus, in so far as the relationship between the real interest rate and saving is concerned, these results give no strong support to the financial liberalization theory. The real interest rate does not play a significant role in the determination of neither financial saving nor total saving. Real income is found to be the most important determinant of total saving while the activities of the informal market is found to be the most important determinant of financial saving.

Empirical Results: Investment Model

The investment model [equations(14) to (16)] is estimated for the 25 African countries in the sample over the period 1971-1992. As in the case of saving equations, a single equation OLS procedure is employed correcting for the presence of autocorrelation through the use of a GAMAECP with a first-order autoregressive and/or a first-order moving average. All the annual data used in the estimation are obtained from World Bank World Tables with the exception of data on nominal interest rate which are obtained from the IMF International Financial Statistics. For the basic investment equation, a similar OLS procedure is used. Preliminary experiments have also shown that investment in some African countries is volatile during years of armed conflicts. To capture this effect, we include in the basic investment equation a War dummy which takes a value of zero in peace years and a value of unity in the years of armed conflicts.[3]

Since the equilibrium interest rate (r^*) is unknown, it is necessary to search for the equilibrium rate which minimizes the sum of square residuals using a trial and error process. That is, different hypothetical values of the real rate of interest that range from +30 percent to -30 percent are initially assumed. Each value is used to calculate the switch variable on the assumption that it represents the equilibrium rate of interest. Each of these variables are then used, together with the other explanatory variables to estimate equation (3). The hypothetical real interest value that yields the estimate with the smallest sum of square residuals is considered as the true equilibrium real interest rate (see Rittenberg 1991).

Table 11.3 shows the values of the real interest rate arrived at in each country's case. The Table reveals that there is a large variations in real interest rate experienced by the African countries in the sample. In 12 countries, the estimated equilibrium real interest rate is positive ranging from a rate of 2.0 percent per annum in the case of Mauritius to a rate of 17 percent in Senegal. In 12 countries, it is negative ranging from a rate of -0.5 percent for Kenya, Gabon, Madagascar, and Zambia and a rate of -27.0 percent for Sierra Leone. This variation is evident even in countries with similar monetary arrangements such as the CFA Franc Zone. For instance, the estimated real interest rate for Côte d'Ivoire is negative amounting to about -7.5 percent, while in Burkina Faso it is positive amounting to 9.0 percent.

Table 11.3. Equilibrium interest rates in 25 African countries, 1970-1992

Country	Equilibrium Interest Rate	Country	Equilibrium Interest Rate
Benin	4.5	Nigeria	3.5
Burkina Faso	9.0	Rwanda	11.5
Côte d' Ivoire	-7.5	Sierra Leone	-27.0
Ghana	8.5	Senegal	17.0
Gabon	-0.5	South Africa	8.5
Gambia	-8.5	Swaziland	-8.5
Kenya	-0.5	Tanzania	0.0
Madagascar	-0.5	Togo	-3.0
Malawi	3.5	Tunisia	4.5
Mali	-7.5	Zaire	2.5
Mauritius	2.0	Zambia	-0.5
Morocco	-8.5	Zimbabwe	3.0
Niger	-9.5		

Table 11.4. Summary of the results of the basic investment (I) in 25 African countries, 1970-1992

Percentage of the total

Independent Variable	Positive and Significant	Positive and Insignificant	Negative and Significant	Negative and Insignificant
B_c	8	64	12	16
r	32	20	40	8
dY	48	32	4	16
[(r-r*)D]	40	8	36	16
WAR	0	40	40	20

It can also be observed that there is no apparent correlation between high inflation rates and negative equilibrium real interest rate. For instance, the equilibrium rate in Zaire (a country with high inflation rates) is 3.0 percent while in Côte d' Ivoire (a country with low inflation rates) is -7.5 percent.

The equilibrium interest rates in Table 11.3 are used to estimate the basic investment equation and the complete results are reported in Table 11.4. Table 11.5, provides a summary of these results and reveals that the supply of bank credit to the private sector is a very important determinant of investment. The coefficient on the supply of credit has the expected positive sign in 19 cases and it is positive and statistically significant in 16 cases. For most of these countries the size of this coefficient is large (in the cases of both elasticity and propensity estimates) indicating a large effect of changes in bank credit on investment. In the case of South Africa and Zimbabwe, for instance a one percent increase in the supply of credit leads to about 0.97 percent increase in investment. In Benin, Burkina Faso, and Côte d' Ivoire a one unit increase in bank credit leads to an increase in investment of about 5 to 7 units.

However, in certain countries, namely, Gabon, Senegal, Tanzania, and Zaire, the coefficient on the supply of credit is negative and highly significant. Although this result is puzzling, but some "African" observations might help to resolve this puzzle. Because of macroeconomic instability, it is observed that in some African countries, investors opt to use the funds supplied by banks for the purchase of foreign exchange which is usually exported and deposited abroad (a form of capital flight) awaiting a large currency devaluation. After devaluation, part of the funds deposited abroad is imported and converted into domestic currency to repay the bank loan and the "profit" is kept abroad. Some investors simply default on the bank loan. If such an operation is in process for sometime and the level of investment is generally falling because of the very macroeconomic instability, it will not be surprising to find a negative and statistically significant relation between the supply of bank credit to the private sector and the level of investment.

Demand factors as approximated by the income accelerator effect proved to be as important as the credit supply factor in the determination of investment. The impact of the change in income on investment is positive in 20 cases and it is positive and significant in 12 cases. However, the impact of real change in income is negative and statistically significant in Rwanda only. Out of the five countries affected by armed conflicts in the sample, the WAR dummy is negative and statistically significant in the case of Kenya and Rwanda. It is negative but not significant in the case of Zimbabwe, positive and statistically insignificant in the cases of Morocco and Tanzania.

We turn now to the subject of primary interest to this study, namely, the effect of the real interest rate on investment. Table 11.5 shows that the effect of the real interest rate on investment has the expected positive sign in 13 cases (52 percent of the total) but it is positive and statistically significant in only 8 cases (32 percent of the total).

Table 11.5. Determinants of investment in Africa, 1970-1992

Country-Period	C	Bc	r	dY	$(r-r^*)D$	WAR	AR(1), MA(1)	R2	D.W.	F(...)
(1) Benin 1971-1991	17.1028 (0.86)	5.7758 (6.53)	-2.2298 (-3.52)	0.3419 (2.58)	-1.0756 (-0.52)		0.94 MA(1) (9.40)	0.83	2.07	14.40
(2) Burkina Faso 1971-1991	36.9152 (2.84)	5.0124 (3.41)	1.2210 (1.94)	0.9726 (3.23)	-15.9355 (-3.54)		0.74 MA(1) (4.16)	0.76	1.73	9.37
(3) Côte d'Ivoire 1972-1992	-248.3500 (-0.76)	7.5372 (2.78)	3.6707 (0.61)	0.1409 (1.79)	-24.2611 (-2.28)		0.66 AR (1) (4.74)	0.87	1.77	20.54
(4) Ghana 1972-1992	114.2210 (1.07)	0.6668 (1.80)	-0.0268 (-0.27)	0.0813 (1.15)	-2.4662 (-1.38)		0.91 AR (1) (5.19)	0.71	1.87	7.18
(5) Gabon 1978-1991	783.8390 (8.54)	-1197.5 (-3.68)	-15.351 (2.99)	0.0162 (0.13)	20.2459 (2.83)		1.01 MA (1) (62.66)	0.80	2.30	6.44
(6) Gambia 1978-1991	0.2891 (3.70)	-0.1347 (0.74)	-0.0015 (-0.88)	-0.0066 (-0.09)	0.0081 (1.75)			0.46	1.92	1.93
(7) Kenya 1971-1991	11.4750 (2.01)	0.8050 (3.69)	-0.5525 (-2.31)	0.2389 (1.57)	-0.6063 (-0.94)	-6.4140 (-2.86)		0.56	1.72	3.86
(8) Madagascar 1972-1992	112.820 (0.75)	368.024 (1.09)	-3.3398 (-1.41)	0.3564 (2.61)	-12.1778 (-0.79)		0.43 AR (1) (1.70)	0.51	1.93	3.06

Table 11.5. Continued... (1)

(9) Malawi 1972-1992	0.4347 (2.44)	0.4900 (0.91)	0.0046 (0.72)	0.2390 (1.15)	0.1205 (2.77)		0.45 AR (1) (1.88)	0.66	2.33	5.72
(10) Mali 1972-1992	-0.8378 (-0.01)	-107.0800 (-0.73)	-1.9870 (-1.37)	0.0307 (0.56)	2.4571 (1.41)		1.06 AR (1) (12.20)	0.92	1.78	34.75
(11) Mauritius 1971-1992	1.5102 (6.76)	0.7536 (16.38)	-0.8215 (-4.33)	0.2347 (2.65)	0.8691 (3.42)			0.95	1.89	85.40
(12) Morocco 1971-1989	32.9784 (3.47)	0.5565 (5.45)	2.1755 (2.30)	0.2384 (1.48)	-3.0686 (-2.87)	2.3769 (1.02)		0.86	1.63	15.78
(13) Niger 1972-1992	201.3980 (4.72)	0.5131 (2.48)	8.3732 (2.52)	0.2997 (3.52)	-12.8648 (-3.47)		-0.39 AR (1) (1.77)	0.72	2.24	7.80
(14) Nigeria 1972-1992	-4.4860 (-0.26)	1.2872 (3.03)	-0.0089 (-1.94)	0.0106 (2.06)	0.1207 (2.94)		0.94 AR (1) (4.23)	0.62	1.49	3.96
(15) Rwanda 1972-1992	2.1360 (1.92)	1.5970 (13.01)	0.0555 (1.10)	-0.2276 (-2.88)	3.1120 (5.41)	-2.0658 (-1.80)		0.97	2.03	91.02
(16) Sierra Leone 1971-1991	3.0493 (11.55)	0.0049 (4.15)	0.0088 (2.50)	0.0131 (2.50)	-0.0516 (-4.48)		0.84 MA (1) (5.69)	0.70	2.04	6.62
(17) Senegal 1972-1992	274.5300 (6.39)	-0.2781 (-2.66)	1.2348 (2.23)	0.0175 (0.22)	-1.2622 (-2.21)		0.49 AR (1) (2.33)	0.67	2.04	6.09

Table 11.5. Continued... (2)

(18) South Africa 1973-1990	10.8158 (0.42)	0.9745 (2.91)	-0.7031 (-1.70)	0.2953 (1.38)	0.8620 (0.40)		1.10 AR (1) (12.04)	0.51	1.40	2.33
(19) Swaziland 1971-1992	0.2932 (3.30)	0.1115 (1.67)	0.0204 (2.10)	-0.0029 (-0.26)	-0.0199 (-1.95)		0.70 AR (1) (3.84)	0.85	1.60	17.02
(20) Tanzania 1971-1989	66.4312 (11.71)	-3.1914 (-5.50)	0.3901 (1.74)	1.1311 (1.77)	1.8756 (1.39)	5.2209 (1.11)		0.76	1.80	8.37
(21) Togo 1971-1992	32.6217 (1.88)	0.4429 (1.76)	-5.8621 (-1.93)	0.0400 (0.33)	5.9957 (1.68)		-0.60 MA (1) (2.79)	0.59	1.76	4.56
(22) Tunisia 1971-1988	1.1612 (6.32)	0.1909 (3.39)	-0.0186 (-1.75)	0.2169 (0.79)	0.0116 (0.09)		-0.92 MA (1) (5.03)	0.76	1.35	7.69
(23) Zaire 1971-1991	13.4447 (9.68)	-0.0735 (-2.61)	0.0075 (0.43)	-0.0028 (-0.23)	-0.9847 (-0.80)		-0.99 MA (1) (-3.59)	0.52	1.76	3.00
(24) Zambia 1971-1991	1.6367 (0.47)	1.2505 (1.12)	0.0022 (0.11)	-0.1411 (0.92)	3.3338 (4.20)		0.67 MA (1) (3.40)	0.72	1.63	7.82
(25) Zimbabwe 1977-1991	0.1178 (0.20)	0.9652 (2.47)	0.0733 (1.50)	0.4775 (2.10)	-0.1054 (-1.53)	-0.0352 (0.18)	0.51 MA (1) (1.62)	0.79	2.08	5.10

Table 11.6. The net effect of real interest rate on investment (in billion of local currency)

Country	α_1	α_2	α_3	Θ_1	Ω_1	Effect of Supply of Credit on Investment	The Net Effect of Real Interest Rate on Investment (in Billions of Local Currency)	
							Below Equilibrium	Above Equilibrium
Benin	5.776	-2.230	0.000	0.000	0.000	0.000	-2.230	-2.230
Burkina Faso	5.012	1.221	-16	0.000	0.000	0.000	1.221	-14.715
Côte d'Ivoire	7.537	0.000	-24	0.000	0.000	0.000	0.000	-24.261
Ghana	0.667	0.000	-2.5	0.000	1.41	0.000	0.000	-2.466
Gabon	-1197.50	-15.4	0.05	0.000	0.000	0.000	-15.351	-15.302
Gambia	0.000	0.000	0	0.000	0.000	0.000	0.000	0.008
Kenya	0.805	-0.55	0.000	0.000	0.000	0.000	-0.553	-0.553
Madagascar	0.000	-3.340	0.000	0.000	0.000	0.000	-3.34	-3.340
Malawi	0.000	0.000	0.12	0.000	0.000	0.000	0.000	0.121
Mali	0.000	-1.99	0.03	0.000	0	0.000	-1.987	-1.954
Mauritius	0.754	-0.82	0.87	0.08	2.3	0.131	-0.690	0.179
Morocco	0.557	2.176	-3.1	0.000	5.15	0.000	2.176	-0.893

Table 11.6 Continued....

Niger	0.513	8.373	-13	0.000	0.000	0.000	8.373	-4.492
Nigeria	1.287	0.000	0.12	0.000	0.000	0.000	-0.009	0.112
Rwanda	1.597	0.000	3.1	0.000	0.000	0.000	0.000	3.101
Sierra Leone	0.005	0.009	0	0.000	0.000	0.000	0.009	-0.043
Senegal	-0.28	1.235	-1.3	0.000	0.000	0.000	1.235	-0.027
South Africa	0.975	-0.7	0.000	0.000	0.41	0.000	-0.703	-0.703
Swaziland	0.115	0.020	0	0.000	0.000	0.000	0.020	-0.027
Tanzania	-3.19	0.390	0.04	-0.3	0.000	0.000	0.390	0.426
Togo	0.443	-5.86	0.06	0.000	-0.130	0.000	-5.861	-5.799
Tunisia	0.191	-0.01	0.000	0	0.000	0.000	-0.011	-0.011
Zaire	-0.07	0.000	0.000	0	0.000	0.000	0.000	0.000
Zambia	0.000	0.000	3.33	0.000	0.000	0.000	0.000	3.334
Zimbabwe	0.965	0.073	-0.1	0	0.000	0.000	0.073	-0.032

These are Burkina Faso, Morocco, Niger, Sierra Leone, Senegal, Swaziland, Tanzania and Zimbabwe. As discussed before, a positive and significant relation between the real interest and investment implies that saving is positively related to the real interest rate. Our previous estimations of the saving functions support this implication in 6 of the 8 cases cited above; the exceptions are Morocco and Senegal. However, the coefficient on the real interest rate is negative in 12 cases and negative and significant in 10 case. These are Benin, Gabon, Kenya, Madagascar, Mali, Mauritius, Nigeria, South Africa, Togo, and Tunisia. The impact of the real interest rate with the switching dummy variable, has the expected negative sign in 13 cases. But it is positive and statistically significant in 10 cases including Benin, Burkina Faso, Côte d' Ivoire, Ghana, Morocco, Niger, Sierra Leone, Senegal, Swaziland, and Zimbabwe. However, the real interest rate and the real interest rate with the switching dummy variable, have the correct signs and are statistically significant in only six cases (Burkina Faso, Morocco, Niger, Sierra Leone, Senegal, and Zimbabwe).

As for the estimates of structural relationship between financial saving and the supply of bank credit [equation (15)], it is found that the relationship is positive in 64% of the sample, but is positive and statistically significant in only four countries (Ghana, Mauritius, Morocco, and South Africa). Its effect is negative and statistically significant in Mali and Togo, while the impact is insignificant in the majority of countries included in the sample. Thus, there appears to be no strong positive relation between financial assets held in the financial system and credit given by the banking sector to the private sector. This, however, might be explained by the fact that credit given to the private sector depends also on the reserve requirements of the commercial banks, credit allocation policies and the availability of profitable investment opportunities and credible investors.

Having estimated the basic parameters of the investment equations, it is now possible to calculate the *net* effect of the change in the real interest rate. As it has been shown before, this net effect depends on the direct effect of the real interest rate on investment and its indirect effect through the supply of credit. Table 11.7 shows the basic estimated parameters and the computation of the net effect when the real interest rate is above equilibrium (given by $\alpha_2 + \alpha_1 \Theta_1 \Omega_1 + \alpha_3$) and the net effect when the real interest rate is above equilibrium (given by $\alpha_2 + \alpha_1 \Theta_1 \Omega_1$. The parameters that were used in the computation are obtained directly from the estimates of equations (14) to (16). Before interpreting the results, it must be noted that a value of zero is given for all those parameters which were not statistically significant.

Table 11.7. Summary of the results of the impact of financial savings (F_s) on bank credit (B_c) in 25 African Countries, 1970-1992

Relationship and Effect	Percentage of the total Positive and Significant	Positive and Insignificant	Negative and Significant	Negative Variable and Insignificant
Effect of F_s on B_c	15	48	8	28

As it can be observed from the table, the only African country where the real interest rate has a positive effect on investment through the supply of credit effect is Mauritius. For this country it is estimated that a one per cent increase in the rate of interest would lead to an increase of financial saving of about 0.07 billions Rubee in constant 1987 prices, (Θ_1). It is also estimated that a one unit increase in financial saving would increase bank credit by about 2.3 units (=Ω_1), while a one unit increase in bank credit would increase investment by 0.75 units (= α_1). The final result of this chain effect (the product $\alpha_1 . \Theta_1 . \Omega_1$), is that a one per cent increase in the real interest rate would lead to an increase of investment of 0.13 billion Rubee in constant 1987 prices. However, contrary to theoretical expectations, the direct effect of an increase in the real interest rate on investment is negative below the equilibrium interest rate and positive above it. Below equilibrium an increase in the real rate of interest would reduce investment by about 0.82 billion Rubee. The total net effect is that an increase in the real rate of interest below equilibrium would reduce investment by about 0.70 billion Rubee (about 3.9 per cent of investment/GDP ratio). In the case where the real interest rate is above equilibrium, the net effect is negative but negligible with a one per cent increase in the real interest rate would increase investment by about 0.18 billion Rubee (or about one per cent of investment/GDP ratio).

For all other countries in the sample, the chain effect that goes from the real interest rate to investment, through the supply of credit is zero. This result follows directly from the insensitivity of financial saving to the real interest rate, indicated by the coefficient (Θ_1) which is statistically not different from zero. In these countries, the net effect of the real interest rate depends on its direct effect on investment below and above equilibrium. Inspecting the results of this net effect, it can be observed that only 6 countries in the sample conform to the theoretical expectation that the effect of the real interest rate on investment would be positive below equilibrium and negative above it. These are Burkina Faso, Ghana, Nigeria, Sierra Leone, Senegal, Swaziland, and Zimbabwe. In the case of Burkina Faso and Swaziland a one per cent increase in the real interest rate would bring about large reductions in investment amounting to 3 and 1.5 percentage points of the investment/GDP ratio. In Côte d' Ivoire and Ghana, movements of the real interest rate below equilibrium seems to have no effect on investment; however, an increase in the interest rate above equilibrium would bring about relatively large reductions in investment. In Zaire, movements in the real rate of interest both below and above equilibrium have no effect on investment. In Gambia, Mali, and Zambia the interest rate has no effect on investment below equilibrium, while its effect above equilibrium is positive but negligible. In Tanzania, the effect is positive both below and above equilibrium. In 7 countries — Gabon, Kenya, Madagascar, Mali, South Africa, Togo, and Tunisia — the net effect of the real interest rate on investment is negative both below and above equilibrium. This ranges from a reduction of 1.7 percentage points in the investment/GDP ratio in the case of Togo to a reduction of 0.12 percent in the case of Kenya.

Conclusion

The empirical results of this chapter do not lend support to the basic assertions of the financial liberalization theory. The real interest rate does not seem to be an important factor in the determination of neither financial saving nor total saving. Real income is found to be the most important determinant of total saving while the activities of the informal market is found to be the most important determinant of financial saving. These findings are not inconsistent with most studies on African as well as other developing countries. In a study of Ghana, Jebuni (1994) found that financial variables were not important in determining private saving, although he found weak support for the real interest rate as a determinant of total gross domestic saving. In this study, the relationship between the real interest rate and total saving for Ghana is positive but not significant. In a study of the BCEAO countries of Benin, Burkina Faso, Côte d' Ivoire and Niger, Senegal and Togo, Leite & Makonnen (1986) estimated three saving models using pooled time series and cross sectional data. They found out that in all cases, real interest rate is positively related to private saving, but this relation is only statistically significant in one of the three models. However, the relation lost its significance when they introduced change in income as a separate independent variable. In a study of 22 Asian and Latin American countries, Gupta (1987) found no support for a positive relation between the real interest rate and saving and he concluded that real income is the dominant determinant of saving. For Asian countries similar conclusions were arrived at by Giovannini (1983) and Cho & Khatkhate (1990).

One study which contradicts these findings is that of Seek & El Nil (1993). Using a sample of 9 countries which have experienced structural adjustment and another sample of 21 countries, they pooled time series and cross sectional data for each of the sample. They found that there is a strong positive relation between the real interest rate and financial saving for both samples, although they concluded that combating inflation is more beneficial than increasing the nominal interest rate.

As for the determinants of investment, it can be concluded that demand factors as approximated by the income accelerator effect appear to be among the most important determinant of investment in the sample of African countries under consideration. The supply of credit to the private sector proved to be equally important as a determinant of investment. But, the supply of credit, itself, appears to have a very weak relationship with financial saving and the latter is not responsive to movements in the real rate of interest. These results should not be surprising in view of the argument that the supply of credit is endogenous and not exogenous as assumed by the financial liberalization theory. That is, if loans given by banks end up deposited in banks, it is not the amount of deposits that determines loans, but the other way round; and that the latter is determined by decisions to invest. As Keynes (1939) once put it "prior saving has no more tendency to release funds available for investment than prior spending has." All that is needed to initiate additional real investment is finance provided by an increase in total bank loans with no need for increased savings as long as the banks can create new finance via acceptable accounting practices. Saving *funds* investment, but does not *finance* it; prior saving is not necessary for investment [African Development Bank (1994)]. As indicated by these results, the real interest rate has no effect on investment through the supply of credit, but it appears to affect investment directly, probably, through its bearing on the cost of investment. However, the direction of this effect is not uniform and seems to depend on each country's peculiar case.

In the light of the foregoing conclusions, it is not difficult to argue that manipulation of the real interest rate is not a reliable policy instrument for resource mobilization in the context of African countries. Both total and financial saving are not responsive to movements in the real interest rate and that its effect on investment is uncertain. Demand factors and the supply of credit which is largely determined by the decisions to invest are important determinants of investment. This being the case, an environment conducive for investment, the reputation of borrowers and incentives for investment, are more important than incentives for saving.

Notes

1. With only a ceiling on deposit rates and no ceiling on loan rates, the banks could charge r_2 to investors. The difference between r_2 and r_1 will represent a profit margin which the banks may use, say, for non-price competition, such as advertising, the expansion of branches, etc. But at r_2 there is no unsatisfied demand for investment.
2. In practice, these components are interdependent and interrelated. This classification is for theoretical convenience.
3. Experiments with the War dummy in all other saving and financial equation indicate that unlike investment, the behavior of saving and bank credit to the private sector were not significantly affected by armed conflicts.

Works Cited

Adams, D. W. "Taking A Fresh Look at Informal Finance.» In Callier,-Philippe, ed. *Financial systems and development in Africa:* Collected papers from an EDI Policy Seminar held in Nairobi, Kenya, from January 29 to February 1, 1990. Economic Development Institute Seminar Series, Washington, D.C.: World Bank, 1991, pages 29-42.

Adera, A. "Instituting Effective Linkages Between the Formal and the Informal Financial Sectors in Africa." *Savings and Development,* 11, 4, 1995.

African Development Bank. *African Development Report.* African Development Bank, Abidjan, Côte d' Ivoire, 1994.

Aredo, D. "The Informal and Semi-Formal Financial Sectors in Ethiopia: A Case Study of the Iqqip, Iddir, and Savings and Credit Cooperatives". *AERC Research Paper no. 21,* Nairobi, 1993.

Aryeetey, E. "Informal Financial Markets in Africa." Paper presented at the Senior Policy Seminar of the AERC, Nairobi, 1995.

Aryeetey, E. "The Structure of Informal Financial Markets in Africa." Paper presented at the Biennial Conference on the African Studies Associations of the United Kingdom, Lancaster University, 1994.

Aryeetey, E. "The Relationship Between Formal and Informal Segments of the Financial Sectors in Ghana." *AERC Research Paper no. 10,* Nairobi, 1992.

Aryeetey, E. and Gockel F. "Mobilizing Domestic Resources for Capital Formation in Ghana: The Role of the Informal Financial Sector." *AERC Research Paper no. 3,* Nairobi, 1991.

Bagashaw, G. "The Economic Role of Traditional Savings and Credit Institutions in Ethiopia. *Savings and Development* 4, 1978.

Benbahmed, R. "Le Systeme Financier Informel dans le Pays Africains de la Zone Franc." *Economic Research Papers No. 25,* African Development Bank, Abidjan, Côte d' Ivoire, 1996.

Bouman, F. "Indigenous Savings and Credit Societies in the Third World: A Message." *Savings and Development,* 4, 1977.

Bratton, M. "Financing Smallholder Production: A Comparison of Individual and Group

Credit Schemes in Zimbabwe." *Public Administration and Development*, 21, 1990.

Callier, P. (ed.) *Financial Systems and Development in Africa*. Washington D.C., EDI Seminar Series, the World Bank, 1991.

Castello, C., Stearns, K. and Christen, R. *Exposing Interest Rates: Their True Significance for Macroenterprises and Credit Programmes*, Discussion Papers Series No.6, Colombia, ACCION, 1991.

Chandavaskar, A. "The Non-Institutional Financial Sector in Developing Countries: Macro-Economic Implications for Saving Policies." *Savings and Development*, IX, 2, 1985.

Chipeta, C. *Indigenous Economics*. New York, Exposition Press, 1981.

Chipeta, C. and Mkandawire, M. The Informal Financial Sector and Macroeconomic Adjustment in Malawi, Nairobi, AERC Research Paper 4, 1991.

Cho, Yoon Che & Khatkhate, D. "Financial Liberalization: Issues and Evidence", *Economic and Political Weekly*, May, 1990.

Delancy. V. "Women at the Cameroon Development Corporation: How Their Money Works." *Rural Financial Markets in Developing Countries*. Baltimore, Johns Hopkins University Press, 1983.

Dornbusch, R. & Reynoso, A. Financial Factors in Economic Development, *American Economic Review Papers and Proceedings*, May 1989.

Giovannini, A. "The Interest Elasticity of Savings in Developing Countries: The Existing Evidence." *World Development*, 11, 7, 1983: 601-607.

Graham. D., Cuevas, E. and Negash, K. "Informal Finance in Rural Niger: Scope, Magnitude and Organization." Economics and Sociology Occasional Paper No. 1472, The Ohio State University, Department of Agricultural Economics and Rural Sociology, 1988.

Granger, C. "Investigating Causal Relations by Econometric Models and Cross Spectral Methods." *Econometrica* 37, 3, 1969: 424-438.

Gupta, Kanhaya L. "Aggregate Savings, Financial Intermediation, and Interest Rate." *Review of Economics and Statistics* 69, May, 1987: 303-311.

Holst, G. "The Role of Informal Financial Institutions in the Mobilization of Savings." *Savings and Development*. Proceedings of a Colloquium. Paris (28-30 May 1984), 1985.

Holst, G. "Savings and Credit for Development: The UN Secretariat's International Integrated Programmes." Paper presented at the Programme for International Economic Cooperation for Nigerian Government Officials, Nigeria. (Photocopy), 1987.

Holt, S. and Ribe, H. "Developing Financial Institutions for the Poor: A Focus on Gender Issues. Paper prepared for PHRWD, 1990.

Hyuha, M., Ndanshau, M, and Kipokola, J. *Scope, Structure and Policy Implications of Informal Financial Markets in Tanzania*, Nairobi, AERC Research Paper 18, 1993.

Jebuni, C. "Financial Structure, Reforms and Economic Development: Ghana." Background paper for the African Development Report 1994, African Development Bank, Abidjan, Côte d'Ivoire, 1994.

Kapur, B. K. "Alternative Stabilization Policies for Less Developed Countries." *Journal of Political Economy* 84, 4, 1976.

Keynes, J.M. "The Process of Capital Formation", *Economic Journal*, September, 1939.

Leite, S. P. & Makonnen, D., "Saving and Interest Rate in the BECAO Countries: An

Empirical Analysis." *Saving and Development* 3, 1986.
Levine, D. *Wax and Gold: Tradition and Innovation in Ethiopian Culture*, University of Chicago Press, 1972.
March, K. and Taqqu, R. *Women's Informal Associations and the Organizational capacity for Development*, Cornell University Monograph Series, No. 5, 1982.
Mauri, A. "The Role of Financial Intermediation in the Mobilization and Allocation of Household Saving in Developing Countries: Interlinks between Organized and Informal Circuits: The Case of Ethiopia." International Experts Meeting on Domestic Savings Mobilization, East-West Centre, Honolulu, 1987.
Mckinnon, R. *Money and Capital in Economic Development* (Washington: Brookings Institution), 1973.
Mckinnon, R. "Financial Repression and Liberalization Problem Within Less Developed Countries." in *The World Economic Order, Past and Prospects*, edited by Grassman S. and Lundberg E. (New York: St. Martins Press), 1981.
Mckinnon, R. (1982), "The Order of Economic Liberalization lessons from Chile and Argentina." *Economic Policy in a World of Change* 17, edited by Brunner K. & Meltzer A. Carnegie-Rochester Conference Series on Public Policy, 1981.
Mckinnon R. "Financial Liberalization in Retrospect: Interest Rate Policies in LDCS." in *The State of Development Economics: Progress and Prospective*, edited by Rains G. & Schultz, London, Basil Blackwell, 1988.
Mckinnon R. "Macroeconomic Instability and Moral Hazard in a Liberalizing Economy." in *Latin American Debt and Adjustment*, edited by Brock et al., New York: Praeger, 1989.
Miller, L. *Agricultural Credit and Finance in Africa*, New York, August, 1977.
Miracle, M, Miracle, D. and Cohen, L. "Informal Savings Mobilization in Africa", *Economic Development and Cultural Change* 28, no. 4 pp. 701-24, 1980.
Mohammed, Nadir A. "Economic Growth and Defence Spending in Sub-Saharan Africa: Beniot and Joerding Revisited." *Journal of African Economies* 2, 2, 1993: 145-156.
Nissanke. M. "Policies for Domestic Resource Mobilization and Financial Reforms." Abidjan, Background Paper, *African Development Report*, ADB, 1996.
Nwanna. G. "Financial Accessibility and Rural Sector Development." *Savings and Development*, XIX, 4, 1995.
Okuda, H. "Financial Factors in Economic Development: A Study of the Financial Liberalization Policy in the Philippines." *The Developing Economies*, XXVIII-3, September, 1990.
Osuntogun, A. and Adeyemo, R." Mobilization of Rural Savings and Credit Extension by Pre-Cooperative Organizations in South Western Nigeria." *Savings and Development* 4, 1981.
Otero, M. *A Handful of Rice: Savings Mobilization by Microenterprises Programmes and Perspectives for the Future*, Monograph Series, 3, Colombia, ACCION International, 1989.
Platteau, J. and Abraham, A. "An Inquiry into Quasi-Credit Contracts: The Role of Reciprocal Credit in Interlinked Deals in Small-Scale Fishing Communities." *The Journal of Development Studies*, 23, no 4 pp. 461-90, 1987.
Quandt, Richard. "The Estimation of the Parameters of a Linear Regression System Obey-

ing Two Separate Regimes" *Journal of the American Statistical Association* 53, 1958: 251-259.
Ramirez-Rojas, C.L. "Currency Substitution in Argentina, Mexico, and Uruguay." *IMF Staff Papers* 32(4), December, 1985, pp. 629-67
Rittenberg, L. "Investment Spending and Interest Rate Policy: The case of Financial Liberalization in Turkey." *Journal of Development Studies*, 27(2), January 1991, pp. 151-67.
Seck, D. & El Nil, Y. M. "Financial Liberalization in Africa." *World Development*, 11), November 1993, pp. 1867-81.
Serieux, J. "Deposit Rates, Money and Price Stabilization Under Structural Adjustment: Theory, and Evidence From Ghana and Kenya." Draft, University of Toronto, 1993.
Shaw, E. *Financial Deepening in Economic Development*, Oxford University Press: Oxford, 1973.
Shrieder, G. and Heidhues, F. "Rural Financial Markets and Food Security of the Poor: The Case of Cameroon." *African Review of Monetary Finance and Baking*, 1-2/1995 (Supplementary Issue of Savings and Development).
Stiglitz, J. E. & Weiss, A. "Credit Rationing in Markets with Imperfect Information." *American Economic Review* 71, 3, June, 1981.
Thirlwall, A. P. & Hussain, M. N. The Balance of Payment Constraint, Capital Inflows and Growth Rate Differences Between Developing Countries, *Oxford Economic Papers* 3, November, 1982.
Van Wijnbergen, S. "Interest Rate Management in LDCS." *Journal of Monetary Economics* 12, September, 1983.
Villanueva, D. & Mirakhor, A. «Strategies for Financial Reforms: Interest Rate Policies, Stabilization, and Bank Supervision in Developing Countries.» *IMF Staff Papers* 37, 3, September, 1990.
Von Pischke, J. *Finance at the Frontier: Debt Capacity and the Role of Credit in Developing the Private Economy*, Economic Development Institute of the World Bank, Washington, The World Bank, 1989.
Von Pischke, J. and Rouse, J. «Selected Successful Experiences in Agricultural Credit and Rural Finance in Africa.» *Savings and Development* 1, 1983.
Wai, (U.T.) "Interest Rate Outside the Organized Money Markets of Underdeveloped Countries." *IMF Staff Papers* 6, Washington, D.C, International Monetary Fund, 1957.
___ " A Revisit to Interest Rate Outside the Organized Money Markets of Underdeveloped Countries." *Banca Nazionale del Lauoro Quarterly Review* 122, 1977.
___ "The Role of Unorganized Financial Markets in Economic Development the Formulation of Monetary Policy." *Savings and Development*, IV, 4, 1980.
Warman, F & Thirlwall, A. P. "Interest Rate, Saving, Investment and Growth in Mexico 1960-90: Tests of Financial Liberalization hypothesis." *Journal of Development Studies* 30(3), April 1994, pp. 629-49
World Bank. *Egypt: Financial Policy for Adjustment and Growth*, The World Bank Washington D.C. 1992.
World Bank. *World Development Report: Financial Systems and Development*, Oxford University Press, 1989.

Chapter 12

Finance Liberalization in Africa:
Legal and Institutional Framework and Lessons from Other Less Developed Countries

Eno L. Inanga & David B. Ekpenyong

Introduction

Financial systems play an important role in economic development. The financial sector forms an important link between a country's macroeconomic policy and the rest of the economy. Its basic role in development is resource mobilization and allocation among productive sectors through financial intermediation, a large scale specialized function performed by specialized financial institutions and their agents. They attract funds from savers in the surplus sector and channel these to borrowers for purposes of profitable investment. A repressed financial system fragments domestic capital market with adverse effects on the quality and quantity of real capital accumulation. The adoption of financial liberalization under these circumstances has been suggested in order to enhance economic growth, a suggestion which many African countries have implemented in various degrees. This chapter analyses and assesses financial liberalization in Africa in the context of legal and institutional framework, and the lessons that can be drawn from the experience of other less developed countries.

Legal and Institutional Framework for Financial Liberalization

Legal Framework
Liberalized financial systems cannot be effective without sound legal and institutional framework. Evolutionary and proactive strategies are possible approaches to financial system development (Popiel 1994). In the evolutionary strategy, financial markets are allowed to develop gradually with the economy. As major distortions or bottlenecks emerge, government intervenes through improvements or changes in laws or regulations. In this strategy, financial deepening and financial system development are basically market driven within an adaptable legal, regulatory and prudential framework.

The pro-active strategy provides legal, regulatory and prudential framework which accelerates financial market development through mechanisms, institutions and financial instruments set up for this purpose. Popiel (1994) sees this as the appropriate strategy for African and other developing countries for three main reasons:

- Inadequate neutral incentive environment and market forces that are insufficiently strong for financial markets to develop by themselves.
- Lack of institution-building capacity to determine the pace and strength of financial markets development.
- Need for flexibility to allow for the use of the most efficient institutional set-up, required training infrastructure and choice of technology that is most suited to the local conditions and level of development.

The proactive approach suggested by Popiel seems to agree with the views of the World Bank (1989), which sees the legal and institutional framework of most developing countries as inadequate to support modern financial processes. Examples of such inadequacy include outdated legal systems leading to poor enforcement of laws concerning collateral and foreclosure. This often creates difficulty in debt collection, monitoring and control, resulting in unwillingness of lenders to enter into certain types of financial contract.

A cohesive and comprehensive legal framework is required under the proactive approach in order to use the contracts that clearly define the rights and obligations of contracting parties. Such a framework should encourage discipline and timely enforcement of contracts, fostering responsibility and prudent behavior on both sides of the financial transaction. Prudent and efficient financial intermediation cannot operate without reliable information on borrowers, and some legislation on accounting and auditing standards, which also ensures honesty on the part of financial institutions, especially if they take deposits from the public (World Bank 1990). Similarly, for a country's equity markets to develop and operate efficiently, legislation should fully incorporate rules of trading, intermediation, information disclosure, take-overs and mergers (Popiel 1988).

Company law is an example of the kind of legislation needed. It not only governs the operations of business enterprises but also protects the interests of company stakeholders. Thus, public disclosure of information on the company's activities should be made mandatory on company management in the appropriate section of the Companies Act. Such information, especially that relating to finance and accounting, should also be statutorily required to be subsequently verified and attested to by auditors.

Because of the role of financial institutions and markets in the development of a sound financial system, additional legislation is normally needed for their operations to complement company law. These are prudential regulations, especially for banks and similar financial institutions that hold an important part of the money supply, create money and intermediate between savings and investment. Prudential regulations cover such issues as criteria for entry, capital adequacy standard, asset diversification, limits on loans to individuals, permissible range of activities, asset classification and provisioning, portfolio concentration and enforcement powers, special accounting, auditing and disclosure standards adapted to the needs of the banks to ensure timely availability of accurate financial information and transparency. The objective is to enhance the safety and soundness of the banking system.

Another legislation relates to financial markets which require not only favorable policies but also legal and institutional infrastructure to support their operations, prevent abuses and protect investors. Investors' confidence is critical to the development of the markets. Brokers, underwriters, and other intermediaries who operate in these markets therefore have to follow laid down professional codes of conduct embodied in the legislation applicable to such institutions as finance and insurance companies, mutual funds and pension funds.

Institutional Framework

The legal, regulatory and prudential framework discussed in the preceding section is essential for fostering financial market functions and promoting and anchoring its institutional framework. The ultimate function of financial markets, as earlier indicated, is to mobilize and allocate resources through financial intermediation in order to accelerate the process of economic growth. The function is performed through two distinct but interrelated components.

One of the components is the money market. It is essentially a market which trades in short-term debt instruments to meet short-term needs of bulk users of funds such as governments, banks and similar institutions. Government treasury bills and similar securities, as well as company commercial bills, are examples of instruments traded in the money market which, in essence, provides a mechanism for effective monetary management and the formulation and implementation of monetary policy. It also includes the inter bank market. Merchant banks, commercial banks, the central bank and other dealers are among the institutions which operate in the money market.

Through the financial instruments, public and private sector operators are able to raise and invest short term funds which, if need be, can be quickly liquidated to satisfy short-term needs. The money market, as an institution, thus provides a framework for liquidity management by monetary authorities. The instruments traded are characterized by short-term maturity, high liquidity, and reliability. Furthermore, as they are traded in large denominations and increasing range, they contribute to increased competition and reduce the financial costs of intermediation (Popiel 1990). Money markets also provide market-based reference point for setting other interest rates and monetary policy implementation through open market operations.

The second component of the financial market is the capital market. The capital market mobilizes long-term debt and equity finance and channels them into productive investments in long-term assets. Capital markets also help to strengthen corporate financial structure and improve the general solvency of the financial system. In addition, they mediate between the conflicting maturity preferences of lenders and borrowers and financial resource reallocation among companies and industries.

The money market and the capital market are inter-related. First, the development of the money market usually precedes capital market development. Second, the same institutions may operate actively in both markets. Hence, the money market serves as a source of liquidity for the long-term investment needs of operators in the capital market.

The capital market may be segmented into monetary intermediaries, non-monetary intermediaries, and securities markets. The monetary intermediaries consist of institutions of a banking nature—central and commercial banks. The non-monetary intermediaries are characterized by a wide range of specialized institutions which trade in debt-related instruments such as term loans, mortgage and leases. Insurance companies, savings and loans institutions and investment trusts are examples of non-monetary intermediary components of the capital market. They play important role in the process of financial deepening. Both monetary and non-monetary segments are non-securities components of the capital market.

The securities segment of the capital market complement traditional lending institutions by providing risk capital (equity) and loan capital (debt). By means of these instruments, the market is able to mobilize long-term savings and provide capital to investors to finance long-term investments thereby broadening ownership of productive assets. Dealers in the securities segment of the capital market include banking institutions, stockbrokers, investment and merchant bankers and venture capitalists which intermediate between the market and the public.

Horch (1989) has argued that in order to develop and achieve the objective of supporting economic growth, the capital market requires environment in which government policies are generally favorable to economic growth. In such environment, resources are allocated in accordance with market forces rather than government directives. Firms and other market operators respond freely to undistorted market signals. In addition, prices of goods and services traded in the market are allowed to reflect their opportunity costs instead of government controls and discriminatory policies. At maturity, a capital market with these characteristics is easily integrated into the international financial market.

Besides conducive macroeconomic environment, the existence of market intermediaries, well developed accounting, auditing, and financial disclosure standards, together with enforced legal and regulatory framework for investor protection are absolutely critical to the effective development of the capital market.

What is the nature of African financial systems that necessitated a call for financial liberalization? To what extent has liberalization achieved its intended objectives? An attempt is made in the next section to find answers to these questions.

African Financial Systems and Financial Liberalization

Financial systems in Africa are noted for their marked variations. Some systems, such as those in Mozambique, Angola, Tanzania, and Guinea are dominantly government-owned, consisting mostly of the central bank and very few commercial banks. Other systems have mixed ownership comprising central banks, public, domestic, private and foreign private financial institutions. These can be further sub-divided into those with rich varieties of institutions such as are found in Nigeria, Zimbabwe, and Kenya, and others with limited varieties of institutions as are found in Malawi, Uganda, Ghana, and other Sub-Saharan African countries (Soyibo 1994).

African governments have, over the years, adopted the policy of financial sector intervention in the hope of promoting economic development (World Bank 1989; Elbadawi et al. 1992). Interest rate controls, directed credit to priority sectors, and securing bank loans at below market interest rates to finance their activities, later turned out to undermine the financial system instead of promoting economic growth. For example, low lending rates encouraged less productive investments and discouraged savers from holding domestic financial assets. Directed credits to priority sectors often resulted in deliberate defaults on the belief that no court action could be taken against the defaulters. In some cases, subsidized credit hardly ever reached their intended beneficiaries.

There was also tendency to concentrate formal financial institutions in urban areas thereby making it difficult to provide credit to people in the rural areas. In some countries, private sector borrowing was largely crowded-out by public sector borrowing. Small firms often had much difficulty in obtaining funds from formal financial institutions to finance businesses. Finally, the tendency of governments of the region to finance public sector deficits through money creation resulted not only in inflation but also in negative real interest rates on deposits. These had adverse consequences for the financial sector. First, savers found it unrewarding to invest in financial assets. Second, it generated capital flight among those unable or unwilling to invest in real assets thereby limiting financial resources that would have been made available for financial intermediation. Coupled with this was the declining inflow of resources to African countries since the 1980s. For example, from a positive figure of US $35.2 billion in 1981, foreign capital inflows to Sub-Saharan African countries declined to a negative figure of US $50.1 billion in 1988 (Roe 1990). It became clear that African countries could no longer depend on foreign resource inflows but have to rely on their domestic resources to finance economic growth and development. The importance of evolving sound macroeconomic policies for building efficient financial systems then became clearly evident if the above objectives were to be achieved. Hence the call for financial liberalization. Generally, the blanket financial reforms were packaged by the World Bank and handed down to African countries.

Villanueva (1988) and El-Nil (1990) have categorized African countries with financial reforms into three in relation to intended objectives. The first category comprises those countries whose objective was to improve the monetary control system. Botswana and Mauritius are examples of countries in this category. The measures adopted to achieve intended objectives included ensuring the adequacy of monetary instruments and replacing direct with indirect controls.

The second category comprises countries that introduce financial sector reforms in order to improve the mobilization and allocation of domestic savings, such as Zaire and Kenya. The Zairian reform involved adopting measures directed at developing and sustaining money and government securities markets. Kenya's concern, on the other hand, was about the adequacy of banking regulation and legislation a concern that led to the introduction of new financial institutions, instruments and measures (Seck and El-Nil 1993).

The final category is made up of countries whose financial sector reform objective was to improve the banking system and the level and structure of interest rates. Mauritania and Senegal pursued this objective through reduction of interest rate subsidies and of bad debts. Burundi, The Gambia, and Sierra Leone, did so by liberalizing interest rates and introducing prime rate or a base lending system. However, there are other countries whose objectives of financial liberalization span across the three categories. Nigeria is an example which, together with Kenya and Zimbabwe, has one of the most developed and diversified financial systems in Sub-Saharan Africa. The economic and financial conditions of the economies of individual African countries, no doubt, have played significant roles in shaping the outcome of the reforms.

In Nigeria and Kenya, for example, the reforms were carried out in conditions of severe macroeconomic imbalances and instability in the financial system (Soyibo 1994). In Ghana and Tanzania, reforms were introduced in periods of both macroeconomic instability and financial distress. In Malawi and Botswana, the banking systems were quite healthy at the commencement of the reforms (Caskey 1992). Zambia and Zaire also had healthy banking systems but macroeconomic instability. In Mauritania, the banks were already bankrupt when reforms were initiated (Paulson 1993). Benin Republic, Niger, Senegal, Côte d'Ivoire, Burkina Faso, Togo, and Mali, former French West African colonies now under the Union Economique et Monétaire Ouest-africanine-UEMOA (West African Economic Monetary Union - WAEMU), were all facing balance of payments problems, internal political crises, and internally inconsistent economic policies before financial sector reform were introduced (Plane 1993). The study by Seck and El-Nil (1993) provided empirical evidence which tended to suggest that African countries stand to gain from financial liberalization due to positive impact of real deposit rates on financial savings and the level of investment.

Sub-Saharan African countries that carried out financial liberalization tended to focus largely on banking. Four types of reforms are typical. These are interest rate liberalization, bank restructuring, privatization of banks, and bank liquidation. Table 12.1 summarizes some of the African countries that carried out financial sector reforms in these areas during the adjustment period. The reforms achieved limited success for reasons of policy credibility, time consistency of the adjustment policies, and lack of independence of the central bank.

The financial reforms also tended to concentrate in the formal sector to the exclusion of long-established informal sector that is quite active in the provision of financial services to households, small farmers and small businesses. Both sectors have been shown to play complementary roles in Malawi (Chipeta and Mkandawire 1992), Tanzania (Hyuha, Ndanshau, and Kipokola 1993), Ethiopia (Aredo 1993), and Ghana (Aryeetey 1991, 1994).

Results within the banking system itself have been mixed. For example, while the reform policies in some countries have led to increase in the number and varieties of banking and other financial institutions, there have been no improvements in the maturity structure of deposit liabilities of banks and non-banks leading to asset-liability mismatch . Yet the demand for long term investment funds never diminished, nor have the reforms brought about any improvement in the access to credit by small and medium-scale enterprises and rural dwellers.

Financial liberalization in many African countries was carried out without explicit study of the initial conditions of the economy. This was so because the blanket financial reform package handed down to them gave them hardly any opportunity for such analysis. The creditors were in a hurry to push the reforms through. It was a matter of "take it or leave it."   As the World Bank (1994) said, it was time Africa began to adjust.

Another factor that accounted for the non-realization of the initial optimism of financial sector reforms in Africa relates to failure to take sequencing into consideration. The experience of Southern Cone countries of Latin America, namely, Argentina, Uruguay, and Chile, with financial liberalization, particularly in relation to high real interest rates and inflation, emphasizes the need to ensure macroeconomic stability before embarking on financial liberalization. Financial liberalization in an inflationary economy will send inappropriate signals, resulting in adverse consequences for the economy. This is illustrated by Adam (1994) in his study of financial liberalization in Zambia under dynamics of inflation between 1992 and 1993. He showed how financial liberalization measures during the period not only had direct fiscal costs in their own right, but by removing the controls which supported the real monetary base, also served to reduce the demand for real domestic currency balances and the seignorage revenue capacity of the economy. Consequently, the fiscal costs of ensuring non-inflationary domestic deficit financing turned out to significantly exceed what would have been necessary without financial liberalization.

Political, administrative, and policy credibility issues might also have led to the dismal performance of financial liberalization in Africa. When economic agents foresee that government persistently reverses policies or adopts "stop-go" approaches in implementing liberalization policies, they may not be willing to commit resources in line with the requirements of liberalization measures, and hence policy credibility becomes low. Policy compatibility issues are also important in the sequencing of economic reforms. For example, in the pursuit of policies of monetary restraint directed at ensuring macroeconomic stability, government can reduce the liquidity of the banking system by many measures (e.g., issuing stabilization securities by the central bank). If, at the same time, the central bank unreservedly underwrites government budget deficits, then policy incompatibility would ensue and this will weaken the sequencing of economic reforms and their implementation.

Case Studies

In this section, we examine the experiences of selected African countries within the framework of institutional and legal reforms that have taken place in the financial sector before and after the reforms with emphasis on the banking sub-sector. Montiel (1995) has identified three conditions relating to institutional framework and three associated with macroeconomic environment which government must put in place for efficient function of a liberalized financial system. The three conditions relating to institutional framework are:

- The existence of appropriate legal framework, well-established property rights and efficient judicial system.
- A financial safety net (deposit guarantees) to avoid liquidity crises.

- Adequate regulatory and monitoring framework to prevent collusion and avoid excessive risk-taking due to moral hazard problems.

Five countries have been selected for study for their distinguishing characteristics. These are: Nigeria and Ghana in West Africa for their common colonial heritage and the nature and size of their financial sector; Kenya in Eastern Africa for its similarities with Nigeria and Ghana and its developed and sophisticated financial system; Tanzania, also a former British colony in East Africa for its unique socialist ideology; and Senegal, a former French colony and a member of the CFA zone.

Nigeria

During the period 1970-1985, Nigeria's financial sector was characterized by financial repression. macroeconomic imbalances and instability (Soyibo 1994). Prior to 1970, banking regulations were largely prudential, aimed mainly at ensuring sound banking practices and protection. From the early 1970s, the aims remained broadly the same as in the previous years, but the control instruments became rather restrictive. The system was so regulated that by the mid-1970s, the Central Bank could stipulate what loans and advances each commercial bank should make to each of the sixteen different priority sectors of the economy, as well as maximum interest ceilings for agricultural and other priority areas (Killick 1993). Government controlled 60 percent of commercial bank share capital while the Central Bank controlled 33 percent of the financial assets.

Prior to the implementation of financial liberalization, government took no serious measures to establish appropriate legal framework under which the financial system would operate. No appropriate safety nets were established to safeguard against liquidity crises and no adequate regulatory and monitoring framework to prevent collusion and excessive risk-taking was put in place.

Financial liberalization brought in a lot of changes in the financial sector. The institutional and legal framework was overhauled. The Central Bank of Nigeria (CBN) Decree No. 24 and the Banks and Other Financial Institutions Decree (BOFID) No. 25 both of 1991 granted autonomy to the apex bank and gave it wider powers to oversee the entire financial system. All financial sector operations formerly within the informal sector were brought under the control and supervision of the CBN.

The Exchange Control Act of 1962 was repeated to remove the restrictions on the amount of foreign exchange which can be imported into or exported from Nigeria. A Foreign Exchange Market was introduced in 1986 to replace the system of administrative controls which consisted of foreign exchange rationing, import licensing and arbitrary determination of the exchange rate. The financial markets were equally liberalized. A Second-Tier Securities Market (SSM) was introduced in 1985 and a complete revitalization of the capital market announced in the 1996 Federal Government budget (*The Guardian*, March 29, 1996:10). Prudential guidelines aimed at bridging the capital adequacy ratio to conform to international standards and to provide for non-performing assets by banks were introduced toward the end of 1990.

Certain important legislations were also introduced on bank supervision aimed at ensuring banks' compliance with public policy along the lines recommended by Polizatto (1992). In that process, Nigeria adopted a complementary model of off-site surveillance and on-site inspection which emphasizes submission of regular reports, assessment of financial position, performance and peer views and detailed periodic examination of banks' records and policy statements.

Nigeria adopted the explicit insurance scheme by establishing the Nigeria Deposit Insurance Corporation (NDIC) by Decree No. 22 of 1988 to insure the deposit liabilities of licensed banks in the country, provide financial and technical assistance to the banks and contribute to the quest for a safe and sound banking environment. The scheme currently indemnifies a depositor up to N50,000 in the event of a bank failure.

Ghana

Prior to financial liberalization, the financial sector was characterized by financial repression which, coupled with balance of payments deficits, overvalued exchange rate, inflationary pressure, and macroeconomic disequilibrium, almost led to the collapse of the country's financial system. Negative real interest rates and budget deficit rose in 1981 to 6.5 percentage as a percentage of GDP (Soyibo, 1994).

The existing banking law did not provide sound prudential and regulatory guidelines for the banking system. Laws governing the operations of the non-bank financial institutions and some informal operators were not formalized. As was the case in Nigeria, interest rates were administratively determined and credit ceilings were also imposed. Maximum and minimum deposit and lending rates were imposed. No appropriate safety nets were established to safeguard against liquidity crises and no adequate regulatory and monitoring framework was put in place to prevent collusion and excessive risk-taking.

With the introduction of financial sector reform in 1983, several drastic liberalization measures were taken. The Banking Law was amended to provide sound prudential and regulatory guidelines for the banking system similar to those adopted in Nigeria and other African countries. Areas covered by the amendment included specification of minimum capital base, limits to risk exposure and improved accounting, auditing and financial reporting standards.

A securities industry law was enacted to support the emerging Ghana capital market. Non-bank financial institutions law was promulgated to formalize the activities of some informed operators.

From 1985, the Bank of Ghana started a phased withdrawal of administered interest rates and credit ceilings. Sectoral lending requirements were abolished. By March 1989, commercial banks were allowed to determine their interest rates. Maximum and minimum deposit on lending rates were replaced with discount rates as the main instrument to influence the overall level of interest rates. To improve liquidity management, treasury bills and bonds were introduced by the Bank of Ghana in 1990 for trading through open market operations which were fully operational in 1992.

Kenya

The financial system of Kenya, though regarded as the most developing in Sub-Saharan Africa, was seriously repressed, highly segmented, with poor licensing and regulatory framework. The financial system was heavily controlled by foreigners. Anomalies in the banking law were rampant thus making it possible for commercial banks to set up near-bank financial intermediaries and circumvent restrictions on banking activities and interest rate ceilings. Near-bank financial intermediaries were not subject to statutory reserve requirements. There were no prudential regulation and regulations on capital and reserve requirements were poorly enforced.

The credit market was greatly segmented given the number and range of financial institutions. The influential large farmers were greatly subsidized by the budget. Credit to the private sector was limited by the government' s pre-emption of credit for financing its budget deficits and those of its parastatals. As was the case in Nigeria and other African countries, there were rigidities in the allocation of credit and ceilings were imposed on interest rates.

Financial liberalization which began in 1989 brought about many changes particularly in the banking sub-sector. The government amended the banking laws thus narrowing the regulatory gap between classes of institutions, imposing more stringent licensing requirements on banks and near-bank financial intermediaries and increasing minimum capital requirements. The Central

Bank's technical and managerial capacity to inspect, monitor and supervise the financial system was strengthened (Husain and Faruqee 1992).

For the first time, guidelines were set for both loan provisioning and minimum financial disclosure requirements. Penalties for violating any of these regulations and guidelines were increased. By July 23, 1993, interest rate controls were completely removed and interest rates on treasury bills fully liberalized. A Deposit Insurance Fund was also established.

Tanzania

The case of Tanzania is different in some ways from the case of Nigeria, Ghana and Kenya which we have already discussed. The Tanzanian economy was based on socialist ideology that preferred central planning to a system that operates on the basis of market forces. As a result, the financial system of the country was, until 1991, state-owned and state-controlled. The banks and financial institutions had been nationalized in 1967 thus turning them into complete monopolies. Competition in the financial sector was precluded by sectoral specialization within the commercial banks (Husain and Faruqee 1994). There was prohibition of entry of new financial institutions into the banking sub-sector. Interest rates were strictly controlled and credit allocation was very ineffective. The capital base of most financial institution was described as extremely low. No safety nets were put in place and the Bank of Tanzania lacked the appropriate institutional and legal framework to sanitize the sector. The entire financial system was noted for poor performance and insolvency of most of the financial institutions whose loans were largely non-performing.

Worried by the poor performance of the financial sector, the government set up a Presidential Commission of Inquiry into the Monetary and Banking System in 1988. The findings of the Commission published in July 1990 recommended far reaching reforms which included interest rate liberalization, strengthening of prudential banking regulations and central bank supervision, restructuring both existing banks and their operating environment, relaxing of entry conditions to facilitate the establishment of domestic and foreign-owned private banks to encourage competition.

By the time the liberalization process was set in motion in 1991, Tanzania had already established a comprehensive reform program. The Banking and Financial Institutions Act was passed in 1991 aimed at creating an autonomous and independent banking system supervised by the Central Bank. The establishment of the Loans and Advances Realization Trust Act in 1991 dealt with the problems of non-performing loans, new requirements and prudential guidelines on capital adequacy to guard against future mismanagement and financial distress. Banks must now diversify in order to spread risks and guard against losses and take necessary steps to strengthen internal controls and improve profitability. There has been a significant adjustment in the exchange rate resulting in the unification of the foreign exchange market.

Senegal

Senegal and its financial system are distinctly different from the countries discussed in this section in terms of evolution which reflects the country's political history as a former French African Colony. The financial system is heavily influenced by financial policies of the two monetary unions that form the CFA Zone. These are the UEMOA (Union Economique Monetaire Ouest-africaine) in West Africa, and the BEAC (Banque des Etats de l' Afrique Centrale) in Central Africa. The two unions evolved progressively from the Bank of Senegal (effectively a currency board) founded in 1853 (Azam 1995).

Although the CFA is guaranteed convertibility thereby having a semblance of "hard currency", member countries of the zone lack the independence of Nigeria, Ghana, Kenya in economic and monetary policy formulation. However, similar to the other countries already examined, Senegal prior to financial liberalization, faced the problems of financial repression as reflected in administratively-determined interest rates, direct government control of bank lending and the use of state-owned banks and central bank to secure disproportionate share of total credit for the public sector.

The operational legal framework prior to financial liberalization was so complex that it was difficult for banks to collect debts which frequently turned into bad debts. Government often interferred in the National Credit Committee's discretionary power to fix bank-by-bank credit ceiling. Changes in the bank supervisory framework were worked out in consultation with the IMF and bilateral donors (France and the United States).

Financial liberalization was concentrated in the banking sub-sector as was the case in Nigeria, Ghana, Kenya, Tanzania, as already discussed. Substantial reforms were carried out in credit policies. Bank-by-bank credit ceilings were restructured. Market-determined interest rates, patterned after Paris rates were introduced along with banking margins to improve the profitability of banks. In 1989, the Council of Ministers of the UEMOA decided that measures should be introduced to make the bank-by-bank ceiling allocations more flexible.

The system of targeting credits to priority areas was eliminated. The operations of the money market was restructured. Entry into the banking sub-sector was restricted. Government's capital share in all banks was reduced to a maximum of 25 percent (except where no suitable buyer of the government's shareholding could be found).

Shareholders undertook to rehabilitate banks under their control through equity or quasi-equity contributions to offset losses and recapitalize the banks to a level compatible with sound banking standards.

Banking inspection and supervision of the banking system was greatly reinforced with the creation of the Commission Bancaire in 1990.

Lessons from Other Less Developed Countries

Less developed countries are far from homogenous. Consequently, one country's experience leading to financial liberalization should be expected to be largely country-specific and therefore different from another.

Thus, in the 1970s and 1980s, several developing countries liberalized their financial markets with mixed results. In some Latin American countries, for example, the reforms initially produced chaos which prevented realization of expected gains. On the other hand, Asian countries such as Korea and Malaysia which also liberalized fared better (Gertler and Rose 1994). The failure of financial liberalization to meet expectations in many Latin American countries was attributed to:

- The rise in interest rates for a substantial class of borrowers leading to a drop in borrower net worth;
- Bad timing of the financial liberalization which coincided with aggregate economic declines, notably falling commodity export prices;
- Inadequate coordination of liberalization with the design of financial safety net.

Avoidance of these mistakes might have resulted in a well-functioning financial system which could, under normal circumstances, contribute significantly to the stimulation of economic growth as suggested by theory and evidence.

Lessons that African countries can learn from other developing countries may thus be classified into general and specific. General lessons are those applicable to all developing countries embarking on financial sector reforms regardless of the uniqueness of their economy (World Bank 1988). Specific lessons relate to individual developing countries whose economy and initial pre-reform conditions and liberalization programmes are comparable to those of individual African countries drawing lessons from their experiences.

General Lessons

(i) African countries must avoid pitfalls of the other developing countries such as those of Latin America identified earlier. They have to recognize the importance of borrower net worth to the sound functioning of financial markets. Consequently, financial liberalization should be pursued simultaneously with macroeconomic policies directly aimed at promoting the growth and stability of the real sector.
(ii) Financial liberalization should not be implemented in an unstable macroeconomic environment. It can exacerbate macroeconomic instability, especially if pursued unilaterally.
(iii) Financial liberalization under price distortions resulting from protection or controls can worsen, rather than improve, resource allocation by inducing the financial system to respond to wrong signals.
(iv) A liberalized financial system must be adequately supported by a legislative and institutional framework with adequate provisions for prudential regulation and banking supervision to enhance financial solvency.
(v) Considerations of equity, sequencing and financial feasibility, should be incorporated in financial liberalization since resulting income and wealth transfers will affect different groups in different ways in the society. The ideal sequence could be:
- Establishment of macroeconomic stability in order to get fiscal deficit under control.
- Development of financial markets and institutions to foster competition.
- Deregulation of interest rates and elimination of controls.

The sequencing has to be gradual, maintaining appropriate balance between excessive speed and undue delay.

Lessons from Country-Specific Experiences

African countries can learn a lot of lessons from financial liberalization experiences of other developing countries often cited as having fared better. The countries selected for this purpose are Chile, Indonesia, Korea, Malaysia, and Turkey where the initial conditions of their financial systems were quite similar to those of many African countries (Atiyas, Caprio, and Hanson 1994). Our analysis focuses on their reform strategies, errors inadvertently made and lessons which African countries can learn from their experiences.

Chile

According to Vittas (1992), the central guidelines for financial reform in Chile were essentially the same as the reforms in other markets which sought reliance mainly on the private sector, fostering of competitive markets, elimination of discrimination among economic agents, and elimination of controls and distortions. Selective credit controls, credit ceilings and interest rate controls were eliminated. By 1974 reserve requirements were reduced and by 1978 all banks were privatized. Capital controls on nonbanks were lifted within 2 years and on banks within 5 years of liberalization. All financial institutions were given "universal" banking powers, allowing free entry into most financial operations to all intermediaries and imposing similar regulation procedures on all financial institutions.

Although the free banking approach predominated in the crucial years of the reform, contradictory approaches were adopted in the process of financial liberalization (de la Cuadra and Valdes 1989). The most important errors were, the absence of prudential regulation, inconsistencies in financial policies, and implicit subsidy to foreign exchange risk generated by the repeated pronouncements by the authorities about the nominal fixation of the exchange rate in spite of the evident symptoms of cumulative financial and macroeconomic disequilibrium (Vittas 1995).

Indonesia

The financial liberalization program which was introduced in March 1983 had both institutional and technology components. The institutional components involved identification of post-deregulation business opportunities; bank reorganization to reinforce risk management and speed management decisions; labour force development aimed at improving staff evaluation, deployment and motivation, and new bank procedures for assets and liabilities management. The technology component involved complete automation of bank functions and procedures (World Bank 1990).

The 1983 financial liberalization removed all credit controls and freed interest rates on all but directed credits. This was accompanied with stabilization programme. The second stage of reforms started in 1988 with the liberalization of the banking system through lowering of entry barriers in order to foster competition between banks and other financial intermediaries. The process of prudential regulation which began in 1989 was further strengthened in 1991 (Hanna 1994).

According to Vittas (1992), financial development in Indonesia has been driven mainly by changes in government policies rather than changes in the rates of economic growth. These policies were effectively directed to ensure appropriate sequencing in the reform process, absence of domestic debt and the absence of foreign exchange controls and openness of foreign exchange markets, which in the end contributed to the success of the financial liberalization programme.

Korea
Financial liberalization in Korea, as in the other countries so far examined, led to the freeing of deposit and lending rates, removal of some of the entry barriers, reduction of control over financial institutions, and the provision of greater autonomy in the day-to-day bank operations and asset management. Although the liberalization of the financial sector remained limited, some specific measures were taken in line with the limited strategy. For example, to prevent any indiscriminate transfer of funds to liberalized financial assets, restrictions were imposed in the form of a minimum transaction unit and ceilings on the handling of some specific businesses. Bank lending rates and most rates in the primary market are still very rigid and unresponsive to market conditions indicating that the Korean financial system is still far from being fully integrated and operating purely on competitive basis. Concerned with a drastic rise in interest rates, the government has also tended to encourage collusion on interest rates by financial institutions (Caprio et al. 1994).

There are certain observed aspects of the Korean financial system which make it different in many ways from the ones earlier discussed. Korean firms rely heavily on external financing. The existence of a dual financial system where the informal financial market has played an important role (despite the government's efforts to suppress it) has been known to be beneficial to the economy. Government control of the banking institutions has been a principal instrument for guiding and regulating private firms. There is a strong growth of nonbank financial institutions which are not heavily restricted in the allocation of their funds and are freer than the banks in circumventing interest rate ceilings on both the sources and uses of funds. There is lack of competition in the banking system which has hampered growth, led to accumulation of non-performing assets and delays in financial liberalization.

Despite these unique features which some may consider problematic, studies have shown that Korean financial market did not experience any major instabilities such as, bankruptcy of financial intermediaries, undesirable shifts in bank portfolios, a large jump in real interest rates or destabilizing capital flows.

The strategy of the approach could have been responsible for this stable state of affairs. The approach to the liberalization process has been very cautious, slow and an-on-going process aimed at avoiding complications which could have resulted from opening up the capital market or concentrating ownership in the major banks. However (Caprio et al. 1994), pointed out that the absence of side effects which often accompany financial liberalization in other countries does not necessarily mean that Korea's financial liberalization is a success. One cannot but wonder what success then means.

Malaysia
The focus of the financial liberalization in Malaysia was on the Central Bank. Prior to this, the various pieces of legislation did not provide the Central Bank with effective overall power to regulate the activities of the various non-bank financial intermediaries that operated at the fringe of the regulated banking sector. The Banking and Financial Institutions Act of 1989 granted the Central Bank sufficient powers to evolve an effective integrated supervisory and regulatory framework for the financial system.

The financial liberalization programme aimed primarily at promoting a competitive free market-oriented financial system, encouraging the growth of the financial system and increasing the response of bank interest rates to market forces.

The Malaysian experience in financial liberalization which has been described as successful followed certain steps which Yusof et al. (1994) recommend strongly. The first step, according to them, should be to raise interest rates administratively, then sort out the health of the banks and simultaneously build the regulatory and monetary strength of the central bank.

Lessons which have been drawn from the Malaysian experience are that:
- financial liberalization should not be seen as a uniformly linear process which, once set in motion, cannot be redirected. The course and pace of the process should be influenced by macroeconomic considerations;
- while a gradualist approach should be adopted in the sequencing and speed of financial reforms, the timing should take into consideration the state of the economy;
- government must be committed to introduce policy changes; and
- good management must be in place and caution exercised in the liberalization process. As Yan (1993) emphasized, good management cannot be legislated, and legislation, no matter how compre-

hensively conceived, cannot bring about good behaviour in bank operators. Ethical standards need moral responsibility to nurture it to maturity. The financial system is as sound as those who manage it, and who should be honest, professional and dedicated.

Turkey

The major objective of the financial sector liberalization in Turkey was to promote financial market development through deregulation and inducing competition by opening up the banking sector to foreigners. This resulted in expansion as new banks entered the industry while existing ones adopted expansionary policies to secure larger market shares. Activities of banks were broadened and consolidated while existing universal banking system was strengthened. A legal and institutional framework was developed for the functioning of financial markets and for effective supervision of financial institutions. The financial reforms were, however, short-lived as the macroeconomic changes induced by the reforms hit corporate profits seriously and left business struggling to adjust. Financial problems in the corporate sector caused distress in the banking sector that called for government intervention, by reimposing among others, ceilings on bank deposits (World Bank 1990a). Government itself had a large share of the blame through its acts of financial indiscipline, for example, through the frequent use of the financial system to finance budget deficits and in the process crowded out financial flows to the private sector.

The following lessons can be drawn from the financial liberalization experience in Turkey:

- The need to maintain a financially strong corporate sector in order to absorb any potential interest rate shocks that follow financial liberalization.
- To understand fully the extent to which macroeconomic uncertainty can hinder the potential positive impact of financial liberalization by placing the burden of public sector borrowing on financial markets.
- The implications of the financial reform' s overemphasis on the development of the banking system at the expense of non-bank financial institutions and instruments. This has been known to slow down financial innovation and the development of the capital market.
- The importance of the judicial system in generating and expanding the benefits of financial liberalization. When the judicial system is efficient, it commands and strengthens the confidence of financial intermediaries in the system. They become innovative and are able and willing to explore new business areas and financial contracts.

Lessons of Experience

In discussing the lessons which African countries can learn from the experiences of other developing countries that undertook financial liberalization, a few comments are necessary.

African countries embarked on financial liberalization out of pressure by their foreign creditors without proper evaluation of the possible implications of the experiments they were asked to embarked upon. The justification for the experiment was the McKinnon and Shaw mythical hypothesis, later debunked by McKinnon himself, that high interest rate promotes savings and economic growth. Ostry and Reinhart (1994) have argued that financial liberalizations which generate higher interest rates will result in savings by households only if households decide to defer consumption. This would happen only if consumption and saving are significantly sensitive to higher interest rates.

However, in many developing African countries, consumption choices are heavily influenced by subsistence considerations. Consequently, raising domestic interest rates is unlikely to stimulate savings. Ostry and Reinhart (1994) have, however, demonstrated in their study that interest rate sensitivity of saving to real rates of return rises with a country's income level as shown in Table 12.4. Thus the same policies in different countries could produce different results depending on the level of development.

The study provides evidence of some reform programmes in Africa causing real interest rates to move from sharply negative to mildly positive levels; but in spite of this, saving failed to respond to changes in real interest rates in many low income countries. This is a lesson which African countries might not have learnt when they embarked on financial liberalization.

The financial liberalization program (and indeed the entire structural reform programme) did not take into account the political, cultural and social peculiarities of the different countries. Full liberalization of the financial system which is what the reform asked for, does not bring about positive results in all nations. In Korea, for instance, government's control of the banking institutions has been a principal instrument for guiding and regulating private firms. As Stiglitz (1996) has emphasized, the earlier U.S. and more recent East Asian experiences suggest, however, that government can play a beneficial economic role in such key policy areas as, education, financial regulation, environmental protection and social welfare. In calling for a more balanced perspective in assessing the role of government in the modern economy, Stiglitz cautions against "rehashing old arguments about the superiority of private over public enterprise." Policy makers should instead, identify areas where the activities of governments and private markets can most effectively complement each other.

The timing of the reform process in relationship to the peculiarities of each reforming country was not given any consideration whatsoever. Consequently, the "Afro-pessimists" as Camdessus (1996) calls them, were quick to rate Africa as non-reform-performers. With regard to the question of sequencing and pacing of financial reform, Cole (1996) advises that policy makers should consider a more gradual market-building approach, one that operates on a time-table of one or two decades rather than a few years, instead of pushing the system ahead too rapidly and in the process slow down the entire liberalization process. This has been the problem which many of the African countries have had to face.

Another key issue is the fact that for any reform to be successful, it should come in stages. Many of those who have assessed the reform process in Africa have tended to miss this important point. Gladly, as Boorman (1996) has pointed out, many African countries have been reasonably successful in implementing "early-stage" structural reforms, for example, reforming exchange rate systems, opening trade and payments systems, removing price controls, and liberalizing production and marketing systems, particularly in the agricultural sector. In the more difficult aspects of reforms which include revenue mobilization, public enterprises, privatization and the financial sector, Boorman (1996) evaluates the performance as uneven. Even at that, Camdessus (1996) confirms that increasing number of African countries have undertaken courageous adjustment and structural reform programs.

After evaluating the entire financial liberalization process as carried out by the other non-African developing countries, we are persuaded to say that in terms of implementing the various financial reforms, African countries have done well. In all the African cases that we have reviewed, changes (in some cases drastic) in the major components of financial liberalization have been effected. Appropriate legal and institutional framework has been established. Financial safety nets to avoid liquidity crises have been put in place and regulatory and monitoring framework has also been established to check collusion and minimize risks.

In accepting to experiment with the reform, African countries seem to have made two major mistakes. The first is to have accepted the reform package without first analyzing how their political and cultural environment would react to the changes which the reforms would bring about. The second is, their failure to have made sufficient allowance in terms of the time frame within which the results of the reform will begin to manifest. The non-African countries took these issues into account seriously.

It is also important to point out that the rather negative reform outcome reported on African countries can be attributed to two major factors: the unduly critical comments from those Camdessus (1996) calls "Afro-Pessimists" and from the fact that most of those comments were based on the first stage of the reform.

Conclusion

In pushing ahead with financial liberalization, African countries and researchers would need to note the following:
- No single formula will work for every country, but what is important is the formulation of country-specific and region-specific policies and cross-regional experiences that can provide the needed experiences. There is need to ask and to provide a satisfactory answer as to whether the relative cost and benefits for Africa of opening up its markets only slowly were the same as those faced in Asia when those countries made their decisions regarding liberalization (Hamanaka 1996; Downes 1996).
- There is need to carry out in-depth investigation as to the connection between culture and economic growth with a view to answering the question, "how receptive is culture to change?" Many of the so-called successful reformers were able to implement the financial reforms within the context of their cultural setting. African countries have not done this.
- As the Korean example reveals, the role of government and its ability to administer any reform is critical. As Stiglitz (1996) advised since government can play beneficial economic role in some key economic sectors, it is therefore important for policy makers to identify areas of cooperation between government and the private sector.
- Experience has shown that reform programmes have been most successful where governments have made them their own and are allowed to experiment with alternative strategies (Zenawi 1996). What has prompted many countries to liberalize is a strong desire to get their economies moving again for the good of their own people coupled with a strong sense of pragmatism regarding what it takes to achieve that. Unfortunately, African countries have been denied this choice. When Nigeria tried to assert this freedom, she was accused of abandoning ownership of the reform which she did not

actively initiate in the first place.
- Financial reform (or any other reform) will thrive in an environment of permanent discipline that ensures, competition, efficiency and transparency.
- Parallel macroeconomic and industrial policy framework which is conducive to efficient investment must be put in place.
- Well-qualified bank staff and a well-established legal and supervisory framework to minimize bank inefficiency must also be put in place and there must be abundance of entrepreneurial skill (Cho 1986).
- There is the need to maintain a financially strong corporate sector in order to absorb any potential interest rate shocks that follow financial liberalization.
- It is important to have an efficient judicial system in generating and expanding the benefits of financial liberalization.

The structural adjustment programme was not African in origin and design. The peculiarity of each country's situation was not taken into consideration in the package proposed for uniform implementation. To guard against recurrence of problems encountered by different African countries in the course of implementing financial liberalization package, we recommend the establishment of an African Bureau of Financial and Economic Research. The main focus of the Bureau would be on research directed at defining and evaluating, with a good degree of concern and commitment, the financial and economic problems and other related issues that are unique to African countries.

Table 12.1: Financial sector reforms undertaken in Africa during the adjustment period

Country	Liberalization & rationalization of interest rates	Restructuring of banks	Privatization of banks	Liquidation of banks
Benin	X			X
Burundi	X			
Congo	X			
Côte d'Ivoire	X	X	X	X
The Gambia	X			
Ghana	X	X		
Kenya	X	X		
Madagascar	X	X	X	
Malawi	X			
Mauritania	X	X	X	
Mozambique	X			
Rwanda	X	X		X
Tanzania	X	X		
Nigeria	X		X	X
Cameroon		X	X	
Guinea		X		X
Mali		X		X
Senegal		X	X	X
Uganda		X		
Guinea-Bissau			X	
Niger				X

Source: Adapted from World Bank 1994 and author's investigations.
Note: The table is not intended to be a comprehensive list of all the financial sector reforms undertaken.
X = applicable

Table 12.2: Financial liberalization in sub-Saharan Africa

The Gambia (September 1985)	Ceilings on interest rates were removed in September 1985, an auction system for issuing Treasury bills was introduced in July 1985, and quantitative controls on credit were removed in September 1990.
Nigeria (July 1987)	Directed credit restrictions were relaxed over the period 1983-1987 by increasing the sectoral aggregation of directed credit allocations. On July 31, 1987, the central bank removed interest rate controls, and raised both the treasury bill and rediscount rates by 4% points, to 15 and 14%. In November 1989, an auction system was instituted for treasury bills and certificates but, the central bank retained a reservation price.
Ghana (September 1987)	Ceilings on interest rates were removed, while the removal of quantitative credit controls was scheduled for 1992.
Malawi (April 1988)	Ceilings on interest rates were removed, and quantitative credit ceilings were eliminated in January 1991.
Uganda (July 1988)	In July 1, 1988, and increase of 10% points was announced on most interest rates.
Benin and Côte d'Ivoire (October 1990)	The BCEAO abolished its preferential discount rate, but bank interest rates remained subject to regulation. Côte d'Ivoire is also a member of the BCEAO, so it was affected by these liberalizing measures.
Cameroon (October 1990)	The BEAC eliminated its preferential lending rates, simplified its interest rate structure, and increased its power to determine interest rate policy with the intention to move toward greater flexibility in rates.
Tanzania (July 1991)	The system of fixed interest rates and fixed differentials was replaced by a single maximum lending rate of 31% on July 25.
Kenya (July 1991)	Interest rate ceilings were removed.

Source: Montiel 1995.

Table 12.3: Financial sector reforms in Nigeria, 1986-1995

1986 Establishment of first and second-tier foreign exchange market.
1987 (i) Removal of interest rate controls
 (ii) Removal of restrictions on licensing of banks
 (iii) Merging of first and second-tier foreign exchange markets.
1988 (i) Establishment of foreign exchange bureaux
 (ii) Relaxation of restrictions on bank portfolio
 (iii) Nigerian Deposit Insurance Corporation established.
1989 (i) Banks permitted to pay interest on demand deposits
 (ii) Auction markets for government securities established
 (iii) Upwards review of capital adequacy standards
 (iv) Raising of cash requirements to mop up excess liquidity.
1990 (i) Increase in required paid-up capital for banks
 (ii) Uniform accounting standards for banks introduced
 (iii) Introduction of stabilization securities to mop up excess liquidity in the financial system.
1991 (i) Imposition of embargo on bank licensing
 (ii) Central bank empowered to regulate and supervise all financial institutions
 (iii) Re-administration of interest rates.
1992 (i) Removal, once again, of interest rate controls
 (ii) Beginning of privatization of government-owned banks
 (iii) Capital market de-regulation
 (iv) Re-organisation of foreign exchange market
 (v) Dismantling of credit controls.
1993 (i) Introduction of indirect monetary instruments
 (ii) Taking over of five commercial banks for purposes of restructuring.
1994 Re-imposition of interest and exchange rate controls.
1995 (i) Modification of interest rate controls
 (ii) Abolition of Exchange Control Act 1962
 (iii) Liberalization of foreign exchange and financial markets
 (iv) Removal of restrictions on repatriation of dividends, profits, loan servicing payments and remittance of proceeds on sale of investments
 (v) Repeal of the Nigerian Enterprises Promotion Decree

Sources: Ikhide and Alawode 1994 and authors' investigations.

Table 12.4: Income & personal saving rates

Country	GDP per Adult in 1995 dollars (US) Average	Personal Saving As Percentage of GDP 1988-1993 Average
Loan-Income	1,380	11.2
Low-Middle Income	2,806	16.7
Upper Middle-Income	4,896	19.5
High Income	16,161	20.0

Source: *IMF World Economic Outlook*, Oct. 1994.
Note:
Low Income Countries in the Sample: Egypt, Ghana, India, Pakistan, Sri Lanka
Low Middle Income Countries in the Sample: Colombia, Costa Rica, Côte d' Ivoire, Morocco, the Philippines.
Upper Middle Income Countries in the Sample: Brazil, Korea, Mexico.

Works Cited

Adam, C. "Financial liberalization and inflation dynamics: some evidence from Zambia". WPS/94.14, Oxford: Centre for the Study of African Economies 1994.

African Development Bank, African Development Report, Abidjan, 1994.

Allechi, M. and M.A. Niamkey, "Evaluating net gains from the CFA Franc zone membership: a different perspective." *World Development* 22, 1994.

Ani, A.A. "Press briefing on the 1995 budget", Federal Republic of Nigeria, Abuja, (mimeo), January 16, 1995a.

Ani, A.A. "Press briefing on the first quarter performance of the 1995 budget", Federal Republic of Nigeria, Abuja, (mimeo.), April 21, 1995b,.

Aredo, D. The Informal and Semi-formal Financial Sectors in Ethiopia: A Study of the IQQUB, IDDIR, and Savings and Credit Co-operatives, AERC Research Paper Twenty One, Nairobi, 1993.

Aryeetey, E. "The relationship between the formal and informal sectors of the financial markets in Ghana". paper presented at the AERC workshop, Nairobi,

May, 1991.

Aryeetey, E. "Financial integration and development in Sub-Saharan Africa: a study of informal finance in Ghana". Overseas Development Institute Working Paper 78, London, 1994.

Atiyas, I. and H. Ersel (1994), "The impact of financial reform: the Turkish experience." in G. Caprio, I. Atiyas and J.A. Hanson (ed.) (1994), *Financial Reform: Theory and Experience*, Cambridge University Press, Cambridge, 1994.

Atse, D. and G. Achiepo "Capital formation in a period of macroeconomic adjustment in the franc zone: the case of Cote d' Ivoire." Final Report, AERC Workshop, Nairobi, (mimeo.), May, 1995.

Azam, J.P. "Macroeconomic reforms in the CFA zone.", paper presented at the Eleventh World Congress of the International Economic Association, Tunis, December, 1995.

Boorman, Jack "Deepening Structural Reforms in Africa.", IMF Survey, IMF, Washington, D.C., 1996.

Callier, P. (ed.), Financial Systems and Development in Africa. *World Bank*, Washington, D.C., June 17, 1990.

Camdessus Michel "African prospects tied to courageous adjustment Efforts." *IMF Survey*, IMF, Washington, D.C. July 29, 1996.

Caprio, G., I. Atiyas and J.A. Hanson (ed.) Financial Reform: Theory and Experience, Cambridge University Press, Cambridge, 1994.

Caskey, J.P. "Macroeconomic implications of financial sector reform programs in Sub-Saharan Africa." Oberlin College, (mimeo.), 1992.

Chant, J. and M. Pangestu "An assessment of financial reforms in Indonesia, 1983-90." in Caprio, G. *et al.* (op. cit.), 1994.

Chipeta, C. and M.L.C. Mkandawire, *Links Between Informal and Formal/Semi-Formal Financial Sectors in Malawi*, AERC Research Paper Fourteen, Nairobi, 1992.

Cho, Y.J. "Inefficiencies from financial liberalization in the absence of well-functioning equity markets." *Journal of Money, Banking and Credit*, May, 1986.

Cho, Y.J. and D. Khatkhate "Lessons of financial liberalization in Asia: a comparative study." *World Bank Discussion*, Paper No. 50, Washington, D.C., 1989.

Cole, David "Deepening structural reform in Africa." *IMF Survey*, IMF, Washington, D.C., June 17, 1996.

Cole, D.C. and B.F. Slade, "Indonesian financial development: a different sequencing." in D. Vittas (ed.) (1992), Financial Regulation, *EDI Development Studies*, Washington, D.C., 1990.

Collier, P. and C. Mayer, "The assessment: financial liberalization, financial systems, and economic growth." Oxford *Review of Economic Policy* 5, 4, Winter, 1989.

Cortes-Douglas, H. "Financial reform in Chile: lessons in regulation and deregulation." in D. Vittas (op. cit.), 1990.

de la Cuadra, S. and S. Valdes, "Myths and facts about instability in financial liberalization in Chile: 1974-1983." Catholic University of Chile, 1989.

Dordunno, C.K. "Exchange rate reforms in Sub-Saharan Africa: some lessons for policy management." paper presented at the Eleventh World Congress of the International Economic Association, Tunis, December, 1995.

Downes, Patrick, "Deepening structural reform in Africa." *IMF Survey*, IMF, Washington, D.C., June 17, 1996.

Edwards, S. "Structural adjustment and stabilization issues on sequencing and speed." EDI Working Paper, *World Bank*, Washington, D.C., 1992.

Ekpenyong, D.B. "Financing distress banks: nature and implications." The Nigerian Banker, The Chartered Institute of Bankers of Nigeria, Lagos, April - June, 1994.

El-Nil, Y.H. "The pre-requisites for a successful financial reform." in P. Callier (ed.), (op. cit.), 1990.

Emenuga, C.E. "Alternative financing and Nigeria's external debt." in *African Debt Burden and Economic Development*, Proceedings of the 1994 Annual Conference of the Nigerian Economic Society, 1994.

Faruqi, S. (ed.), Financial Sector Reforms in Asian and Latin American Countries. EDI Seminar Series, *World Bank*, Washington, D.C., 1993.

Fry, M. "Financial development: theories and recent experience." *Oxford Review of Economic Policy* 5, 4, Winter, 1989.

Gertler, M., and A. Rose, "Finance, public policy and growth." in G. Caprio et al. (op. cit.), 1994.

Glen, J. "An introduction to the microstructure of emerging Markets." International Finance Corporation Discussion Paper 24, Washington, D.C., 1994.

The Guardian, March 29, 1996.

Hamanaka, Hideichiro, "Deepening structural reform in Africa." *IMF Survey*, IMF, Washington, D.C., June 17, 1996.

Hanna, D.P. "Indonesian experience with financial sector reform." *World Bank Discussion Paper 237*, Washington, D.C., 1994.

Harberger, A. "Observations on the Chilean economy, 1973-83." *Economic Development and Cultural Change* 33, 3 (April), 1985.

Horch, H. "Policies for developing financial markets." *EDI Working Papers*, Washington, D.C., 1989.

Husain, I. and R. Faruqee (ed.), "Adjustment in Africa: Lessons from Country Case Studies." *World Regional and Sectoral Studies*, Washington, D.C., 1994.

Hyuha, M., M.O. Ndanshau and J.P. Kipokola, Scope, *Structure and Policy Implications of Informal Financial Markets in Tanzania*, AERC Research Paper Eighteen, 1993.

Ikhide, S.I. "Positive interest rates: financial deepening and the mobilization of savings in Africa." *Development Policy Review* 11, 4, 1993.

Ikhide, S.I. and A.A. Alawode, "Financial sector reforms, macroeconomic stability and the order of economic liberalization: the evidence from Nigeria." paper presented at the AERC Workshop, Nairobi, May, 1994.

Inanga, E.L. "Financial sector reforms in Sub-Saharan Africa." paper presented at the Eleventh World Congress of the International Economic Association, Tunis, December, 1995.

Inanga, E.L. "Financial sector liberalization: curse or blessing?" *IMF Newsletter* 9, August, 1995.

Kariuki, P. "Interest rate liberalization and the allocative efficiency of credit: some evidence from small and medium scale industry in Kenya" in C. Harvey (1994) (ed.), *Constraints on the Success of Structural Adjustment Programme in Africa*, Macmillan Press Ltd., London, 1966.

Killick, Tony, The Adaptive Economy, *EDI Development Studies*, World Bank, Washington, D.C., 1993.

Koo, B.H. "Industrial policy and financial reforms in Korea." in S. Faruqi (ed.) (op. cit.), 1993.

McKinnon, R.I. *Money and Capital in Economic Development*, Brookings Institution, Washington, D.C., 1973.

McKinnon, R.I. "Financial liberalization and economic development: a reassessment of interest rate policies in Asia and Latin America." *Oxford Review of Economic Policy* 5, 4 Winter, 1989.

Mans, D. "Tanzania: resolute action." in Husain, I. and R. Faruqee (op. cit.), 1994.

de Melo, J. and J. Tybout "The effects of financial liberalization on savings and investment in Uruguay." *World Bank Report* No. DRD 129, Washington, D.C., 1985.

Ministry of Finance, Central Bank of Nigeria Decree No. 24, Lagos, 1991.

Ministry of Finance, Bank and Other Financial Institutions Decree No. 25, Lagos, 1991.

Montiel, P.J. Financial Policies and Economic Growth: Theory, Evidence and Country-Specific Experience from Sub-Saharan Africa, *AERC Special Paper Eighteen*, Nairobi, 1995.

Mwega, F.M., S.M. Ngola and N. Mwangi, Real Interest Rates and the Mobilization of Private Savings in Africa, *AERC Research Paper 2*, Nairobi 1990.

Nam, S.W. "Korea's financial reform since the early 1980s." in G. Caprio *et al.* (op. cit.), 1994.

Nelson, E.R. "Monetary management in Sub-Saharan Africa: Senegal." Bethesda, Maryland, DAI (mimeo.), 1991.

Ngugi, R.W. and J.W. Kabubo (1995), "Financial sector reforms and interest rate liberalization: the Kenyan experience." Final research report presented at AERC Workshop, Nairobi, May, 1991.

Nigerian Deposit Insurance Corporation, Annual Report and Statement of Accounts, Lagos, 1989.

Nissanke, M. "Liberalization experience and structural impediments to savings mobilization and financial intermediation." International Development Centre, University of Oxford (mimeo.), 1991.

Ojo, M.O. "The economics of controls and deregulation: the Nigerian case study." Research Occasional Paper 10, Central Bank of Nigeria, Lagos, 1994.

Oresotu, F.O. 'Interest rates behaviour under a programme of financial reform: the Nigerian case', Central Bank of Nigeria, *Economic and Financial Review* 30, 2, 1992.

Oshikoya, T.W. "Interest liberalization, savings, investment and growth: the case of Kenya." *Savings and Development* 26, 1992.

Ostry, J. and C.M. Reinhart, *IMF World Economic Outlook*, Washington, D.C., October, 1994.

Paulson, J.A. "Some unresolved issues in Africa financial reforms." in Lawrence H. White (ed.) African Finance, Institute of Contemporary Studies Press, San Francisco, 1993.

Plane, P. "Financial crises and the process of adjustment in the Franc zone: the experience of the West African Monetary Union." in Lawrence H. White (ed.), (op. cit.), 1993.

Polizatto, V.P. "Prudential regulation and banking supervision." in D. Vittas (op. cit.), 1992.

Popiel, Paul A. "Development of money and capital markets." *EDI Working Papers*, Washington, D.C., 1988.

Popiel, Paul A. "Recent developments and innovations in international financial markets." *EDI Working Paper*, Washington, D.C., 1989.

Popiel, Paul A. "Developing financial markets in Sub-Saharan Africa." in P. Callier (op. cit.), 1990.

Popiel, Paul A. "Financial systems in Sub-Saharan Africa." *World Bank Discussion Papers 260*, Washington, D.C., 1994.

Rodrik, D. (1990), "How should structural adjustment programs be designed?" *World Development* 18, 7, 1990.

Seck, D. and Y.H. El-Nil "Financial liberalization in Africa." *World Development* 21, 11, 1993.

Shaw, E.S. *Financial Deepening in Economic Development*, New York, Oxford University Press, 1973.

Sowa, N.K. "Financial liberalization and macroeconomic stability in Ghana." A paper presented at the AERC Workshop, Nairobi, December 4 - 9, 1994.

Sowa, N.K. and T.O. Antwi-Asare "Financial reforms and monetary policy in Ghana." AERC Research Proposal, June, 1995.

Soyıbo, A. "Financial liberalization and bank restructuring in Sub-Saharan Africa." Paper presented at the plenary session of the AERC Workshop, Nairobi, December 4 - 9, 1994.

Stiglitz, J.E. "Financial markets and development." ' , Oxford *Review of Economic Policy* 5, 4, 1989.

Stiglitz, J.E. "The role of the State in the financial markets." Proceedings of the 1993 Annual World Bank Conference on Development Economics, *World Bank*, Washington, D.C., 1994.

Stiglitz, J.E. "Bank conference highlights development's dynamic nature." *IMF Survey*, IMF, Washington, D.C. June 17, 1996.

Swamy, G. "Kenya: patchy, intermittent commitment." in i. Husain and R. Faruqee, op. cit., 1994.

Turtelboom, B. Interest Rate Liberalization: Some Lessons from Africa, *IMF WP/91/121*, Washington, D.C., 1991.

Umo, B.O. "Empirical tests for some savings hypothesis for African countries." *Financial Journal* 2(2), 1981.

Villanueva, Delano, "Issues in financial sector reform." Finance and Development 25, March, 1988.

Villanueva, Delano and Mirakhor, "Strategies for financial reforms." *IMF Staff Papers*, September, 1990.

Wong, C. and A. Mirakhor, "Strategies for financial reforms: interest rate policies, stabilization and bank supervision in developing countries." *IMF Staff Papers* 37, 3, 1990.

Wong, C. "Market-based systems of monetary control in developing countries: operating procedures and related issues." *IMF Working Paper*, WP/91/40, 1991.

World Bank, *World Development Report*, Washington, D.C., 1989.

World Bank, *World Development Report*, Washington, D.C., 1990a.

World Bank, *Financial Systems and Development*, Policy Research and External Affairs, Washington, D.C., 1990b.

World Bank, Global Economic Prospects and Developing Countries, Washington, D.C., 1993.

World Bank, Adjustment in Africa, Policy Research Report, Washington, D.C., 1994.

Yan, L.S. "The institutional perspective of financial market reform: the Malaysian experience." in S. Faruqi (op. cit.), 1993.

Yusof, Z.A.; A.A. Hussain; I. Alowi; L.C. Sing and S. Singh (1994), "Financial reform in Malaysia." in G. Caprio et al. (op. cit.), 1989.

Zenawi, Melles, "Deepening structural reform in Africa." *IMF Survey*, IMF, Washington, D.C., June 17, 1996.

Chapter 13

Financial Liberalization, Emerging Stock Markets and Economic Developments in Africa

Temitope W. Oshikoya and Osita Ogbu

Introduction

Since the mid-1980s, financial liberalization in several African countries has been implemented largely through on-going structural adjustment programs. As a prerequisite for the financial liberalization programs, stabilization policies have been designed to ensure macroeconomic stability, low inflation and reduced budget deficits. The focus has been on liberalizing interest rates, deregulation of the financial sector, strengthening the banking system, introduction of new financial instruments, and development of securities markets, in particular the stock market. Stock market is viewed as a medium to encourage savings, help channel savings into productive investment, improve the efficiency and productivity of investments. The emphasis on the growth of stock markets for domestic resource mobilization have also been strengthened by the need to attract foreign capital in non-debt creating forms.

The trends towards promoting stock markets in African countries in recent years may also be linked to developments in the world financial markets which have been characterized by increased securitization, liberalization and integration. These features have been encouraged by several factors including the rapid technological development, the deregulation of financial markets in industrial countries, the globalization of these markets, the introduction of new financial instruments, and the increasing role of the institutional pension fund and mutual fund investors. It is increasingly viewed that emerging markets offer significant diversification potential for global investment portfolios. Harvey (1995) has shown that while individually, emerging stock markets have been quite volatile, as a group their risk-adjusted return has been higher than that of developed markets, and they have shown low and negative correlations with the more developed financial markets and among themselves.

However, despite the recent enthusiasm and drive towards the establishment of stock markets in African countries, it is important to examine their role in the context of the continent's structural adjustment programs and economic development. This chapter reviews: (1) the broader issue of financial liberalization and their impacts on the savings-investment process in Africa and; (2) the channels through which the emerging stock markets in Africa may foster or obstructs the economic development process.

The study is divided into five sections. (i) Introduction; (ii) an outline of the theoretical perspectives on the role of finance in economic growth; (iii) an examination of the experiences of several African countries with financial liberalization within the context of the on-going structural adjustment programs; (iv) synthesis of relevant policy and analytical issues in emerging stock markets in Africa; and (v) summary.

Theoretical Perspectives on Finance and Economic Development[1]

The theoretical perspectives on the role of finance in economic growth can be broadly divided into three areas: (1) the Keynesian view of investment determinants; (2) the Modigliani-Miller irrelevance propositions; and (3) the financial repression hypothesis.

The Keynesian View of Finance
The Keynesian perspective on the role of finance in economic growth portends that investment decisions are primarily determined by the level of confidence, expected demand and the "animal spirits" of the private investors. Underlying the Keynesian view is the fundamental message that it is investment that determines saving, and not vice versa. Although, in principle, the rate of interest matters, in practice it is regarded as being relatively insignificant compared to demand factors. High real interest rates may stifle investment and growth. The disequilibrium approach within the context of the Keynesian tradition implies that investment depends on prospects for profits and the binding constraints on firm's sales (Sneessens 1987; Malinvaund 1980). Moreover, it is not necessarily the case that a perfect capital market will lead to an optimal allocation of investment. Indeed Keynes in the General Theory likens the stock market to a gambling casino dominated by speculators and investors with short-term outlook.

The Modigliani-Miller Propositions

The Modigliani-Miller (1958, 1961) irrelevance propositions have dominated modern neo-classical theory of finance and investment until recently. The M-M irrelevance propositions state that in fully developed capital markets, with perfect competition, no transaction costs and no taxation, and with full and symmetric information among all investors, the stock-market valuation of the firm is independent of its financing or dividend pay-out decisions. The market value of a firm will be determined by earnings prospects and risk of its underlying real assets and would be invariant to its capital structure or the division between internal or external sources for financing its investment plans. At the macroeconomic level, the propositions imply a dichotomy between finance and the real economy. Corporate growth and investment decisions are determined by the real economic factors such as productivity, demand for output, technical progress and relative factor prices of capital and labour.

More recent theoretical developments have invalidated the Modigliani-Miller propositions and produced an optimal capital structure which maximizes its stock-market valuation by relaxing some of the underlying assumptions. The introduction of corporate tax incentives that allow interest to be deducted as costs would favour debt finance. However, a high level of debt may increase bankruptcy and financial distress during economic recession. Some literature on the capital structure of firms and investment decisions (see Mayer 1989; Fazzari, Hubbard, and Petersen 1988) has emphasized the lack of perfect substitution between internal finance from retained earnings and external finance from bonds, equity or bank credit. In this framework, investment is affected by financial factors and the availability of the appropriate financial instruments could constrain a firm's investment plans and growth. From a macroeconomic perspective, the sensitivity of investment to liquidity and corporate retained earnings which in turn are dependent on the business cycle, macroeconomic instability may affect firms that rely heavily on internal finance. Other models (see Stiglitz and Weiss 1981) involving asymmetric information, adverse selection, moral hazard, agency and transaction costs and incentive effects point to the importance of financial constraints on investment decisions and the links between corporate capital structures and the financial decisions of the real economy. Asymmetric information may make interest rate policy inefficient in discriminating between good and bad borrowers. Under such conditions, firms may face binding financial constraints in the capital markets as credit rationing and quantitative constraints are imposed by creditors.

The Financial Repression Hypothesis

The theories of finance outlined above have been mainly concerned with advanced economies where the capital markets are well developed. In the case of underdeveloped capital markets, the McKinnon-Shaw framework has explicitly sought to relate capital-market developments to long-term economic growth in the developing countries (McKinnon 1973; Shaw 1973). The McKinnon-Shaw proposition is that a repressed financial sector interferes with development in several ways: savings vehicles are not well developed; financial intermediaries that collect savings do not allocate them efficiently among competing uses; and firms are discouraged from investing because of financial policies repression that reduce the returns to investment or make them uncertain; as a result growth is retarded. Thus financial liberalization theory argues for improved growth through financial deepening and financial sector reform. The key relations of financial liberalization paradigm are: positive real deposit rates raise the saving rate; a positive correlation between the degree of financial deepening and the growth rate; increased real rates raise the level of investment; and increased real deposit rates promote economic growth (Dornbusch and Reynoso 1988; Oshikoya 1992).

The McKinnon-Shaw proposition is based on the underlying classical assumption that savings determine investment and that a full utilization of resources is always guaranteed. This is contrary to the fundamental Keynesian framework which suggests that it is investment that determines savings and that the supply of loans is endogenous through the money multiplier process by which the banking system could create additional credit without increasing the deposit base. The structuralist models developed by Taylor (1983) and Van Wijnbergen (1983) also points out that the financial liberalization framework ignores the several channels through which high real interest rates could adversely affect costs, investment and the level of demand in the economy. An increase in nominal interest rates will raise financing costs to the firms which may lead to rising prices, a fall in real wages and a reduction in aggregate demand and capacity utilization. High interest rates will reduce capital accumulation and growth as investment demand is related to capacity utilization. Furthermore, empirical evidence from many countries which have liberalized their credit markets and increased real interest rates does not indicate a systematic rise in aggregate savings. Apart from the well documented evidence on the experiences of the Southern Cone countries in the 1970s, the Cho and Khatkhate (1989) study of the financial liberalization experience of five Asian countries Indonesia, Malaysia, Philippines, Republic of Korea, and Sri Lanka suggests that financial reforms do not seem to have made any significant difference to the saving and investment activities in the liberalized economies. The experience of African countries with financial liberalization is discussed in the third section of this chapter, see also ADR (1994).

Although, the McKinnon-Shaw framework focuses on the financial repression in developing countries, it does not examine the role of the stockmarket. The framework also ignores the institutional aspects of the financial system. As Chandavarkar (1991) has pointed out eliminating financial repression through a positive real interest rates may not be sufficient in mobilizing and allocating domestic resources efficiently. The institutional aspects of the liberalization process, restructuring of financial institutions, effective regulation and supervision of the banking system are important factors to be considered. The appropriate sequencing of the overall stabilization and structural reforms including financial liberalization need to be considered. For example, the likely impact of simultaneous liberalization of interest rates and the exchange rates and a deregulation of capital controls on external capital mobility, exports and imports, the domestic goods and services market should be taken into consideration.

The existence of informal and curb market, moral hazard, adverse selection effects, and asymmetric information affect the behaviour of financial markets in developing countries. As outlined above, Stiglitz and Weiss (1981) has shown that there will still be credit rationing, even with financial liberalization, because of imperfect information on the part of borrowers and lenders. To reduce the inefficiencies associated with weak credit markets, Cho (1986) argued that credit markets need to be complemented by a well-functioning equity market in developing countries. The underlying assumption in this analysis is that both potential shareholders and risk-neutral lenders have the same level of information on firms with which to judge the expected productivities of their investment projects. It has been suggested that Cho's analysis ignores information costs and the informational requirements it imposes on the individual equity investor as well as agency problem in management-controlled corporations.

The theory relating to the determinants of total saving is different from that relating to financial saving. The decision to save in financial form is a portfolio choice among assets; the decision to save itself is an intertemporal choice between consumption today and consumption tomorrow. Total saving consists of public and private saving, of which financial saving is a part. Total domestic saving is expected to be positively related to income (the Keynesian absolute income hypothesis), but the sign and significance of the coefficient on the real interest variable will depend on the relative strength of the income and substitution effects of the relative price change between consumption today and consumption tomorrow. The substitution effect of a higher interest rate is to encourage agents to sacrifice current consumption for future consumption, but the income effect is to discourage current saving by giving agents more income in the present, and the two effects may cancel each other out. The same argument would apply to private saving alone. This being so, it is perhaps surprising, as Dornbusch and Reynoso (1989) have remarked, "to find so strong a belief in the ability of higher interest rates to mobilize saving."

Financial saving measures the amount of total saving that is channelled via financial assets. Since saving is a flow, it is measured by the change in the stock of monetary assets issued by the banks and financial intermediaries—hence the importance of financial intermediation in encouraging financial saving. Full financial deepening describes a situation in which all saving is done in financial form; and full financial augmentation would occur when all new saving flows into financial assets. The ratio of financial saving to total saving will depend on the proportion of total saving that goes into non-financial assets, including real assets, hoarding, and capital flight; the portion that goes to the informal financial sector; and the portion that goes into the underground economy and the parallel market for foreign exchange. Even if total saving is stagnant, financial saving can increase if the leakages into non-financial saving decrease. The financial liberalization school argues that a rise in the real deposit rate of interest will cause a switching to financial assets which will result in a greater supply of credit to finance real investment, and to a superior allocation of investment resources.

Financial Liberalization and the Saving–Investment Process in Africa

In this section we review the empirical literature on the saving-investment process. This review is based in part on the work of the authors, the work of other colleagues in the field, and the summary found in the ADR (1994).

When countries embark on financial reforms, they usually have two objectives in mind: (a) to raise the level of saving and investment in an economy; and (b) to improve the allocation of investment resources consistent with certain economic and social objectives. It is envisaged that increasing the quantity, and improving the quality, of investment will raise the growth rate and improve living standards in the countries concerned. Our objective here would be to examine the results of liberalization with reference to the impact of the real interest rate on financial saving, private saving, total saving and investment, and also on the growth rate.

Under the auspices of the Structural Adjustment Program, a number of African countries have introduced financial liberalization programs. A summary of measures undertaken in nine countries, surveyed by Seck and El Nil (1993) is shown in Table 13.1. Most of the countries placed emphasis on full or partial liberalization of interest rates, and on partial lifting of restrictions on the allocation of credit. ADR (1994) reports on the empirical findings on the relation between financial saving and movements in real interest rates and other variables in several African countries including Egypt, Nigeria, Côte d' Ivoire, and Ghana. In Egypt, there is a significant relationship between financial saving and the real interest rate lagged one period using data for 1966-90 period. Financial saving is also affected by the level of real income, and the inflation differential between Egypt and the United States. For Nigeria, there also appears to be strong support for the interest rate liberalization hypothesis over the period 1960-91 as both real interest rates and real income are significant determinant of financial saving. Seck and El Nil (1993) use pool time series and cross-section data over the years 1974-89 for a sample of 9 countries which have experienced structural adjustment programs and another sample of 21 African countries. They find a strong positive relation between financial saving and the real deposit rate of interest for both samples, although they conclude from their results that curbing inflation is more beneficial than raising the nominal rate of interest.

However, evidence on the impact of interest rates on the level of private or total saving in several African countries is weak. Leite and Makonnen (1986), in a study of the BCEAO countries of Benin, Burkina Faso, Côte d' Ivoire, Niger, Senegal, and Togo, test three models of private saving behaviour using pooled time series and cross-section data over the period 1967-80—the first, relating private saving to income and real interest rates; the second, adding lagged saving as a separate independent variable; and the third, adding the change in income. In all cases, the interest rate variable is positive, but is only significant in the lagged savings model. Jebuni (1994) does not find financial variables important in determining the private savings ratio, although he does find weak support for the real interest rate as a determinant of the total gross domestic savings ratio over the period 1970-90 in Ghana. For Egypt, on the other hand, Hussain (1994) finds that total saving is determined positively by real income and negatively by the inflation differential between Egypt and the United States and by the real interest rate. These empirical results suggest strong substitution between financial and non-financial assets, but no substitution between present and future consumption.

These results for some African countries are consistent with those found for other developing countries as well. Gupta (1987) study of financial liberalization in twenty-two Asian and Latin American countries over the period 1967-1976 shows that there is little support for the hypothesis that the positive substitution effect of real interest rates on savings dominates the negative income effect as the most important determinant of saving is real income. Cho and Khatkhate (1990) review of the financial liberalization experience of five Asian countries also concludes that financial sector reform does not seem to have made any significant difference to the savings and investment activities in the liberalized countries. There are no systematic trend or pattern in regard to saving as the decisions to save are determined by several factors while the relationship between savings and real interest rates remains ambiguous.

From a theoretical stand point, the rate of interest may affect investment in two opposite directions. On the one hand, interest rates may affect investment positively through the effect on financial saving and the supply of credit to the private sector. On the other hand, interest rates may be expected to affect investment negatively, holding the supply of credit constant, if the rate of interest is considered as a proxy for the price of credit, and the rate of interest is above its equilibrium value. The classical investment function suggests that if the interest rate is below equilibrium, investment is constrained by saving, and a rise in interest rates should raise investment. If the interest rate is above equilibrium, however, a rise in interest rates will discourage investment. In order to determine the overall net effect of interest rates on investment, it is necessary to calculate the magnitude of the two separate effects, distinguishing at the same time between interest rate regimes below and above equilibrium. This suggests the use of a switching empirical investment function in which the sign of the interest rate variable may change according to the relationship between the interest rate and its equilibrium value (see Warman and Thirlwall 1994).

Table 13.1: Summary of Financial Reform Measures by Selected African Countries

Deregulation §		Legal Reserve Public Sector Share of Domestic credit * Ratio + Securities μ		Requirements		Real Interest Deposit Rate Rate ±		Competitively Priced-financial	
Deposit	Lending	Before	After	Before	After	Before	After	Before	After
Botswana F.P	-117 S.BP		-122	N/A	N/A	-1.35	-2.95		F
Egypt F.P	66.9 S.TB.CD.GB		71.7	25/15	15/15	-1.6	-8.95		F
Gambia F	60.9 TB		-223	15	15	-2.62	1.05		F

Ghana F	90.6 TB.GB	59	25	15	-56.4	-9.05	F
Kenya F	31.2 S.TB.GB.FXC	31.2	20	24	11	0.55	F
Malawi F	63.2 TB.GB.BP	40.8	10	20	-5.1	-2.95	F
Nigeria F.C	51.7 S.TB.GB	52	N/A	30	4.05	-9.4	F.M
Tunisia F.C.P	13.9 S.TB.CD.CP	11.9	0	2	-3.25	0.15	F.M
Zimbabwe R	49.4 S.TB.CD	41.2	N/A	10	0	N/A	R

Source: Seek and El Nil African Development Report (1991).

* Public sector's share of domestic credit, before refers to the starting year of the reform program, after refers to end of 1991.

+ Legal reserve requirements ratio before refers to the starting year of the reform program, after refers to the situation as of June 1991. Ratio applicable to commercial banks are considered.

± Real deposit rates, before refers to the average of the two years prior to program start, after refers to the average of the two most recent years for which data are available.

§ Interest rate deregulation: F = fully liberalized. P = directed credit to priority sectors at below-market rates. M = minimum deposit rate enforced by Central Bank. C = maximum lending rate enforced by Central Bank. R = fully or partially regulated.

ꟾ Competitively priced financial securities. S = stocks. TB = Treasury bills. CD = certificates of deposit. CB = Government banks. EXC = foreign exchange certificates. BP = Central Bank paper. CP = Commercial paper.

The results of empirical findings of the impact of interest rate and the supply of credit from the banking system on investment using the switching investment function are also reported in ADR (1994). The empirical work on Egypt and Nigeria referred to earlier estimate such a 'switching' model, and also the relationship between interest rates and the supply of credit from the banking system. For Egypt, there is no evidence that real interest rates have been below equilibrium, constraining savings and investment. The interest rate effect on investment is to reduce investment by £E 92 million for every one percentage point rise in the real interest rate. The supply of credit effect of higher interest rates is to increase investment by £E 41 million, giving a net negative effect of £E 51 million. For Nigeria, a net negative effect of interest rates on investment is also estimated. The equilibrium real interest rate is estimated at 3.5 per cent, but, overall, the negative effect of the cost of capital outweighs the positive effect of interest rates on the supply of credit. Other studies such as that of Seck and El Nil (1993) find a positive relation between the investment/GDP ratio and the real deposit rate, but the relation is unstable when other independent variables are added. For Ghana, it is found that the interest rate is not important in determining the investment ratio, but the flow of credit to the private sector is.

The channel through which interest rate affects economic growth rate is more complex as the savings/investment ratio and the productivity of capital also determine the growth of output in a given economy. There is an interdependence between savings and growth as a higher growth rate requires a higher savings ratio in order for the ratio of money balances to income to remain at a constant level while at the same time the savings ratio depends on the rate of interest, growth of output itself and other variables. The growth effect on the propensity to save depends, in turn, on the financial conditions in the economy. The financial assets-income ratio is likely to be higher, the more developed the financial system thus resulting in a higher propensity to save. The interdependence nature of saving and growth suggests the use of a virtuous circle model framework in which growth is a function of the rate of interest and other variables such as government saving, foreign saving, inflation and export growth; and from this, the structural parameters of the savings function may also be obtained (see Warman and Thirlwall 1994). This model which includes the interest rate, export growth, the government savings ratio and the foreign savings ratio is tested for Egypt by Hussain (1994). The export growth is the principal determinant of output growth. Even though the interest rate variable is positive but neither it nor the foreign savings variable is statistically significant. Government saving, however, is demonstrated to have a significant positive effect on growth. Similarly, Oshikoya finds export growth to be the most important determinant of output growth in Nigeria. Growth is negatively related to the rate of inflation, as well as being negatively and significantly related to the rate of interest. Neither government saving nor foreign saving appears to be significant. Seck and El Nil (1993), find a positive relation between the real deposit rate of interest and growth in the sample of nine countries referred to earlier, but the relation appears to be indirect.

These empirical results suggest that financial saving appears to be favourably influenced by real interest rates, but not total saving. There appears to be a positive relation between the stock of financial assets held in the financial system, banking sector credit to the private sector and the level of investment. It is the positive relationship between the stock of **financial** assets and the level of investment and (consequently), growth that underscores the importance of the stock market during financial reform. Stock markets provide non-interest elastic savings instruments (equities) which could contribute to higher rates of saving even in financial structures where saving do not respond sufficiently to interest rates as has been noted in the foregoing discussion for several African countries. The particular attraction of stock markets also relate the their potential to motivate external financial inflows. Further, while higher bank-based financial savings are often crowded out of productive investments through the overbearing influence of the public sector in Africa, equity flows through the stock markets are directly tied to real investments. The structure of the financial sector that emerges from the reforms is therefore central in linking the savings and investment process to growth in Africa. Improving the scope and performance of stock markets are thus useful means through which financial liberalization is intended to raise the level of savings and investment as well as improve the allocation of resources, and hence generate a higher productivity of capital in Africa.

Africa's Emerging Securities Markets

Professional investment managers seeking to diversify their global portfolio have rekindled interest in emerging markets in developing countries in general and in African countries in particular. In this context, Miles Morland has provided a useful classification of world stock markets into three: mature, growth and emerging.[2] The mature financial markets, such as the US, UK, France, Japan, and Australia, can offer good opportunities but are cyclical. The growth financial markets are countries such as Turkey, Taiwan, Argentina, Mexico, Brazil, Malaysia, Korea, South Africa, and India. The process of emergence in these markets were accompanied by spectacular performances in their equity markets. Over the last five and ten years, some of the best performing offshore funds were those invested in these countries. Over five years, the value of $100 would have increased in the following funds as shown in brackets JF Indian Pacific ($1188), Genesis Chile ($701), Thorton New

Tiger Hong Kong ($434), Deltec Latin America ($416). Over ten years, J.F Philippine ($2428), Colonial Securities Hong Kong ($1089), and Baring IUF Malaysia and Singapore ($873).[3] Once they have emerged, Growth markets are influenced strongly by hot foreign money and, consequently, behave in a different way to emerging markets. In 1993, the growth markets increased by 63.7 percent, measured by IFC emerging market index (EMI), as record $47 billion of foreign money flowed into them. However, over the long term they should do well due mainly to the underlying growth of their economies.

According to Morland, these markets have emerged and those investing in them will not benefit a second time in stock prices that can accompany the process of emergence. There are now only three major geographic areas where investors will find true emerging markets: Africa, Arabia, and the countries of the old communist world. There are odd pockets in Latin America and the Far East, such as Paraguay and Burma, but nearly all the markets in these areas have now emerged. Investing in true emerging markets is hard but not impossible. Often, by the time a country fund is available, the opportunity has passed. Indeed, the launching of the first New York Stock Exchange listed country fund is usually a sure signal that a market has finished its first big upward move and is about to collapse. Meanwhile, the true emerging markets, too small or too difficult to attract the foreign capital did very little in 1993. In 1994, the reverse happened. The IFC Composite was down 2.2 per cent while the real emerging markets picked up. In 1994, African stock markets were reported to have posted the biggest gains in U.S. dollar terms among all markets worldwide — Kenya (175%), Ghanaian stocks (70%), Zimbabwe (30%), Egypt (167%), and Tunisia (114%). In 1995, African stock exchanges gained about 40 per cent, with the value of stocks on the Nigerian Stock Markets and Côte d' Ivoire' s bourse registering over 100 per cent increase in dollar terms.[4] As a result of such performances and in the context of broad financial reforms, several countries including Tanzania, Uganda, Sudan, and Zambia have also put in place mechanisms for establishing stock markets.

Up to the end of 1994, there were 14 stock exchanges in the continent. These were Cairo (Egypt), Casablanca (Morocco), Tunis (Tunisia) in North Africa; Abidjan (Côte d' Ivoire), Accra (Ghana), and Lagos (Nigeria) in West Africa; Nairobi (Kenya), Windhoeck (Namibia), Gaborone (Botswana), Johannesburg (South Africa), Port Louis (Mauritius), Lusaka (Zambia), Harare (Zimbabwe) and Swaziland in Eastern and Southern Africa. With the exception of the Johannesburg Stock Exchange, these markets are small in comparison to other emerging markets in Asia and Latin America. At the end of 1994 there were about 1150 listed companies in the Africa markets put together. The market capitalization of the listed companies amounted to $240 billion for South Africa and about $25 billion for other African countries. On the other hand, in 1992 the market capitalizations for Brazil, Chile, India and Indonesia were respectively US$ 45 billion, US$ 30 billion, US$ 65 billion, and US$ 12 billion.

African stock markets are also small in comparison with their economies—with the ratio of market capitalization to GDP averaging 17.3 per cent (Table 13.2). The limited supply of securities in the markets and the prevailing buy and hold attitudes of most investors have also contributed to low trading volume and turnover ratio. Turnover is poor with less than 10 percent of market capitalization traded annually on most stock exchanges. The low capitalization, low trading volume and turnover would suggest the embryonic nature of most African stock markets.

Table 13.2: Capitalization of Africa's Stock Markets

Countries Ranked	No. of Listed Companies	Market Capitalization ($ billion)	GDP ($billion)	% of GDP
1. South Africa	645	240	106	226.4
2. Egypt	23	7	36	19.4
3. Morocco	52	5	27	18.5
4. Tunisia	23	2.5	13	19.2
5. Zimbabwe	65	2	5	40.0
6. Ghana	18	2	6.1	32.8
7. Kenya	52	1.9	4.7	40.4
8. Mauritius	36	1.7	2.8	60.7
9. Nigeria	180	1.3	31.3	4.2
10. Zambia	7	0.6	3.7	16.2
11. Côte d'Ivoire	26	0.52	9.3	5.6
12. Botswana	11	0.35	3.9	9.0
13. Swaziland	4	0.33	2.3	14.3
14. Namibia	13	0.21	2.1	10.0
Total	1155	265.41	253.2	
South Africa	645	240	106	226.4
Rest of Africa	510	25.41	147.2	17.3

Source: Focus on Africa—Africa on the Threshold of a New Era (GT Management PLC).

The channels through which these stock markets could affect the financial sector deepening and economic development process are outlined below.

Competitive and Solvent Financial System and Corporate Sector

The development of securities market could help to strengthen corporate capital structure and efficient and competitive financial system. The capital structure of firms in African countries where there are no viable equity markets are generally characterized by heavy reliance on internal finance and bank borrowings which tend to raise the debt/equity ratios. The undercapitalization of firms with high debt/equity ratios tends to lower the viability and solvency of both the corporate sector and the banking system especially during economic downturn. There are interest rate risk and mismatch of maturities on the balance sheet of firms implied by the heavy reliance on bank overdrafts and other short term credits to finance fixed assets and working capital. During a period of monetary tightening and credit squeezes, banks usually limit overdraft lines, which tend to aggravate the problems of companies without sufficient long term equity funding. In a few countries where domestic bond markets exist, these are generally dominated by government treasury funding which crowds out the private sector needs for fixed interest rate funding. With minor exceptions, the international fixed rate bond markets have been closed to African corporations. Thus the development of an active market for equities could provide an alternative to the banking system for both savers and users of fund by allowing savers to compare the yield on equity or bond with the interest rate on deposits, and companies can compare the cost of various sources of finance.

The financial systems of many developing countries are characterized by high ownership structure resulting in oligopolistic practices which creates privileged access to credit for large companies but limited access to smaller and emerging companies. A viable equity market can serve to make the financial system more competitive and efficient. Without equity markets, companies have to rely on internal finance through retained earnings. Large and well established enterprises are in a privileged position because they can make investments from retained earnings and bank borrowing while new companies do not have easy access to finance. Without being subjected to the scrutiny of the marketplace, big firms get bigger. Ownership of corporations and thus decision making remains in the hands of a few. For the emerging smaller companies, retained earnings and fresh cash injections from the controlling shareholders may not be able to keep pace with the needs for more equity financing which only an organized marketplace could provide. Banking institutions are reluctant to provide longer term finance which such companies need to expand existing business, purchase new equipment, and penetrate new markets. The corporate sector would also be strengthened by the requirements of equity markets for the development of widely accepted accounting standards, disclosure of regular, adequate and reliable information. While closely held companies can camouflage poor investment decisions and low profitability at least for a while, publicly held companies cannot afford this luxury. The availability of reliable information would help investors to make comparisons of the performance and long term prospects of companies; corporations to make better investments and strategic decisions; and provide better statistics for economic policy makers. Although efficient equity markets force corporations to compete on an equal basis for the funds of investors, they can be blamed for favouring large firms, suffer from high volatility, and focus on short term financial return rather than long-term economic return.

The Intermediation Role of Financial Savings Mobilization and the Allocative Efficiency of Investment

The stock-market is supposed to encourage savings by providing households with an additional instrument which may better meet their risk preferences and liquidity needs. In well-developed capital markets, share-ownership provides individuals with a relatively liquid means of sharing risk in investment projects. To the extent that securities and bonds are a viable and relatively secure form of investment with an attractive long term return, they serve two functions: Stocks provide an incentive to save and invest rather than to consume, buy land and real estate (thereby fuelling speculation in this sector), or seek more profitable investment alternatives abroad. Financial savings are promoted over non-financial savings and the domestic savings rate as a whole may increase. They compete with bank deposits etc. which may be subject to interest rate controls. This exerts pressure to keep the controlled interest rates closer to market rates which are more likely to reflect inflation and scarcity of funds. Liquid and active bond markets also help to mitigate one of the most critical problems of finance in the developing world—the availability of long-term funds. While investors want liquidity, corporations and the government need to be assured of long term credits to match their long term assets or, in the case of the government, to finance development projects. To reconcile these conflicting concerns, secondary markets are required. These can enable new issues in the primary markets to be successful.

As discussed in the last section, financial sector reforms in Africa, specifically interest rate liberalization, encouraged the substitution of other form of savings for financial savings with very little effect on total savings. While the development of the stock-markets may encourage financial savings, it may not necessarily lead to an increase in aggregate savings if financial savings simply represent the substitution of one form of saving for another, for example, bank savings or government bonds for corporate shares in the stock-market. In some developing countries where the stock market activities increased substantially in the 1980s, there is little or no evidence of an increase in aggregate savings. Increased market activities in the Latin America region outside of Chile was not accompanied by increased total savings. Indeed, Mexico's private savings rate fell by more than 50 percent from 18.8 percent of GDP in 1988 to 9.1 percent in 1994.[5]

The recent development of interest in African stock markets is also predicated on the notion that they be used for efficient allocation of savings to productive investment. However, the evidence on the contribution of equity markets to corporate finance in both advanced and emerging securities markets suggests that the role of the stock markets in providing capital for corporate expansion is mixed. Taggart (1985) showed that internal funds financed 52 percent of expansion in the United States over 1970-79 while new stock issues contributed only 3 percent to corporate growth. Based on a flow of funds accounts analysis, Mayer (1989) found that the stock market contributed only a small proportion of corporate finance of firms in eight industrial countries over 1970-85. Indeed, in both Britain and the USA, firms bought back more equity than they had issued. Singh and Hamid (1992) examined financing patterns of the top listed companies manufacturing firms quoted on the stock markets in each of ten developing countries—Brazil, India, Jordan, Malaysia, Mexico, Pakistan, Republic of Korea, Thailand, Turkey, and Zimbabwe and found that firms rely more on external resources for corporate expansion. The proportion of corporate growth financed by internal finance was about 16 percent for Korean firms, 30 percent in Malaysia, 57 percent in Zimbabwe, and 67 percent in Pakistan. For the 10 sample developing countries, equity finance contributed about 40 percent of net asset growth. However, the relative importance of external debt and equity varied widely between countries. New equity contributed only 16 percent of new capital for Indian firms, but 66 percent for Turkish firms, and about 44 percent for firms in Zimbabwe.

Macroeconomic and Financial Effects

The significant size of capital inflows that some developing countries have recently attracted caused concern about the macroeconomic impact of these flows and the increased risks they might entail for domestic capital markets (Calvo, Leiderman and Reinhart 1993; Corbo and Hernandez 1996). Evidence that the recent surge of capital inflows has resulted in an appreciation of real exchange rates thus threatening the competitiveness of recipient countries is reported from a number of Latin American countries while the increase in capital inflows in recent years in some fast-growing Asian countries has not been accompanied by an appreciation of real exchange rates (Reisen 1993a and 1993b). The latter countries have successfully reacted by using a combination of policies, including complete liberalization of the current account, fiscal restraint, sterilization of operations and other measures. As the size of FPEI inflows in individual countries is still a relatively small fraction of total capital inflows their impact on real exchange rates might have been negligible. However, short-term speculative flows and their size relative to the size of emerging markets may have increased market volatility and made management of monetary policy more difficult. In addition, it remains the threat of a massive withdrawal of cumulative equity investment which could result in severe balance of payments difficulties and/or in a sharp drop in stock prices. Both effects could be equally detrimental to the economy of the host country. It should not be assumed that it makes sense to establish an organized securities exchange until a careful study has been made of the political, economic, financial, legal, tax and institutional environment; and until any necessary changes have been made to create a conducive environment. Even when conditions are favourable for the creation of an organized market, certain inherent potential problems cannot be avoided.

Although markets react quickly to many events, the cause and effect relationship is not always clear because markets tend to over-react. The resulting market cycles are particularly pronounced during the early stages of equity markets, when the supply of stocks is limited, manipulation relatively easy, investors unsophisticated, underwriters and brokers inexperienced, and securities legislation often inadequate. Major booms and busts in the secondary market undermine the confidence of investors and affect the ability of companies to raise new funds in the primary market. Moreover, the securities market is not isolated from the rest of the financial sector. A major crash in the equity market or dramatic changes in the interest rates and thus prices in the bond market may undermine public confidence in a country's financial sector as a whole, at least momentarily.

Volatility

Stock-market prices tend to fluctuate more than other economic variables, even in fully developed markets. However, the capital markets of developing countries exhibit much greater volatility than those of advanced economies. As Table 13.4 indicates, the standard deviations of equity returns in the emerging markets tended to be considerably higher than those of the developed markets of Europe, United States, and Japan. The dollar value of listed equities in emerging markets has also fluctuated considerably due to developments in exchange rate and other macroeconomic factors. For example, the Nigerian market capitalization which stood at $3.2 billion in 1983 declined to $960 in 1988 on account of the local currency devaluation. Following the devaluation of the Mexican peso in 1994, the Mexican stock market fell from about $175 billion to $85 billion. In 1995, the Mexican Bolsa index fell by 44 percent after falling by about 40 percent in 1994. Between 1982 and 1985 the Brazilian stock-market index rose five-fold to $43 billion, but by 1987 they had fallen to $17 billion or about 28 percent of their 1985 value. The Brazilian Market which also rose by 65 percent in 1994 was down by 28 percent in 1995. The Korean market's dollar capitalization increased from $14 billion in 1986 through $94 billion in 1988 to $141 billion in 1989, but was down to $96 billion by end-1991.

The high degree of volatility in these markets create negative shocks which could undermine the financial system as a whole. The considerable uncertainty involves could make share prices and the stock markets much less useful as a guide to the allocation of resources. The cost of capital to corporations could increase when high volatility due to market fluctuations discourage risk-averse investors.

Table 13.4: Means, Standard Deviations and Autocorrelation of International Equity Returns, 1976-92

Market	Starting & month	Arithmetic mean	Geometry mean	Standard deviation	Autocorrelation p1	p2	p12
Industrial markets							
Australia	1976.01	15.95	12.17	26.34	0.02	40.13	-0.10
Austria	1976.01	15.20	12.31	24.21	0.14	0.02	0.01
Belgium	1976.01	18.03	15.80	20.97	0.07	0.07	-0.01
Canada	1976.01	12.44	10.39	19.93	-0.02	-0.07	-0.11
Denmark	1976.01	14.98	13.13	19.08	-0.07	0.06	-0.18
Finland	1988.01	-9.66	-12.17	22.15	0.09	-0.33	0.03
France	1976.01	17.78	14.51	25.26	0.02	-0.02	-0.10
Germany	1976.01	15.17	12.73	21.81	-0.04	-0.01	-0.08
Hong Kong	1976.01	25.45	19.25	33.88	0.02	-0.05	-0.06
Ireland	1988.01	12.61	9.72	24.25	-0.19	-0.11	-0.25
Italy	1976.01	14.68	11.11	26.84	0.18	-0.03	0.07
Japan	1976.01	17.97	15.20	23.38	0.01	-0.03	0.12
Netherlands	1976.01	18.95	17.30	17.53	-0.06	-0.09	0.01
New Zealand	1988.01	-1.98	-5.18	26.12	-0.04	-0.09	-0.10
Norway	1976.01	16.60	12.49	28.41	0.12	-0.04	-0.02
Singapore and Malaysia	1976.01	16.72	13.05	26.21	0.03	0.02	-0.05
Spain	1976.01	10.32	7.32	24.47	0.11	0.00	-0.03
Sweden	1976.01	18.65	15.87	23.24	0.08	0.00	0.01
Switzerland	1976.01	14.18	12.37	18.74	0.05	0.00	-0.03
United Kingdom	1976.01	19.20	16.50	22.90	-0.01	-0.09	-0.14
United States	1976.01	14.27	13.00	15.46	-0.01	-0.06	-0.02

Emerging markets

Argentina	1976.01	71.66	27.02	105.06	0.05	0.06	-0.10
Brazil	1976.01	22.69	4.71	60.83	0.03	-0.04	0.03
Chile	1976.01	38.65	30.90	39.84	0.17	0.26	0.09
Colombia	1985.01	45.60	40.27	32.57	0.49	0.16	0.03
Greece	1976.01	9.75	3.82	36.27	0.12	0.18	-0.05
India	1976.01	21.45	17.88	26.87	0.09	-0.10	-0.09
Indonesia	1990.01	-6.29	-12.35	34.95	0.30	0.24	0.19
Jordan	1979.01	10.14	8.53	18.04	0.00	0.02	-0.02
Korea Republic of	1976.01	20.02	15.15	31.97	0.01	0.07	0.12
Malaysia	1985.01	13.56	9.81	26.90	0.05	0.08	-0.10
Mexico	1976.01	30.44	19.02	45.00	0.25	-0.08	0.01
Nigeria	1985.01	2.18	-6.36	37.20	0.09	-0.13	-0.08
Pakistan	1985.01	25.65	32.21	22.38	0.27	-0.24	0.13
Philippines	1985.01	51.16	43.23	38.79	0.33	0.02	0.06
Portugal	1986.01	40.85	29.00	51.43	0.27	0.03	0.03
Taiwan (China)	1985.01	39.93	25.37	54.06	0.06	0.04	0.13
Thailand	1976.01	21.55	18.11	25.69	0.12	0.16	0.05
Turkey	1978.01	47.89	22.04	76.71	0.24	0.10	-0.16
Venezuela	1985.01	37.92	26.23	47.52	0.27	0.18	-0.06
Zimbabwe	1976.01	10.16	4.33	34.30	0.13	0.15	-0.04

Source: Campbell R. Harvey. The World Bank Economic Review—January 1995 (Vol. 9, No. 1).
Note: Values are based on U.S dollar returns from monthly data from January 1976 to June 1992.
Pj denotes the jth autocorrelation coefficient.

Institutional Factors

An organized securities requires a securities exchange, a securities commission or other regulatory agency, and intermediaries such as underwriters, dealers, brokers, investment managers and securities analysts. This can be costly, although virtually all costs are borne directly by those who benefit. The intermediaries receive their fees from the issuers or investors to whom they provide a service; the stock exchange is usually funded through fees paid by investors and issuers; even the expenses of the securities commission may be partially paid for by registration fees rather than being a major burden on the government budget. Companies which go public not only pay the initial underwriting expenses but are also subject to the continuous costs of providing financial information, transferring shares, paying dividends and other aspects of shareholder relations. However, this relatively small cost can be set against the public relations value of being a widely recognized name. Regulators will pay continuous attention to the securities market. This may divert attention from initiatives in other parts of the financial sector, especially if regulatory resources are limited; but the exposure to market forces and the experience gained in contacts with private corporations may make those involved in regulation more aware of markets outside the country or company.

One way of overcoming some of the institutional impediments is to foster the creation of regional stock markets which would permit cross-border trading of securities and allow investors to diversify portfolio risk across national boundaries. To facilitate the establishment of regional securities exchange, it will be important to improve regional telecommunication facilities; regional clearing houses in the absence of regional convertible currencies; and harmonize securities market regulations, information and listing requirements, accounting and auditing standards.

Sustainability of Foreign Portfolio Equity Flows

Equity investments in Africa's emerging markets are perceived to have future growth potential as they offer diversification benefits because of their impressive returns and the low monthly correlation between their returns and equity returns in developed countries. The diversification benefits, however, are likely to diminish as these economies become closely integrated with the global economy and markets. Furthermore, in light of the recent experiences with the financial crisis in Mexico and Latin America, concerns have been raised about the sustainability of foreign portfolio equity flows in fragile stock markets such as in Africa. As George Soros once remarked:

> Foreign investing usually involves boom/bust sequences. I have been in the foreign investment business from the early part of my career. I have seen many cycles. And I concluded early on that foreign investors acting as a herd always prove to be wrong... And it is certainly true of this emerging-markets investing mania that reached its climax at the end of [1994]. It's the biggest foreign investment boom that I have ever seen, and the corresponding bust, which started with the devaluation of the Mexican peso in December 1994, is likely to be equally significant (*Fortune*, September 4, 1995: 64).

Although capital controls in emerging stock markets have been able to insulate the domestic market from price movements, country-specific factors relating to credible economic reforms, enabling environment, favourable institutional and regulatory framework would be more important in attracting and sustaining foreign portfolio equity flows in Africa.

In this respect, the lessons of Chile's experience with developing its stock market and attracting equity flows are instructive. The Chilean experience since early 1980s suggests that a gradual accumulation of privately-managed pension funds can encourage the development of capital markets. Following the boom-and-bust episode of the early 1980s, Chile implemented a series of measures that guards against hot money flows from abroad while mobilizing domestic savings. First, it deters speculative inflows by requiring foreign investors to keep capital in Chile for one year; and requiring deposits of 30 percent of foreign loans at the central bank at no interest for one year. These limited and selective capital controls were strong deterrents to short-term speculative investors which prevented domestic fiscal and monetary policies from being jeopardized by huge short-term inflows of foreign currency. Second, it encourages exports competitiveness through a managed float which keeps the peso's exchange rate price-competitive against a basket currencies and by rebating the 18 percent value-added tax to exporters. Third, it spurs savings by taxing corporate dividends more than profits; and it maintains a fiscal budget surplus which lowers public sector borrowing requirements from the financial markets. Fourth, it encourages pension savings through privately managed funds. The effectiveness of these measures in Chile were borne out by its financial markets resilience in the face of the fragility of the wave of stock market and exchange control reforms in Latin America revealed by the devaluation of the Mexican Peso. While Mexico, Argentina and other Latin American countries were severely affected, Chile is witnessing impressive gains. Stock market capitalization has increased from about 20 percent of GDP in 1981 to about 90 percent in the 1990s as closed companies have shown a growing tendency to go public and to accept standard record-keeping and auditing practices, encouraged by better access to pension funding financing. The size of Chile's private pension fund is estimated at $28 billion compared to $1.4 billion and $400 million in Argentina and Peru. In 1995, the growth rate of gross domestic product is estimated at 6 percent as in 1990-1994; with inflation declining from 27 percent in 1990 to 8.3 per cent. Trade surplus has widened and total investment increased to 28 percent of GDP in 1994, mainly financed by domestic savings which averaged 25.4 percent of GDP.

Summary and Conclusion

This chapter surveys the analytical and empirical issues relating to financial sector liberalization in Africa and policy issues relating to the region's emerging stock markets and their role in economic development.

The analytical underpinning for financial sector liberalization in Africa has been based generally on the financial repression hypothesis which implies that a repressed financial sector interferes with development as savings vehicles are not well developed and financial intermediaries are prevented from performing their allocative role in the saving-investment process efficiently. The hypothesis therefore argues for improved growth through financial deepening and liberalization. The concept of total saving must, however, be separated from financial savings which measures the amount of total saving that is channelled via financial assets. Even if total saving is constant, financial saving can increase if leakages into non-financial saving decline. The policy implication of financial liberalization theory is that an increase in the real deposit ratio will lead to a substitution into financial assets resulting in a greater supply of credit to finance real investment for economic growth

These empirical results suggest that financial saving appears to be favourably influenced by real interest rates. There appears to be a positive relation between the stock of financial assets held in the financial system, bank sector credit to the private sector and the level of investment. On the other hand, higher interest rates discourage investment, and the net effect of interest rates on investment may turn out to be negative. Since total savings and investment are not positively related to interest rates, it is not surprising that the growth rates of the African countries surveyed also appear to be invariant with respect to the real rate of interest. There are more important factors at work such as the growth of exports.

An important aspect of the financial sector reforms in Africa has been the emphasis in the growth of stock markets for domestic resource mobilization. Although they remain small by most widely used capitalization ratios and standards, interest in Africa's emerging stock markets have been influenced on the domestic side by implementation of macroeconomic policy reforms and structural adjustment programs, and by developments in the world financial markets favouring diversification and need to generate high equity portfolio returns.

The chapter outlines several development issues relating to the influence of stock markets on the economic development process. Among them are: competitive and solvent financial system and corporate sector; the intermediation role of financial savings mobilization and the allocative efficiency of investment; macroeconomic and financial effects; volatility and sustainability of foreign portfolio equity flows.

While the development of securities market could help to strengthen the corporate sector because of the requirements for the development of widely accepted accounting standards and disclosure of reliable information, they can also suffer from high volatility and focus on short-term financial return rather than on long-term economic return. Market volatility associated with high speculative equity investments could accentuate market cycles, made management of monetary policy difficult and undermine confidence in the broader banking and financial sector.

The lessons of experience of countries such as Chile suggest that these problems could be dealt with varying degrees of success. Some of the important policy factors that could help to enhance the role of the equity markets in Africa's development process include: an enabling environment further through credible macroeconomic through credible macroeconomic policy; promotion of private sector initiatives; improvement in the regulatory framework; infrastructure and information technology and promotion of regional stock exchanges in Africa.

Notes

1. For more detailed exposition on the theories of finance, investment and the stock market and their relationship to economic growth see Copeland and Weston 1983, Lee 1983, Singh 1993, Fry 1988, and Serven and Solimano 1992.
2. See Miles Morland, "Where Have all the Emerging Markets Gone." *Financial Times*, July 22/23, 1995.
3. "The Money Report: Best Performing Offshore Funds." *International Herald Tribune*, September 23-24, 1995.
4. The emerging stock markets in Africa have also recently been given wide coverage by the leading financial press. See, for example, "Stalking Africa's Fledgling Stock Markets." *The Economist*, June 11, 1994; "African Exchanges Pause to Take A Breath." *The Financial Times*, June 5, 1995; "Africa: Continent of Hazard and Opportunity." *Financial Times*, February 7, 1994; "Emerging Now, It's African Stocks," *International Herald Tribune*, April 5, 1994; "Africa: Take Investment By The Horn." *The International October*, 1994.
5. "A Survey of Latin American Finance." *The Economist*, December 9th, 1995:16.

Works Cited

Abebe, A., "Financial Repression and Its Impact on Financial Development and Economic Growth in the African Least Developed Countries." *Savings and Development* 14, 1, 1990.

African Development Report, An Annual Publication of the African Development Bank, 1994.

Atje, Raymond and Boyan Jovanic, "Stock Markets and Development." *European Economic Review* 37 (June): 632-40, 1993.

Callier, P. (ed.), "Financial Systems and Development in Africa". *Economic Development Institute of the World Bank*, Washington: World Bank, 1991..

Calvo, Guillermo A., Leonardo Leiderman, and Carmen Reinhart, "The Capital Inflows Problem: Concepts and Issues." IMF Paper on Policy Analysis and Assessment 93/10. International Monetary Fund, Washington D.C. Processed 1993.

Chandavarkar, A., "Financial Liberalization: Theory and Evidence." Unpublished paper presented at World Bank Seminar on Financial Liberalization, 16-17 September, 1991.

Cho, Yoon Je, "Inefficiencies from Financial Liberalization in the Absence of Well-functioning Equity Markets." *Journal of Money, Credit and Banking* 18, 2, May 1986.

Cho, Yoon Je and Khatkhate, D. "Financial Liberalization: Issues and Evidence." *Economic and Political Weekly*, 20 May, 1990.

Chuhan, Punan, "Sources of Portfolio Investment in Emerging Markets." Working Pa-

per, World Bank, International Economics Department, Washington D.C. Processed, (1992).
Chuhan, P., Claessens, S. and Mamingi, N., "Equity and Bond Flows to Latin America and Asia: The Role of Global and Country Factors." Working Paper 1160. World Bank, International Economics Department, Washington D.C. Processed.
Copeland, Thomas E. and Watson, J. Fred, "Financial Theory and Corporate Policy", 1983.
Corbo, Vittorio and Hernandez, Leornardo, "Macroeconomic Adjustment to Capital Inflows: Lessons from Recent Latin American and East Asian Experience." *The World Bank Research Observer* 11, 1, 1996.
Dailami, Mansoor, and Michael Atkin, "Stock Markets in Developing Countries: Key Issues and a Research Agenda." Policy Research Working Paper 515. World Bank, Country Economic Department, Washington D.C., 1990.
De Angelo., H. and Masuli, R., "Optimal Capital Structure Under Corporate and Personal Taxation." *Journal of Financial Economics* 8, 1980:3-81.
Dornbusch, R. and Reynoso, A., "Financial Factors in Economic Development." *American Economic Review*, Papers and Proceedings, May, 1989.
El-Erian, Mohamed A. and Manmohen S. Kumar, "Emerging Equity Markets in Middle Eastern Countries." *IMF Staff Papers*, Washington, D.C., 1995.
Erunza, Vihang, Lemma Senbet, and Ishac Diwan, "Country Funds: Theory and Evidence." Paper presented at the World Bank Symposium on Portfolio Investment in Developing Countries. Washington, D.C. September, 1993.
Fazzari, S.M., Hubbard, R.G. and Petersen, B.C., "Financing Constraints and Corporate Investment." *Brookings Paper on Economic Activity* I, pp. 141-95, 1988.
Fischer, B. and Reisen, H. "Towards Capital Account Convertibility." OECD Development Center Policy Brief No. 4. Paris: Organization for Economic Cooperation and Development Processes, 1992.
Frank, J. and Mayer, C. "Capital Markets and Corporate Control: A study of France, Germany and the U.K." *Economic Policy*, 1990.
Frankel, Jeffrey A. "The Internationalization of Equity Markets." *NBER Working Paper 4590*, Cambridge, Mass.: National Bureau of Economic Research, 1993.
Fry, M.J., "Money, Interest and Banking in Economic Development." London, 1988.
Giovanni, A., "The Interest Elasticity of Savings in Developing Countries: The Existing Evidence." *World Development*, vol. 11 no. 7, July, pp. 601-607, 1983.
Greenwald, B., Stiglitz, J.E. and Weiss, A., "Information Imperfection in the Capital Market and Macroeconomic Fluctuations." *American Economic Review* 74, 1984.
Harvey, Campbell R., "The Risk Exposure of Emerging Equity Markets." *The World Bank Economic Review* 9, 1, 1995.
Hussain, M.N., "Savings, Economic Growth and Financial Liberalization: The Case of Egypt.", *African Development Report-1994*, 1992..
IFC, *Emerging Stock Markets Factbook*. Washington, D.C., 1994.
Jensen, M.C. and Meckling, W.M., "Theory of the Firm: Managerial Behaviour, Agency Costs and Ownership Structure." *Journal of Financial Economics*, 1976: 305-360.
Lee, Cheng F., "Financial Analysis and Planning: Theory and Applications." 1983.
Leite, S.P. and Makonnen, D., "Savings and Interest Rates in the BCEAO Countries: An

Empirical Analysis." *Savings and Development*, 3, 1986.

Levine, Ross, and Sara Zervos "Policy, Stock Market Development, and Long-Run Growth." Paper presented at the World Bank Conference on Stock Markets, Corporate Finance and Economic Growth. Washington, D.C 1995..,

Mayer, Colin, "New Issues in Corporate Finance." *European Economic Review*, June, 1988)

Mayer, Collin P., "Myths of the West: Lessons from Developed Countries for Development Finance." *Policy Research Working Paper 301*, World Bank, Washington, D.C. Processed, 1989.

Mckinnon, Ronald, I. "Money and Capital in Economic Development." Washington, Brookings Institution, 1973.

Miller, M.H., "Debt and Taxes." *Journal of Finance* 32, 1977:261-75.

Mirakhor, Abbas, and Delano Villanueva, "Strategies for Financial Reform." *IMF Staff Papers 37* (September), 1990: 509-30.

Modigliani, F. and Miller, M.H., "The Cost of Capital, Corporate Finance and the Theory of Investment." *American Economic Review* 48, 1958:201-297.

Modigliani, F. and Cohen, R., "Inflation, Rational Valuation and the market." *Financial Analyst Journal*, March/April, 1979:24-44.

Mullin, John, "Emerging Equity Markets in the Global Economy", *Federal Reserve Bank of New York Quarterly Review* (Summer), 1993:54-83.

Mullins, M. and Wadhwani, S.B., "The Effects of the Stock-Market on Investment: A Comparative Study." *The European Economic Review*, 33, 1989:939-961.

Myers, S., "Determinants of Corporate Borrowing." *Journal of Financial Economics*, 5, 1977: 147-75.

Myers, S.C., "The Capital Structure Puzzle.", *Journal of Finance*, 39, no. 3 pp. 575-92 1984.

Oshikoya, T.W., "Interest Rate Liberalization, Savings, Investment and Growth: The Case of Kenya." *Savings & Development Quarterly Review* 3-XVI, 1992: 305-320.

Oshikoya, T.W., "Financial Sector Reforms, Interest Rate Liberalization and Economic Growth in Nigeria." *African Development Report-1994*, 1994.

Oshikoya, T.W., "Macroeconomic Determinants of Domestic Private Investment in Africa: An Empirical Analysis." *Economic Development and Cultural Change* 5, 42, pp. 573-96, 1994.

Papaioannou, Michael, and Lawrence Duke, "The Internationalization of Equity Markets." *Finance and Development 30*, 3, September, 1993: 36-39.

Reisen, Helmut, "The Case for Sterilized Intervention in Latin America.", Paper presented at the 6th Annual Inter-American Seminar on Economics, May 28-29, Caracas, Venezuela. Paris: Organization for Economic Cooperation and Development (OECD), Development Center. Processed, 1993.

Schinasi, Garry J., and Monica Hargraves, "Boom and Bust in Asset Markets in the 1980s: Causes and Consequences." *IMF Staff Studies for the World Economic Outlook*. Washington, D.C., 1993.

Schwert, G.W., "Why Does Stock-Market Volatility Change?" 1989.

Seck, D. and El Nil, Y.M., "Financial Liberalization in Africa." *World Development*, vol. 21 no. 11 pp. 1867-81, 1993.

Serven, L. and Soliamano, A., "Adjustment Policies and Investment Performance in Developing Countries: Theory, Country Experiences and Policy Implications." *Policy Planning and Research*, Working Paper Series 606, World Bank, Washington, D.C. 1991.

Shaw, Edward, "Financial Deepening in Economic Development." Oxford University Press, New York, 1988.

Shiller, Robert J., "Do Stock Prices Move Too Much to be Justified by Subsequent Changes in Dividends." *American Economic Review 71*, June, 1981: 421-36.

Singh, Ajit, "The Stock-Market and Economic Development: Should Developing Countries Encourage Stock-Markets?" *UNCTAD REVIEW*, 4, 1993.

Singh, Ajit, "How Do Large Corporations in Developing Countries Finance Their Growth." In *Finance and the International Economy*. Amex Review Prize, Essays. Oxford, U.K.: Oxford University Press, 1994.

Sing, A. and Hamid, J., "Corporate Financial Structures in Developing Countries." Washington D.C., World Bank, IFC Technical Paper 1, 1992.

Stiglitz, Joseph E., and A. Weiss, "Credit Rationing in Markets with Imperfect Information." *American Economic Review 71*, June, 1981: 394-410.

Stiglitz, J.E. and Weiss, A., "Credit Markets and the Control of Capital." *Journal of Money Credit and Banking* 17, 2, 1986.

Taggart, Jr., R.A., "Secular Patterns in the Financing of U.S. Corporations." B.M. Friedman (ed.), *Corporate Capital Structure in the United States*, University of Chicago Press, Chicago, 1985.

Taylor, L., "Structuralist Macroeconomics: Applicable Models for the Third World." New York: Basic Books, 1983.

Thirwal, A.P. and Hussain, M.N., "The Balance of Payments Constraint, Capital Flows and Growth Rate Differences Between Developing Countries." *Oxford Economic Papers*, November, 1982.

Van Agmatel, Antoine W., "Emerging Securities Market: Investment Banking Opportunities in the Developing World." London : Euromoney, 1984.

Van Wijnbergen, Sweeder, "Interest Rate Management in LDCs." *Journal of Monetary Economies*, 1983a.

Van Wijnbergen, Sweeder, Credit Policy, Inflation and Growth in a Financially Repressed Economy. *Journal of Development Economics*, 13 (1-2), August-October 1983, 1983b: 45-65.

Villenueva, Delano, "Issues in Financial Sector Reform." *Finance and Development 25* March, 1988: 14-17.

Warman, F. and Thirlwall, A.P. "Interest Rates, Savings, Investment and Growth in Mexico 1960-90: Tests of the Financial Liberalization Hypothesis." *Journal of Development Studies*, April, 1994.

Yao, K., "Financial Development, Economic Growth and Financial Reforms under Structural Adjustment: The Case of Côte d'Ivoire." *African Development Report-1994*.

Chapter 14

The Outcome of Financial Sector Reforms in West Africa

Chidozie Emenuga

Introduction

Financial systems have long been recognized to play an important role in economic development. This recognition dates back to Goldsmith (1955), Cameron (1967), Mckinnon (1973) and Shaw, (1973), which demonstrated that the financial sector could be a catalyst of economic growth if it is developed and healthy.

The benefits accruable from a healthy and developed financial system relate to savings mobilization and efficient financial intermediation roles (Gibson and Tsakalotos 1994). First, through the financial intermediation functions of the financial institutions savers and borrowers are linked up and this reduces transactions and search costs. Second, they create liquidity in the economy by borrowing short-term and lending long-term. Third, they reduce information costs, provide risk management services and reduce risks involved in financial transactions. Fourth, the intermediaries bring the benefits of asset diversification to the economy. Fifth, they mobilize savings from atomized individuals for investment, thereby solving the problem of indivisibility in financial transactions. Finally, mobilized savings are invested in the most productive ventures irrespective of the source of the savings.

The above benefits of financial intermediation translate into the economy-wide benefits which motivate financial reforms where the system is considered undeveloped. These are to:

(i) increase the size of domestic savings channeled through the formal financial sector;
(ii) improve the efficiency of financial intermediation; and
(iii) enhance the effectiveness of monetary policy.

Based on these expectations, many African countries have implemented financial liberalization as a component of the structural adjustment program under varying financial structures and different macroeconomic conditions. For instance, at the commencement of the reforms, Nigeria already had more advanced financial institutions and assets than Ghana, Sierra Leone and The Gambia. But generally, the reforms commenced in these countries on a background of macroeconomic imbalance and financial distress.

Expected Outcome of Financial Reforms

Through the removal of the elements of financial repression, particularly controlled interest rates, financial sector reform is expected to lead to higher nominal and real interest rates. This is the postulate of the hypothesis (Mckinnon 1973; Shaw 1973). A higher real interest rate encourages people to substitute consumption for savings (the substitution effect). On the other hand, the higher interest income on savings makes savers to achieve their saving targets with lower stock of savings (the wealth or income effect). The two effects operate in opposing directions and the net outcome would depend on which one that dominates. The underlying logic of the Mckinnon-Shaw doctrine is that the substitution effect would outweigh the wealth effect. Financial savings will further be boosted by a shift in the savers wealth portfolio from non-financial assets to financial assets (asset substitution effect).

Contrary to the Mckinnon-Shaw premise, the increased real interest rate may not necessarily lead to improved private savings. In very poor countries for instance, the level of income could be so low that households spend very high proportion of their earnings on basic needs. In such a case, even with high real interest rates, very little (if any) proportion of income could be saved. This implies that the Mckinnon-Shaw proposition would therefore be more relevant in rich nations. A recent study of this proposition (Ogaki et al. 1996) found that a 100% rise in real interest rate leads, in the long run, to a 66.7% rise in savings in high income countries but to only 10% rise in very low income countries. This basic needs explanation and even the tendency for dissaving in Africa might explain the insensitivity of savings to real interest rates in some African countries (see Oshikoya 1992 for the case of Kenya).

Also, in an under-banked economy, where the financial markets are rudimentary, with a large size of financial intermediation taking place in the informal sector, savings may not be sensitive to real interest rates. The informal financial sector is of course large in most of sub-Saharan Africa (Aryeetey and Udry 1994). For the savers who operate in the formal financial sector, a history of government interference in the deposit market or growing incidence of bank distress could scare them from saving in financial instruments despite the lure of rising real interest rate. Also, innovations in financial products following financial liberalization, such as consumer credit facilities could induce a rise in the consumption habit of the people, thereby reducing their savings (a liquidity effect).

Even when financial reform leads to increased savings, it may not promote growth. The use to which savings is committed is an important linkage. Economic activities would be stimulated if more of the growth in savings is channeled to productive activities. On the contrary, the gains to economic growth through increased credit to the private sector would be sidelined if the increased savings is used to finance public sector deficits (Wijnbergen 1983). Also, if the increase in interest rate is excessive, the lending portfolio of banks could become riskier just as firms would face harder times in meeting interest and capital repayment commitments. Baring some of these possible negative developments, financial liberalization would naturally improve the financial intermediation process and lead to more investment, productivity and increased economic growth (Mckinnon 1991)

Measures of financial sector development are used to assess the effectiveness of financial reform. Some of these measures are (Bisat, Johnston, and Sundararajan 1992:

- The growth of private financial assets as measured by the size of currency (M2 and xx), which also indicates the liquidity position of the financial system. Their growth rates and ratios to GDP show the degree of monetization and financial market development.
- The flow of credit to the private sector from the financial institutions. The credit flow from the central bank to financial institutions highlights the role of the public sector in financial intermediation and would diminish as the financial sector develop, with increased mobilization of private savings.
- The growth of financial institutions' credit to the private sector relative to the growth of private sector deposits with financial institutions. A faster growth of the former relative to the latter indicates that there is pressure on domestic resources. This process reflects

poor performance of the financial sector in resource mobilization.

Other general indices used to assess the performance of the financial system and the effectiveness of the liberalization policies include, the trend of real interest rates, real GDP growth rates, number and types of financial institutions, and the spread between deposit and lending rates which show the efficiency of financial intermediation.

In analyzing the effects of financial liberalization it is often useful to appraise the fiscal process. Fiscal stability and the success of monetary reforms are interrelated. Without fiscal balance, the resulting increased budgetary expenditure could motivate authorities to resort to larger deficit (perhaps financed through money creation) which is capable of undermining the gains from the reform. Where the deficits are financed through market instruments, domestic interest rates could rise even in real rates while exchange rate could appreciate. These developments could shift the burden of adjustment to the real sector (Johnston 1994). Hence the controversy as to whether financial reforms are capable of improving the performance of the financial sector and the economy where there is fiscal instability (El Nil 1992). Also depending on its nature, the fiscal process could contribute to higher rate of inflation and where the expansion in credit that follows liberalization is rapid and greater than deposit growth, this could lead to a loss of macroeconomic control, further causing and exacerbating the rate of inflation (Bisat, Johnston, and Sundararajan 1992). This process could spirally cause large increases in interest rates and prices and adversely affect the real sector. The fiscal stance as well as the behavior of the price level (i.e., the rate of inflation) are therefore important variables in assessing the outcome of financial reform.

Reforms and Results

Nigeria

Reform
At the inception of comprehensive financial sector reform in 1987, the sector was highly repressed. Interest rate controls, selective credit guidelines, ceilings on credit expansion and use of reserve requirements and other direct monetary control instruments were typical features of the pre-reform financial system in Nigeria. Entry into banking business was restricted and public sector-owned banks dominated the industry. The reform of the foreign exchange market which hitherto was also controlled began a year ahead of the general financial sector reform.

In context, the financial sector reform was a component of the Structural Adjustment Program (SAP) which kicked off in 1986. Although the policy planks of SAP in Nigeria were the prototype prescriptions of the Bretton Woods institutions, the program was sold to Nigerians by government as Nigeria's alternative to IMF loan-based adjustment. The introduction of the program was on the heels of the rejection of IMF loan package with its conditionalities, a decision that reflected the consensus of a nationwide debate.

The major financial sector reform policies implemented were deregulation of interest rates, exchange rate and entry into banking business. Other measures implemented include, establishment of Nigeria Deposit Insurance Corporation, strengthening the regulatory and supervisory institutions, upward review of capital adequacy standards, capital market deregulation and introduction of indirect monetary policy instruments. Four highly distressed banks were liquidated while the central bank took over the management of others. Government banks were also privatized. The details and the sequencing of the reform measures are contained in the Appendix.

A peculiar feature of the reform program in Nigeria is the associated inconsistency in policy implementation. The reform of the foreign exchange market started in 1986 with the dismantling of exchange controls and establishment of a market-based autonomous foreign exchange market. Bureaux de change were allowed to operate from 1988. However a fixed official exchange rate has continued to exist alongside the autonomous market. In 1994 the gradual market-based depreciation in the official exchange rate was truncated by a sharp devaluation in a bid to close the widening gap between the official and the autonomous exchange rate. Unsatisfied with the observed further widening of the gap between the two exchange rates, government outlawed the autonomous foreign exchange market and reintroduced exchange controls in 1994. But after a full year of exchange controls, the autonomous market was brought back in 1995 to co-exist with the fixed official exchange rate. A foreign exchange subsidy of about 300% still exists for some government favored consumption such as pilgrimage and sporting events for which the official rate applies. The continued operation of the official exchange rate brings with it a great deal of distortions in the domestic allocation of resources within the public sector. This is very pronounced in the vertical distribution of export earnings among the three levels of government. The revenue from this foreign exchange rent to the federal government now constitutes one of its major revenue sources (Emenuga 1996a). Fiscal gains thus appears to be an incentive factor in retaining the current structure of the foreign exchange market.

A similar pattern of policy reversals applies to the reform of interest rates. First introduced in 1987, the market-determined interest rates ruled until 1991 when interest rates were capped. But after only a year of controls, market forces were permitted once more to determine all interest rates in 1992 and 1993. Since 1994, the pre-reform policy of controls has been retained.

While indirect monetary instruments (open market operations) have been initiated since 1993, some measures of controls such as sectoral credit allocation guidelines have continued to be applied.

In the sphere of bank licensing and regulation, the reform was ushered in with deregulation of bank licensing in 1987. This was immediately greeted with the establishment of many new banks, leading to the doubling of the number of operating banks within three years. When the increase in the required banks paid up capital in 1989 and the reform of their accounting procedure (1990) appeared insufficient to curb the "excesses" of the sector, government placed total embargo on bank licensing in 1991. This is yet to be lifted. Privatization of banks was suspended after applying the measure to a few banks. In fact, Government has drawn up plans to buy back its stakes in the major commercial banks, apparently in sympathy with the political class who have been deprived of one of their lucrative "spoils" through the divestiture. The "de-privatization" is already being implemented for one of the three largest commercial banks.

Some of the issues highlighted above point to the disorderly manner in which the reform has been implemented in Nigeria. In effect, the reform has not been a one-shot smooth process. This therefore complicates the task of assessing its outcome.

Results

The liberalization of bank licensing at the onset of the reform resulted in the establishment of many new banks. The number of operating banks almost doubled within three years into the reform and tripled in the fifth year (Table 14.1). It required official re-imposition of embargo on bank licensing in 1991 to halt this growth. Profitability of investment and access to credit and foreign exchange were among the major motives for bank ownership.

The growth in the number of banks appears to have had only a marginal impact on the concentration of the industry. The share of the three largest banks in the total assets of the industry decreased from 37.3% before the reform to an average of 34.6% over the reform period, a reduction of only 2.7% points despite the over 200% growth of operating institutions. The new banks were generally small and undercapitalized, a situation that later led to an upward review of banks minimum operating capital.

The competitiveness that resulted from the entry of new banks and the liberalization of interest rates brought about sharp rise in nominal deposit and lending rates. The increase was however moderated by the occasional reimposition of interest rate ceilings. Nevertheless, the average deposit and lending rates doubled in the third year of the reform (Table 14.1)

Surprisingly, the competition for deposits which drove nominal interest rates up could not ensure a cheaper cost of intermediation. The nominal interest rate spread rather worsened from 2.9% pre-reform average to 4.0%. This is one indication that the reform did not improve the availability of loanable funds. Higher spread was one of the strategies employed by the banks for survival. The banking environment that emerged from the reform is a lot inefficient, undercapitalized, riskier, less liquid and generated lower return on assets relative to the pre-reform period (Sobodu and Akiode 1994).The incidence of fraud, and of non-performing loans also increased with the reform. The quality of management is a major determinant of a bank' s long-term survival (Siems 1992; Pentalone and Platt 1987) and the dearth of qualified management personnel to meet the challenges of sudden growth in the industry contributed to the poor health of the banking industry (Ikhide and Alawode 1994). It was in 1991 that government promulgated the Bank and other financial institutions Decree (No. 24) and the Central Bank of Nigeria Decree (No.25) which spelt out comprehensive guidelines for bank regulation, supervision and liquidation. By this time the incidence of distress in the industry was already rampant.

Although interest rates responded positively to financial liberalization, real rates behaved differently. For most of the reform years, real deposit rate was negative, and averaged -13.5% compared to -7.65 during financial repression (Table 14.1). The high rates of inflation during the reform coupled with reimposition of interest rate ceilings brought about the negative real deposit rate.

If financial savings is interest elastic, the negative real deposit rate would lead to poorer savings mobilization. The relationship between real savings and interest rate was recently explored for Nigeria (Soyibo and Adekanye 1992) and the empirical evidence identified only a weak influence of real interest rate on real savings. It further shows that the era of financial liberalization could not make any difference to the tenuous link. Over the reform period the rate of real resource mobilization declined. The Savings/GDP ratio dropped from 15.4% to 12.4%. The growth of real savings also slowed down by 7.5% points (Table 14.2). These developments are in harmony with the econometric evidence. The decline in real wage income during the reform could have also contributed to the fall in real savings. For instance, the index of real wage income for the middle level public service cadre declined from 100 in 1987 to 40 in 1990 and 34 in 1992 (Federal Ministry of Finance, *Approved Budget*, various issues).

The expected increase in financial savings during the reform was only realized in the rural areas. Measured by the share of deposits mobilized in banks' rural branches in the total deposits, rural savings increased from 1.9% before the reform to 10.8%. This trend nets off the role of community banks and the People's bank in rural credit mobilization. The two set of institutions were established to enable rural dwellers and the poor save and have access to credit at rates lower than the high rates that came with the financial reform. The privatization exercise which required investors to pay through banks contributed much to the growing culture of financial savings in the rural areas. This is a plausible explanation since the rate of rural savings growth only rose significantly in 1989 after the start of privatization in 1988. Good as this growing culture of rural savings may be for breaking financial dualism, the economy would have been worse for it if the informal sector in Nigeria, from where savings now move to the formal sector is more efficient in resource use. This proposition will await empirical investigation for substantiation.

In the first few years of the reform, the share of the banking system's credit to the private sector improved and superseded the flows to government for the first time in five years. Later, government's reliance on Central Bank's financing for the soaring deficits overturned the table to its favor. From 50.7% average before the reform the share of the private sector in the total credit decreased to 49.7% after the reform. In 1993 and 1994, only 34% of the total credit went to the private sector (Table 14.2).

The bulk of the credit that was channeled to the private sector was mainly directed toward short-term investment. Between 1987 and 1994, 50% of the private sector credits went to call money, 32.5% to lending maturing within 12 months, 12% for 1-5 years maturity (medium-term) and only 4.8% for long-term commitments exceeding five years (Central Bank of Nigeria, *Annual Report and Statement of Accounts*, various issues). Again much of the short-term private sector credits was invested in foreign exchange speculation. Through the incentives created by the retention of the overvalued official foreign exchange market together with the parallel market rate, politically "connected" rent-seekers emerged, buying foreign exchange from the official market and reselling in the parallel market at premium that in some cases was above 200% (Emenuga 1996a). Bank loans were largely held in speculative balances (call money) because bid for foreign exchange was required to be backed with Naira cover (Sobodu and Akiode 1994). Other reasons for the short-term lending behavior of banks were the uncertainty that surrounded lending for real investment in the face of the unstable macroeconomic environment (Soyibo 1994b) and the policy uncertainty that characterized the reform period.

The banking system's lending to the public sector was part of the financing facility for fiscal deficits. The Structural Adjustment Program implemented from 1986-1994 did not include fiscal reform. Nor did the financial sector component of SAP include a reform of government borrowing from the financial system. Rather the reform which led to increases in nominal interest rates prompted government to rely more on Central Bank financing than previously. Since 1987 government has abandoned the issue of Federal Government Development Stocks (FDS), government's long-term bond in preference for treasury financing possibly due to the high cost of servicing the debt. The share of Central bank's credit in the total deficit financing has consequently risen from 25.4% in 1987 to 67.9% between 1987 and 1994 (Table 14.3). Also, due to the absence of a stabilization program, the size of government's deficit increased during the reform years. Confronted with increasing costs of its fixed (budgeted) commitments as a result of inflationary forces during the reform, government easily found rescue in the mint (Emenuga 1996c).This operation was made smooth by the absence of central bank independence (Emenuga 1996b). The ratio of deficits to GDP could therefore not improve through the reform but rather increased from 7.0% to 9.2% (Table 14.3). With the deteriorating fiscal balance, and worse still, with a heavy reliance on Central Bank's financed deficit, the explanation for the inflation rates which assumed unprecedentedly high levels during the reform becomes not far fetched. The average rate of inflation doubled over the reform period, rising from 16.0% to 33.6%. It has been shown empirically that the Central Bank-based deficit financing has been a strong causal factor in the domestic price instability. (Ndebbio 1995). Exchange rate depreciation and "appropriate pricing of petroleum products" are additional causes of inflation during the reform.

The depth of the financial sector, measured by the M_2/GDP ratio, contrary to expectations, was not improved by the reform. It declined. The fall in financial deepening from 32.6% to 26% implies that the growth of the financial sector lags behind the tempo of economic activities, suggesting that the financial sector may not have been the source of real GDP growth within the period.

Interestingly, despite the dashed expectations on the developments in the financial sector, the performance of the economy improved over the reform period. Negative trends in real GDP growth were reversed. Except for the first year of financial reform which was also the first full year of implementing the Structural Adjustment Program (1987) there were positive real GDP growth in the rest of the years. Overall, the economy improved from -2.7 average annual growth to 5.0% during the reform but to ascribe this impressive performance to financial liberalization would be dubious. In 1985 and 1986, the immediate years preceding the reform, real GDP grew by 9.4% and 3.45 respectively whereas it dropped to -0.6% in the first year of the reform (Table 14.3). The channels through which the reform would have led to improved growth have been shown to have deteriorated during the reform. Real deposit rate, real savings, efficiency and depth in financial intermediation and credit flow to the private sector, became poorer during the reform. The rate of inflation also worsened. The growth in real GDP must have been influenced then by forces outside the financial sector. Since the period of implementing the Structural Adjustment Program (SAP) encompassed that of the financial sector reform, the positive economic performance might be situated within the wider adjustment package.

Ghana

Reform

The comprehensive economic adjustment program which embodied the financial sector reform started in April 1983 following years of continuous decline in economic performance. The first phase of the Economic Recovery Program (ERP) dated from 1983 to 1986 and focused on stabilization measures. The policies implemented include currency devaluation, tighter fiscal management, and liberalization of prices including interest rates.

Prior to the reform program, the financial sector was repressed, featuring interest rate ceiling, sectoral credit guidelines, credit ceiling and use of direct monetary control tools, including reserve requirements.

The liberalization of interest rates was gradual. It was not until

September 1987 that the prescription of minimum and maximum deposit rates was abolished. The phased transition to a market-determined interest rates was stepped up in 1988 with the introduction of the Financial Sector Structural Adjustment Program (FINSAP). The specific focus on the financial sector was aimed at:

> creating a sound prudential and regulatory framework for banking; strengthening bank supervision, restructuring distressed banks; human resource development in banks; and development of fully liberalized money and capital markets. (Bank of Ghana, *Annual Report*, 1989/90:1)

The thrust of FINSAP was therefore to fully deregulate the financial sector through introduction of market-oriented monetary management instruments and at the same time ensuring the protection of the overall health of financial institutions through adequate regulation and supervision.

With FINSAP, government completely deregulated interest rates, eliminated selective credit guidelines and implemented measures to usher in indirect monetary controls. Open Market Operations was introduced for liquidity management. However, the use of reserve requirement was retained. Other features of the financial sector reform include, licensing of new private banks, establishment of a stock exchange, granting more supervisory powers to the Central Bank and promulgation of laws to formalize the activities of non-bank financial institutions. It may be noted that the implementation of these financial reform programs was done alongside the general Structural Adjustment Program.

The gradual phasing of the program implementation, the consistency in pursuing the reform and absence of any major policy reversal are remarkable features of the financial sector reform in Ghana.

Results

In many respects the outcome of the reform in Ghana reflected the pace of its implementation. The nominal deposit and lending rates increased gradually from 1984 in consonance with the scope of liberalization. After five years, the average nominal lending rate increased by just 3.7% over its level at the start of the program (Table 14.4). Immediately the reform started, interest rate spread dropped considerably but later rose above the pre-reform levels. The higher default risk and its associated cost within the reform period was a major cause of the increasing interest rate spread (Aryeetey and Seini 1992).

Improvement in the real deposit rate has been very significant. In the first year of the reforms (stabilization cum interest rate liberalization), real deposit rate increased by 86.7% points (from -111.5% to -24.8%). Further into the reform positive real deposit rates were achieved. Between the pre-reform and the reform periods real deposit rate increased by 52.2% although it remained negative (Table 14.4). This resulted mainly from the stabilization measures which led to lower rates of inflation. The contribution of nominal interest rate growth was marginal.

The rise in real deposit rates did not however lead to higher rate of savings. The ratio of savings to GDP decreased slightly by 0.6% over the reform period (Table 14.5). This poor performance of savings is believed to have resulted from the public loss of confidence in the banking system as a result of government intervention in 1982 (Aryeetey et al. 1991).1 The mobilization of savings gradually improved with the continued implementation of the reform program especially in the 1990s, when real deposit rate became positive.

The effect of the tight fiscal operations that accompanied the financial sector reform showed up in larger flow of domestic credit to the private sector, resulting in increase in the private sector share of credits from 9.2% to 20.3% (Table 14.5). Real private sector credit also increased by 2.2% points.

The period of adjustment witnessed increased macroeconomic stability. Deficit/GDP ratio dropped drastically from 5.1% average to 0.1%. The rate of inflation declined by 47.0% points, from 73.3% to 26.3% on the average. In 1983 the rate of inflation was as high as 123% whereas since the full implementation of the reform the highest rate has been 39.8%. In some years it was as low as 10%. An impressive growth in real GDP has also been achieved. For the four consecutive years preceding the reform, real GDP growth was negative. But since the adjustment Program, the real GDP growth has ranged from 3.0% to 5.6% (Table 14.6).

The growth of real GDP is linked to the financial sector through increase savings, investment and improved efficiency in financial intermediation. A recent study on Ghana identified several factors that affect private investment. Among these are the growth of credit to the private sector, macroeconomic stability and fiscal stance (deficit\GDP ratio) all of which have improved since the reform. (Asante 1994). Hence, either by mere association or causality, the financial sector reform in Ghana, with the accompanying fiscal and macroeconomic adjustment is identifiable with better economic performance. A possible source of the growth is the increased efficiency in resource allocation arising through larger flows to the private sector. However the contribution of the financial sector to the real GDP growth could only be marginal. For one reason, the share of private sector credits in the total credits still remained small (20%) even after the improvement brought about by the reform. Secondly, the reform did not improve the level of financial savings, a possible source of increased investment. Higher level of investment could have derived from external flows. Further, the depth of the financial system has not improved with the reform. The ratio of M_2/GDP declined slightly from an average 17.0% to 15.5% (Table 14.6). As with savings, the persistence of negative real interest rate for some years into the reform and the loss of public confidence in the financial sector which healed but slowed as the reform progressed impacted adversely on the deepening of the financial sector.

So far the gains from the reform in Ghana are in terms of releasing more resources to the private sector, lower rates of inflation and modest improvement in real interest and rate. The credit for these gains owe much to the consistency in program implementation. This ensured a progressive improvement in the financial and macroeconomic indices. The discipline of government in cutting down the budget deficit and borrowing less from the financial system also helped. A clearer picture of the state of the financial system that resulted from the reform and of its linkage with the real sector will however emerge if the reform is further implemented with the past zeal and zest.

Sierra Leone

Reform

In 1986, Sierra Leone embarked on Structural Adjustment Program in fulfillment of the conditionalities for an IMF loan package. Financial sector reform was part of the comprehensive conditionality. The reform measures implemented include devaluation, privatization, removal of consumer subsidies, price decontrols, liberalization of interest rates, and public expenditure controls.

Results

Nominal deposit and lending rates increased with their upward adjustment at the inception of financial liberalization. Both the deposit and the lending rates doubled within four years of the reform reaching all time levels of 40.5 and 52.3% respectively in 1990. These increases were only in nominal terms as the real interest rates were negative and declining. The post reform average real deposit rate declined by 33.8% having deteriorated from -40.3% to -74.1% (Table 14.7). Interest rate spread also declined by 7.5% points during the reform. The declining real interest rates and the falling efficiency in financial intermediation were also associated with falling real savings. The savings/GDP ratio dropped by about half of its size before the reform. The reform of public expenditure released more resources to the private sector as the private sector share of bank credits increased significantly from the inception of the reform but relative to GDP, credit to the private sector dropped from 5.3% to 3.8%.

The reform program was accompanied by reduction in the level of government deficit. This reflected in the falling deficit/GDP ratio since 1989. But the rate of inflation kept increasing and remained the bane of the reform program. In the first year of the reform the inflation rate was more than double its historical peak and remained high for most of the subsequent years. This unsavory development emanated from supply shocks (decline in mineral and agricultural output and exports) during the reform (Elliot and Braimah 1993) and the sharp rise in interest rates. The high rate of inflation was a major factor in the rising real interest rates.

The depth of the financial sector and the availability of loanable funds performed poorly during the reform. The M_2/GDP ratio declined substantially.

Despite the less than satisfactory outcome in the financial sector, the growth of real GDP was revived and sustained. On the average, it improved from -0.5% to 3.2% (Table 14.9). A major factor in the performance of the economy during the reform was the inflow of foreign capital in both loan and aid forms. Of course this was the motivating factor for the reform. Since the right financial sector for efficient financial intermediation has not emerged from the reform, it becomes suspect how long the achieved growth will last especially if the capital inflow becomes market-based.

Summary and Conclusion

In this chapter we have analyzed the financial sector reform programs in three West African countries—Nigeria, Ghana, and Sierra Leone. The analyses are based on the policies implemented, their effects on the financial sector vis-à-vis the expected effects and the link with the economy.

Before the reforms, the economies experienced various forms of financial repression and macroeconomic problems. The three countries, were characterized by officially fixed low interest rates, high rates of inflation, negative real interest rates, low rate of savings, and dominance of the public sector in the domestic credit market. There were also shallow financial sector, high budget deficit and negative growth of real GDP.

In all the three countries, the financial sector reform was implemented as a component of the Structural Adjustment Program whose content and form were the typical package of Bretton Woods institutions. The menu of the implemented financial sector reform measures include interest rate deregulation, elimination of credit ceiling, decontrol of entry and operation in the banking industry, increased regulatory and supervisory powers of the Central banks over the banking system, and gradual introduction of indirect monetary control instruments. The speed of implementation, the scope and the concomitant policies however differed. In Ghana, the implementation was gradual, consistent and progressed with the pace of cutback in the public sector deficit. In Nigeria, it was with speed but without fiscal reform and marked with several policy reversals.

Through the reform, the countries had hoped to achieve a more competitive, healthier, efficient and deeper financial system. In particular, it was expected that liberalizing interest rates would lead to higher real interest rates and ultimately to more financial savings. Through more efficient intermediation, the higher savings would increase the level of investment, resulting in higher real income.

The outcome of the reform has been mixed, differing from one country to another. In Nigeria, the liberalization of interest rates and entry into banking business gave rise to sharp increases in nominal interest rates. With the additional effects of currency devaluation and higher Central bank-financed public sector deficits, the rate of inflation soared. Consequently, real interest rates remained negative, and further declined. The converse of the McKinnon-Shaw doctrine then held true as real savings and financial depth decreased. The health of the banking industry deteriorated with growing number of distressed banks. The outstanding elements of repression in the system such as the controlled foreign exchange market, and the ensuing macroeconomic instability motivated banks to lend the scarce savings for short-term and speculative activities rather than for real investment.

In Ghana, nominal interest rates increased gradually in line with the slow pace of the reform. With the drastic reduction in the public sector deficits, the inflation rate was significantly reduced. Real interest rates improved rapidly. However, given the high level of the pre-reform negative real interest rate, it took longer time to achieve positive real deposit rate. But real savings did not improve appreciably. What is however not clear is whether the slower rate of savings growth was purely an aftermath of the previous government interference with financial savings or an indication that there were little idle savings to be mobilized.

The fiscal reform that accompanied the financial sector reform in Ghana ensured that an increasing percentage of domestic credits flowed to the private sector, although the public sector still controlled 80% of the credits. The depth of the financial sector however declined slightly after the reform. Given these developments, one is reluctant to attribute the favorable real GDP performance to financial reform. Nevertheless, the reform has given birth to a stronger financial system and with the combination of the larger Structural Adjustment a healthier macroeconomic environment has emerged.

In Sierra Leone, real interest rates, which were negative before the reform, further deteriorated despite the sharp rise in nominal interest rates. This was caused by the escalation in the rate of inflation due to shocks in commodity supplies and rising prices. The ratio of savings to GDP dropped by half just as the ratio of private sector credits to GDP declined substantially in spite of the decreasing flow of credits to the public sector. Similarly, the depth of the financial sector became shallower. The tripling of the interest rate spread is one indication of the fragile health of the financial sector during the reform. While the financial sector reform was fast, fiscal reform was very gradual, achieving only a marginal reduction in fiscal deficits.

The reasons for the outcome of the reforms seem to be evident in their policy content and the implementation process. In Nigeria, the reform proceeded with fast speed or even haste. More banks were established without adequate managerial, regulatory and supervisory capacity. The absence of fiscal reform before and during the reform was a costly policy oversight. The partial nature of the reform was also a source of distortions. For instance, the rent-seeking opportunities introduced by the operation of dual exchange rates seriously distorted allocation of financial assets and discouraged investment. The reform was therefore hasty in the execution of the approved measures, partial in scope, destitute of other concomitant adjustment measures, and inconsistent. Is it a surprise that it has not yielded the desired results.

A similar problem of haste marred the reform in Sierra Leone. Interest rate and commodity prices were freed beyond the supply capacity of the economy. The resulting high rates of inflation became a great stumbling block in the reform process. Relative to the speed of liberalization, the reform of the fiscal process was too slow to give the right effects. The experience of Sierra Leone points to the salient issue of the role of the real sector in financial reform. A shock in the real sector was a single major stroke that undermined the financial reform.

By hindsight, Ghana's approach appears to be the right step in financial reform. It proceeded slowly, consistently and trailed behind the speed of macroeconomic and fiscal adjustment. It also paced with the managerial, regulatory and supervisory capacity and infrastructure in the financial system. But wither the anticipated miracles and the reward for reforming according to the rules?

The outcome of the reform in these countries is a pointer that all is not well with the Mckinnon-Shaw paradigm as interpreted and applied to Africa. The target of achieving positive and rising real deposit rate via increase in nominal interest rate was obviously a wrong approach given the structure of the economies and the financial sectors—weak real sector, fragile and undeveloped financial sector. A more effective means to improve real interest rate would be through macroeconomic stabilization and fiscal reforms. The significant improvement is real interest rate in Ghana was achieved through this and not through unbridled loosening of holds on nominal interest rates as happened in Nigeria and Sierra Leone. In fact it was the failure of this approach that led to reintroduction of controls in Nigeria.

We now seem to know better in terms of reform sequencing (Villanueva and Mirakhor 1990; Diaz-Alejandro 1985; Gibson and Tsakalotos 1994 and World Bank Report 1994). For the reform to succeed, it has to be preceded by macroeconomic and fiscal stability and strengthened capacity of regulatory and supervisory institutions. And this is of such an overwhelming importance that it would pay off better to postpone financial reforms till the conditions are met. Concurrent implementation of financial reforms and the pre-conditional adjustment programs as the World Bank Report would suggest does not appear to be appropriate. In fact financial sector reform should start late in the adjustment timetable.

Also, the expectation of improved savings and investment through increased real interest rates appears exaggerated if not misplaced. In pursuing the reforms, greater importance was paid to interest rate rather than income in savings mobilization. So was the role of the dominant informal sector where savings are mostly in non-financial assets neglected. For the reforms to play a crucial role in economic development, the state needs to play active role not only in terms of setting the preconditions, of macroeconomic cum fiscal stability and institutional infrastructure but also in designing appropriate instruments and mechanisms to target financial transactions in the informal sector.

While we recommend that countries like Nigeria and Sierra Leone which have reformed in error need to follow the rules of the game, while

Ghana and others that have started off well should persevere, we note that the scope of the reforms need to be enlarged to accommodate both the informal sector and programs to raise the level of income. Given these tasks the role of the state is therefore far from gone.

Table 14.1: Nigeria: financial rates, 1980-1994 (%)

Year	Number of banks (units)	Growth rate of banks	Share of 3 largest banks in total assets	Nominal lending rate	Nominal deposit rate	Interest rate spread	Real deposit rate
1980	26	0.0	+	9.5	6.5	3.0	-3.4
1981	26	0.0	+	10.0	6.5	3.5	-14.4
1982	28	7.7	+	11.8	8.0	3.5	-14.4
1983	32	14.3	+	11.5	8.0	3.5	-15.2
1984	36	12.5	+	13.0	10.0	3.0	-29.6
1985	37	2.8	+	11.8	10.0	1.8	4.5
1986	39	5.4	37.3	12.0	10.0	2.0	4.6
1987	45	15.4	36.6	19.2	15.8	3.4	5.6
1988	58	28.9	35.2	17.6	14.3	3.3	-24.0
1989	76	31.0	39.6	24.6	21.2	3.4	-19.7
1990	107	40.8	35.1	27.7	23.0	4.7	15.5
1991	119	11.2	31.6	20.8	20.1	0.7	7.1
1992	119	0.0	32.8	31.2	20.5	10.7	-24.0
1993	119	0.0	31.4	29.2	27.1	2.1	-30.1
1994	118	-0.8	+	23.0	19.0	4.0	-38.0
1980-1986[a]	37	7.1	37.3	11.4	8.4	2.9	-7.6
1987-1994[b]	109	15.8	34.6	24.2	20.1	4.0	-13.5
Change between the two periods	72	8.7	-2.7	12.8	11.7	1.1	-5.9

+ = not available
[a] = pre-reform period
[b] = reform period

Source: Computed from:
(i) Central Bank of Nigeria: *Statistical Bulletin*, Vol.15, Nos. 1 & 2,1994
(ii) Central Bank of Nigeria: *Annual Report and Statement of Accounts*, (various issues).

Table 14.2: Nigeria: financial ratios, 1980-1994 (%)

Year	Savings/GDP	Growth of real savings	Rural savings total savings	Private sector share of total credit	Credit to private sector/ GDP
1980	11.4	25.7	+	66.7	14.2
1981	13.0	-5.8	+	59.5	19.1
1982	14.5	5.4	0.8	53.8	22.1
1983	16.5	1.5	0.7	44.0	21.7
1984	17.3	-15.2	1.8	41.3	20.3
1985	17.3	6.8	1.6	42.0	18.9
1986	19.2	6.4	4.5	47.3	23.8
1987	17.2	21.1	5.0	60.6	23.4
1988	16.0	-20.5	4.1	57.2	20.5
1989	10.6	-33.6	18.4	59.1	13.7
1990	11.4	18.8	18.6	54.6	14.0
1991	11.6	12.9	17.8	54.1	14.0
1992	9.8	-0.9	5.2	43.0	11.1
1993	12.2	0.0	14.0	34.1	13.3
1994	10.4	-30.1	2.9	34.9	13.6
1980-1986[a]	15.5	3.5	1.9	50.7	20.0
1987-1994[b]	12.4	-4.0	10.8	49.7	15.5
change between the two periods	-3.1	-7.5	8.9	-1.0	-4.5

+ = not available
[a] = pre-reform period
[b] = reform period

Source: Computed from:
(i) Central Bank of Nigeria: *Statistical Bulletin*, Vol.15, Nos. 1 & 2, 1994
(ii) Central Bank of Nigeria: *Annual Report and Statement of Accounts*, (various issues).

Table 14.3: Nigeria: macroeconomic indicators, 1980-1994 (%)

Year	Central Bank's credit to Govt	M_2/GDP	Deficit/GDP	Inflation rate	Real growth
1980	+	29.0	6.7	9.9	5.5
1981	+	30.7	7.7	20.9	-26.8
1982	+	32.8	11.9	7.7	-0.3
1983	+	34.2	5.9	23.2	-5.4
1984	+	34.3	4.2	39.6	-5.1
1985	+	33.3	2.4	5.5	9.4
1986	+	34.1	10.3	5.4	3.2
1987	25.4	28.1	5.4	10.2	-0.6
1988	75.4	30.0	8.4	38.2	10.0
1989	67.5	20.8	6.7	40.9	7.3
1990	82.4	25.2	8.5	7.5	8.1
1991	91.9	26.9	11.0	13.0	4.8
1992	80.4	23.5	10.1	44.5	3.6
1993	61.7	24.6	15.4	57.2	2.7
1994	58.3	28.6	7.9	57.0	3.8
1980-1986[a]	+	32.6	7.0	16.0	-2.7
1987-1994[b]	67.9	26.0	9.2	33.6	5.0
change between the two periods		-6.6	2.2	17.6	7.7

+	=	not available
[a]	=	pre-reform period
[b]	=	reform period

Source: Computed from data from:

(I) Central Bank of Nigeria: *Statistical Bulletin*, Vol.15, Nos. 1 & 2,1994
(ii) Central Bank of Nigeria: *Annual Report and Statement of Accounts*, (various issues).

Table 14.4: Ghana: financial rates, 1979-1992 (%)

Year	Nominal lending rate	Nominal deposit rate	Interest rate spread	Real deposit rate
1979	19.0	11.5	7.5	-43.0
1980	19.0	11.5	7.5	-39.7
1981	19.0	11.5	7.5	-104.9
1982	19.0	11.5	7.5	-10.8
1983	19.0	11.5	7.5	-111.5
1984	21.2	15.0	6.3	-24.8
1985	21.2	15.8	5.4	-5.5
1986	20.2	17.0	3.0	-7.6
1987	25.5	17.6	7.9	-22.2
1988	25.6	16.5	9.1	-14.8
1989	+	+	+	+
1990	+	+	+	+
1991	+	21.3	+	3.3
1992	+	16.3	+	6.2
1979-1983a	19.0	11.5	7.5	-61.8
1984-1992b	22.7	17.1	6.3	-9.3
change between the two periods	3.7	5.6	-1.2	52.5

+	=	not available
a	=	pre-reform period
b	=	reform period

Source: Computed from:
(i) IMF, *International Financial Statistics*, (various issues)
(ii) ADB, *African Development Report*, 1995.

Table 14.5: Ghana: financial ratios, 1979-1992 (%)

Year	Savings/GDP	Private sector share of total credit	private sector credit/GDP
1979	4.5	10.8	2.8
1980	4.3	9.5	2.1
1981	3.6	8.4	1.4
1982	4.2	8.5	1.8
1983	2.2	8.6	1.5
1984	1.9	12.3	2.2
1985	2.5	13.8	3.1
1986	2.7	16.0	3.7
1987	2.9	11.7	3.2
1988	3.1	26.8	3.1
1989	3.8	34.2	5.9
1990	3.2	37.5	4.7
1991	4.3	19.3	3.5
1992	5.5	20.9	4.6
1979-1983[a]	3.8	9.2	1.6
1984-1992[b]	3.2	20.3	3.8
Increase between the two periods	-0.6	11.1	2.2

[a] = pre-reform period
[b] = reform period

Source: Computed from:
(i) IMF, *International Financial Statistics*, (various issues)
(ii) ADB, *African Development Report*, 1995.

Table 14.6: Ghana: macroeconomic indicators 1979-1992 (%)

Year	M_2/GDP	Budget deficit /GDP	Inflation rate	Real GDP growth rate
1979	21.1	6.4	54.5	2.5
1980	18.6	4.2	50.2	-0.6
1981	16.6	6.5	116.4	-2.4
1982	17.2	5.6	22.3	-7.5
1983	11.30	3.0	123.0	-0.7
1984	11.81	1.8	39.8	3.4
1985	13.62	2.2	10.3	5.2
1986	17.0	-0.1	24.6	5.0
1987	18.0	-0.5	39.8	4.7
1988	18.0	-2.9	31.3	5.6
1989	17.0	-	25.2	5.3
1990	14.0	-0.7	37.2	3.0
1991	13.0	-1.5	18.0	5.4
1992	17.0	4.8	10.1	3.7
1979-1983[a]	17.0	5.1	73.3	-1.7
1984-1992[b]	15.5	0.1	26.3	4.6
Change between the two periods	-1.5	-5.0	-47.0	6.3

[a] = pre-reform period
[b] = reform period

Source: Computed from:
(i) IMF, *International Financial Statistics*, (various issues)
(ii) ADB, *African Development Report*, 1995.

Table 14.7: Sierra Leone: financial rates, 1982-1990 (%)

Year	Nominal lending rate	Nominal deposit rate	Interest rate spread	Real deposit rate
1982	15.0	10.0	5.0	-18.6
1983	17.3	11.0	6.3	-0.1
1984	18.0	12.0	6.0	-54.7
1985	17.0	13.3	3.7	-62.7
1986	17.2	14.2	3.0	-65.4
1987	28.5	12.7	15.8	-168.3
1988	28.0	16.3	11.7	-14.8
1989	29.7	20.0	9.7	-42.9
1990	52.3	40.5	11.8	-70.5
1982-1986[a]	16.9	12.1	4.8	-40.3
1987-1990[b]	34.6	22.4	12.3	-74.1
Change between the two periods	17.7	10.3	7.5	-33.8

[a] = pre-reform period
[b] = reform period

Source: Computed from:
(i) IMF, *International Financial Statistics*, (various issues)
(ii) ADB, *African Development Report*, 1995.

Table 14.8: Sierra Leone: financial ratios, 1982-1991 (%)

Year	Private sector deposit/GDP	Savings/GDP	Private sector share of total	Credit to private sector/GDP
1982	0.3	10.4	15.4	7.1
1983	0.3	10.2	14.4	6.9
1984	0.1	8.3	9.3	4.1
1985	0.1	6.5	7.9	2.8
1986	0.2	5.8	14.2	5.7
1987	0.2	4.4	14.0	3.1
1988	0.2	3.6	24.8	2.3
1989	0.2	3.6	24.8	2.3
1990	0.1	4.3	21.3	3.9
1991	0.1	4.2	34.5	5.4
1982-1986[a]	0.2	8.2	13.1	5.3
1987-1991[b]	0.2	4.1	22.4	3.8
Change between the two periods	0.0	-4.1	9.3	-1.5

[a] = pre-reform period
[b] = reform period

Source: Computed from :
(i) IMF, *International Financial Statistics*, (various issues)
(ii) ADB, *African Development Report*, 1995.

Table 14.9: Sierra Leone: macroeconomic indicators, 1982-1991 (%)

Year	M2/GDP	Budget Deficit	Inflation	Real
1982	26.2	10.4	28.6	2.7
1983	29.4	14.4	11.1	-2.6
1984	26.0	7.6	66.7	1.4
1985	25.4	7.9	76.0	-2.7
1986	30.5	+	79.6	-1.4
1987	19.0	16.9	181.0	4.2
1988	20.2	7.4	31.1	4.0
1989	23.3	6.5	62.9	0.2
1990	21.5	2.9	111.0	5.1
1991	20.9	6.9	102.7	2.4
1982-1986[a]	27.5	10.1	52.4	-0.5
1987-1991[b]	21.0	8.1	97.7	3.2
change between the two periods	-6.5	-2.0	45.3	3.7

+ = not available
[a] = pre-reform period
[b] = reform period

Source: Computed from:
(i) IMF, *International Financial Statistics*, (various issues)
(ii) ADB, *African Development Report*, 1995.

Notes

1. In 1982, the new military regime which came to power in 1981 confiscated financial assets of politicians and other people suspected to have amassed wealth through public expense. Government also recalled the 5000 Cedis notes in circulation within a very short time.

Works Cited

Aryeetey, E. and C. Udry. "The Characteristics of Informal Financial Markets in Africa." paper presented at the Plenary Session, African Economic Consortium Workshop, Nairobi, Kenya, December 4-9, 1994.

Aryeetey, E. and Seini, W. "An Analysis of the Transaction Costs of Lending in Ghana." Research Report presented at African Economic Research Consortium Workshop, Nairobi, Kenya, December 5-10, 1992.

Aryeetey, E., Asante, Y. and Kyei, A. "Mobilizing Domestic Savings for African Development and Diversification: A Ghanaian Case Study." Mimeo, International Development Centre, Queen Elizabeth House, Oxford University, Oxford, 1991.

Asante, Y. "Determinants of Private Investment Behavior in Ghana." Research Report presented at the African Economic Research Consortium Workshop, Nairobi, Kenya, December 4-9, 1994.

Bisat, A., Johnston, R., and Sundararajan, V. "Issues in Managing and Sequencing Financial Sector Reforms: Lessons from Experiences in Five Developing Countries." International Monetary Fund, Working paper, WP/92/82, 1992.

Cameron, R. *Banking in the Early Stages of Industrialization: A Study in Comparative Economic History*. New York: Oxford University Press, 1967.

Diaz-Alejandro, C. "Goodbye Financial Repression, Hello Financial Crash." *Journal of Development Economics*, 19, 1985:1-24.

Emenuga, C. "Distortions in the Nigerian Financial Markets." In A.Ekpo (ed.), *Fiscal and Monetary Policies During Structural Adjustment in Nigeria*, Proceedings of Senior Policy Seminar, held at the Central Bank of Nigeria, Lagos, March 6 - 7, 1996, 1996a.

___ " Institutional Setting For Post Reform Management of the Nigerian Financial System: A Case For Central Bank Independence." *Post-Reform Management of the Nigerian Economy*, Nigerian Economic Society, Published 1996 Annual Conference Proceedings, 1996b.

___ "Seigniorage Revenue and Financial Reform in Nigeria." Paper Presented at the African Economic Research Consortium Workshop, Nairobi, Kenya, May 24, 30, 1996c.

Elliot, J. and Braima, S. "An Assessment of Budgetary and Fiscal Performance of the Sierra Leone Economy." Research Report presented at the African Economic Research Consortium Workshop, Cape Town, may 29 - June 3, 1993.

Gibson, H. and Tsakalotos, E. "The Scope and Limits of Financial Liberalization in Developing Countries: A Critical Survey." *The Journal of Development Studies*. 30, 3 (April), 1994: 578-628.

Goldsmith, R.W. "Financial Structure and Economic Growth in Advanced Countries." In M. Abramovitz (ed.) *Capital Formation and Economic Growth*, Princeton, NJ: Princeton University Press, 1955.

Ikhide, S. and Alawode, A. "Financial Reforms, Macroeconomic Stability and the Order of Economic Liberalization: The Evidence from Nigeria." Research

Report presented at the African Economic Consortium Workshop, Nairobi, Kenya, December 4-9, 1994.

Johnston, R. "The Speed of Financial Sector Reform: Risks and Strategies." International Monetary Fund, IMF Paper on Policy Analysis and Assessment, 1994.

Johnston, R. and Brekk, O. "Monetary Control Procedures and Financial Reform: Approaches, Issues and Recent Experiences in Developing Countries." *IMF Working Paper*, WP/89/48 (June), 1989.

McMahon and Medhora. "The Process of Financial Liberalization in Developing Countries: Research Directions." Mimeo. Ottawa, Canada: International Development Research Center, 1994.

Mckinnon, R.I. *Money and Capital in Economic Development*, Washington, D.C., The Brookings Institution, 1973.

Mckinnon. "*The Order of Economic: Liberalization.* Baltimore, M.D., John Hopkins University Press, 1991.

Ndebbio, J. "Fiscal Operations, Money Supply and Inflationary Development in Nigeria." African Economic Research Consortium, Forthcoming as a monograph, 1995.

Ogaki, M. Ostry, J. and Reinhart, C. "Saving Behavior in Low and Middle-Income Developing Countries: A Comparison." *IMF Staff Papers*, March, 1996.

Oshikoya, T. "Interest Liberalization, Savings, Investment and Growth: The Case of Kenya." *Savings and Development.* 26, 1992.

Pentalone, C. and Platt, M. "Predicting Commercial Bank Failure Since Deregulation." *Federal Reserve Bank of Boston.* New England Economic Review, July/August, 1987: 37-47.

Siems, T., "Quantifying Management's Role in Bank Survival." *Federal Bank of Dallas Economic Review*, First Quarter, 1987:29-41, 1992.

Wijnbergen, S. "Interest Rate Management in LDCs." *Journal of Monetary Economics*, 1983.

Shaw, E. *Financial Deepening in Economic Development*, New York: Oxford University Press, 1973.

Sobodu, O. and Akiode, P. "Bank Performance, Supervision and Privatization in Nigeria: Analyzing the Transition to a Deregulated Economy." Research Report presented at the African Economic Research Consortium Workshop, Nairobi, Kenya, December 4-9, 1994.

Soyibo, A. "Financial Liberalization and Bank Restructuring in sub-Saharan Africa: Some Lessons for Sequencing and Policy Design." paper presented at the Plenary Session, African Economic Research Consortium Workshop, Nairobi, Kenya, December 4-9, 1994a.

Soyibo, A. "The Savings-Investment Process in Nigeria: An Empirical Study of the Supply Side." *African Economic Research Consortium Research paper.* 12, 1994a.

Soyibo, A. and Adekanye, F. "Financial System Regulation, Deregulation and Savings Mobilization in Nigeria." *African Economic Research Consortium Research Paper* 11, 1992.

Stiglitz, J. "Financial Markets and Development." *Oxford Review of Economic Policy.* 5, 4, 1989.

"The Role of the State in Financial Markets." *Proceedings of the World Bank Annual Conference on Development Economics,* The World Bank, 1993.

Stiglitz, J. and Weiss, A. "Credit Rationing in Markets with Imperfect Information." *American Economic Review.* 71, 1981.

Villanueva D. and Mirakhor A. Strategies for Financial Reforms. *IMF Staff Papers.* 37, 3, 1990.

World Bank. *Adjustment in Africa: Reforms, Results and the Road Ahead,* World Bank Policy Research Report, Oxford University Press, New York, 1994.

Appendix: An Outline of the Financial Sector Reform in Nigeria

A. Exchange Rate

1986   Establishment of the first-tier and second-tier (autonomous) foreign exchange markets.
1988   Bureaux de change established.
1992   Devaluation of the official exchange rate.
1994   Reintroduction of exchange controls and suspension of bureaux de change.
1995   Exchange controls relaxed.
Operation of bureaux de change permitted.
Autonomous foreign exchange market introduced.
1996   Official fixed foreign exchange market operated for Government transactions
Continued operation of the autonomous foreign exchange market.

B. Interest Rate and Monetary Policy

1987   Deregulation of interest rate
1989   Auction market for Government securities introduced.
Continued use of direct monetary policy instruments (cash reserve requirements)
1990   Introduction of stabilization securities for liquidity management.
1991   Reintroduction of interest rate controls.
1992   Removal of interest rate controls.
Liberalization of bank credit market.
1993   Introduction of indirect monetary instruments (open market operations)
1994   Re-imposition of interest controls.
Review of Central Bank operations
1995   Continuation of interest controls
Initiated fiscal reforms.
1996   Retention of interest controls
Continuation of fiscal reforms

C. Banking and Capital Market

1987   Deregulation of bank licensing
1988   Restrictions on bank portfolio relaxed
Deposit guarantee scheme established.
1989   Review of Banks' capital adequacy standards.
1990   Reform of accounting procedure for banks.
1991   Embargo on bank licensing
Strengthening of bank regulation and supervision.
1992   Privatization of banks commenced.

	Deregulation of the capital market.
1993	Restructuring of distressed banks.
1994	Liquidation of banks.
1995	Liberalization of capital flows.
1996	Liberalization of the capital market continues.

Contributors

Adebayo Olukoshi is Professor of International Relations and graduated from Ahmadu Bello University, Zaria and Leeds University, England. He was Director of Research at the Nigerian Institute of International Affairs, Lagos and Programme Officer at the Nordiska Afrikainstitutet, Uppsala. He is currently Executive Secretary of the Council for the Development of Social Science Research in Africa, Dakar. He has published extensively on contemporary politics in Africa.

Ademola Oyejide is Professor of Economics at the University of Ibadan, Nigeria. He has written extensively on trade policies and liberalisation, regional integration, and industrialisation. He has held leading positions in various academic bodies and institutions and is currently one of the Managing Editors of the *Journal of African Economies*.

Ali Abdel Gadir Ali was Professor of Economics at the University of Gezira, Sudan. He also worked as the Director of Research at the Inter Arab Investment Guarantee Corporation, Kuwait and Director of the Economic and Social Policy Division at the UN ECA, Addis Ababa. He is currently Advisor at the Arab Planning Institute, Kuwait. His research interests are in development economics including evaluation of macroeconomic policies, inequality and poverty analysis.

Charles C. Soludo is currently Professor of Economics, University of Nigeria, Nsukka, and the Executive Director, African Institute for Applied Economics, Enugu. He has consulted extensively for various international organizations, including the World Bank, the IMF, UNCTAD, UNECA, IDRC, etc. He was previously Visiting Professor at Swarthmore College, P.A., USA, and Visiting Scholar at the Brookings Institution, Oxford University and Cambridge University.

Chidozie Emenuga received both his master's and doctorate degrees in economics from the University of Ibadan, Nigeria, where he also taught in the Department of Economics. He is currently with the African Development Bank, Abidjan, Côte d'Ivoire. Dr. Emenuga is a member of the governing council of the Nigerian Economic Society and a member of the African Economic Research Consortium. Among many others, his published works include "Export Promotion in Nigeria: Analysis of Policies and Performance," published for the 1995 Annual Conference of the Nigerian Economic Society.

David Ekpenyong teaches in the Department of Economics at the University of Ibadan, Nigeria.

Francis Mwega, PhD, is currently associate professor of economics at the University of Nairobi, where he has taught since the mid-1980s. He attended Makerere University (Uganda) and the Universities of Nairobi (Kenya) and Illinois at Urbana-Champaign (USA). He has written extensively on the African economies, with a particular reference to Kenya.

Elwathig M Kameir holds degrees from universities of Khartoum, Sudan and Hull, England. He was lecturer and associate Professor of Sociology, University of Khartoum between 1980 and 1990 and worked as consultant for several regional and international organisations, including the ECA, ADB, Arab League, ILO, UNICEF, and UNESCO. He has actively participated in both the Arab and African research communities and held leading professional and academic positions in them. He is currently Senior Programme Officer, IDRC Regional Development Research Centre of Canada. He was responsible for the project: The African Perspectives on Structural Adjustment which has led to this volume. He is currently, the Executive Director of the African Technology Policy Studies Network (ATPS).

Samuel Wangwe, is a Professor of Economics and currently the Executive Director Economic and Social Research Foundation (ESRF), Dar es Salaam, Tanzania.

Temitope W. Oshikoya, is the Manager of the Research Division of the African Development Bank He is team leader of the *African Development Report,* the flagship publication of the Bank now published by the Oxford University Press. He is also the Editor of the *African Development Review.*

Dr. Oshikoya has published extensively on development policy issues in such journals as *World Development, Economic Development and Cultural Change*, and *Journal of African Economies*.

Thandika Mkandawire served as Executive Secretary of the Council for the Development of Social Science Research in Africa (CODESRIA) in Dakar Senegal, from 1986 until 1996. He is currently the Director of the United Nations Research Institute for Social Development (UNRISD) in Geneva, Switzerland. He has published extensively on development economics and problems of policy making in Africa.

Tshikala B. Tshibaka is currently Chief of Policy Assistance at the Food and Agricultural Organisation Regional Office, Accra. Professor Tshibaka has taught in various universities in Africa and the United States. He has written widely in the area of Macroeconomics and Agriculture. Among his publications include the edited collection *Structural Adjustment and Agriculture in West Africa* (CODESRIA: Dakar, 1998). His areas of research interest include, among others, Trade and Globalisation Issues, Regional Integration and Cooperation, Agricultural Policies, Public Assistance to Agriculture, Food Security and Poverty, and Youth and Women in Agricultural Development.